The Phantom God

What Neuroscience Reveals about the Compulsion to Believe

JOHN C. WATHEY

Prometheus Books

Essex, Connecticut

Prometheus Books

An imprint of Globe Pequot, the trade division of
The Rowman & Littlefield Publishing Group, Inc.
4501 Forbes Blvd., Ste. 200
Lanham, MD 20706
www.rowman.com

Distributed by NATIONAL BOOK NETWORK

British Library Cataloguing in Publication Information Available

Library of Congress Cataloging-in-Publication Data Available

ISBN 978-1-63388-806-7 (cloth : alk. paper) | ISBN 978-1-63388-807-4 (ebook)

∞™ The paper used in this publication meets the minimum requirements of American
National Standard for Information Sciences—Permanence of Paper for Printed Library
Materials, ANSI/NISO Z39.48-1992

In memory of
M. Michael Huntley

I became my own test subject. "Look at that! I'm talking with God. It sure feels real, but it must be a trick of the brain." It *had* to be a trick of the brain, since it was beginning to look like a personal god probably did not exist. What a strange and wonderful thing to realize.—Dan Barker[1]

Mythology is not invented rationally; mythology cannot be rationally understood. Theological interpreters render it ridiculous. Literary criticism reduces it to metaphor. A new and very promising approach is opened, however, when it is viewed in the light of biological psychology as a function of the human nervous system, precisely homologous to the innate and learned sign stimuli that release and direct the energies of nature—of which our brain itself is but the most amazing flower.—Joseph Campbell[2]

Contents

Prologue

It was not neuroscience that caused former Christian evangelist Dan Barker to lose his belief in God. It was more a gradually escalating discomfort with the unfulfilled promises of the religion, the empirically testable aspects of Scripture that seemed not to hold up to experiment, and the growing realization that its most fundamental ideas no longer made sense. As with many believers, his personal experience of the presence of God had once seemed compelling evidence of the divine, and to this day he marvels that he can still summon those feelings. His flash of insight that appears as the first epigraph to this book beautifully expresses both the joy of seeing through an illusion and the deep questions that immediately follow. *Why* are humans prone to this illusion? *How* does the brain trick us in this way? In my first book, *The Illusion of God's Presence*, I tried to answer the *why* question. Here I deal mainly with *how*, and that takes us into the brain—a journey that a prescient Joseph Campbell could only imagine in 1959.

Since the advent of functional neuroimaging, the temptation to peek into the brains of people praying, meditating, or speaking in tongues has been irresistible. It has spawned a new discipline, or at least a new genre in popular science writing: *neurotheology*. Most of the results, however, have been disappointing. There is no consensus about what kind of mystical experience should be studied, little consistency in findings across studies, a dearth of clear hypotheses to test, and ample room for unconstrained speculation. Understandably, neurotheology has its detractors. Yet religion really is a profoundly important aspect of human behavior, one that merits scientific scrutiny. The neuroscience of religious experience is in its infancy, its techniques are rapidly improving, and what it most needs now are specific and testable hypotheses that lead to good experimental questions.

I did my PhD in the lab of Ted Bullock, one of the founding fathers of neu-ro*ethology* (not to be confused with the almost identically spelled neuro*theology*).

Ethology is the study of animal behavior in its natural context and emphasizes the evolutionary origin of the behavior and its role in the animal's reproductive success. Neuroethology extends this exploration into the brain. It seeks to understand the brain by applying the methods of neuroscience to the natural behaviors of highly specialized animals. The essence of the idea is that the complexity of the brain is overwhelming, and the only hope of understanding it is to start with a part of the brain where we completely understand the problem it is trying to solve. We can best do this in the brain of an animal that is highly specialized for a distinct and easily understood behavior—preferably one that has already been well studied by ethologists. Famous examples include infrared vision in rattlesnakes, communication and navigation by electric fields in weakly electric fish, echolocation in bats, sexual pair-bonding in prairie voles, the development of singing in songbirds, and the ability of barn owls to catch mice in total darkness using only sound cues.

With each of these examples, understanding the behavior had to come first. It took decades of behavioral study to understand that birds sing to establish territories and attract mates, that their song is partly innate but mostly learned, and that chicks need to hear the normal song during a critical period that precedes their first attempts at singing. Only then did it make sense to explore the neural circuitry in the brain that makes this possible. The behavioral studies mainly address "why" questions: Why do birds sing? Why is their song abnormal if they are raised in isolation? The neurobiological studies address mainly "how" questions: How do birds sing? How does the brain encode the innate aspects of the song? How does the brain of a chick learn the fine details of the song from its father? What neurons, fiber tracts, and neurotransmitters are involved? How do these change during development?

In my previous book, *The Illusion of God's Presence*, I tried to do mainly the first part—the ethological or behavioral part—for the scientific study of religion. The intense and highly emotional longing for a loving God had been largely ignored in most of the scientific literature on religion, and my background in ethology suggested a new answer to this puzzle. That's what most of my first book was about, and the questions it addresses are mainly "why" questions. I did, however, include a few chapters that touched on neurobiology, and I confessed to my readers that these were mere snippets of a second book that had taken shape, almost by accident, as I was writing the first. Although the second book was unfinished at the time, I knew what it would cover, and I referred to it as "the sequel" throughout *The Illusion of God's Presence*.

This book, *The Phantom God*, is that second book. I wrote it to try to understand the circuitry in the brain that underlies the feeling of God's presence—in other words, to address the "how" questions. More precisely, I try to make sense of experiments that others have done that bear on the problem, and I suggest promising directions for future research. Part of what's new and special about *The Phantom God* is that it draws connections to fields of neuroscientific research previously considered unrelated to human religiousness, like mother-infant bonding in nonhuman mammals and language acquisition in human infants. There are many surprising connections and insights of this kind that can be seen only in the brain. For example, part of the circuitry of mother-infant bonding is also the epicenter of drug addiction—an observation that may explain a lot about the compulsive and tenacious aspects of religious belief. Similarly, the orbitofrontal cortex of an infant specifically responds to the mother's voice and, in adult Carmelite nuns, to their recalling the feeling of God's presence. Damage to this part of the brain causes acquired sociopathy, and the most severe cases are those in which the damage occurred in infancy; here we see the intersection in the brain of the social and neonatal roots of religion I emphasized in my first book.

Although *The Phantom God* clearly grew out of my work on *The Illusion of God's Presence*, it stands on its own. You can understand *The Phantom God* without reading its predecessor because part I, comprising two chapters, summarizes the essential ideas and evidence from the first book that form the behavioral foundations of this one. The title of the first chapter is an all-encompassing "why" question: "Why Is God Two-Faced?" Why do believers tell you that their God will damn you to hell yet in the same breath insist that he loves you unconditionally? The answer, in my view, is the key to an ethological understanding of human religiousness. Equipped with that answer, we embark on an expedition deep into the human brain.

In part II, I lay the foundation for some promising new mechanistic questions, mainly by addressing this fundamental one: where in the brain is the circuitry that produces the illusion of a divine presence? Along the way I describe evidence from neuroscience that bears on the central hypothesis I lay out in chapter 1. My broader goal, however, is to explain as best I can the neural mechanisms behind that illusion and the emotions of spiritual experience.

The overarching hypothesis is that the illusion of God's presence comes from an innate neural model of mother, the purpose of which is to establish the infant's part of the mother-infant bond. Anomalous activation of this circuitry in the mature brain gives rise to a specific kind of spiritual longing and the expectation that a primordial savior exists. If this idea is right, then some of our most powerful spiritual feelings and experiences arise from the neural circuitry of neonatal emotion and expectation, and this insight leads to testable predictions. The desperate helplessness that precipitates the comforting sensation of God's presence and the crying of a newborn for the comforting arms and breast of its mother should share a common neural mechanism. Similarly, the satisfaction and enfolding love experienced by an infant at its mother's breast and the feelings that typically constitute spiritual experience—unconditional love and ecstatic union with a divine presence—should also have a common neural substrate. The correspondence of these infantile and adult mental states should be evident in similarities of localization in the brain, of neuronal activity, and of sensitivity to drugs. If that correspondence cannot be found, then the hypothesis is wrong.

One strength of the hypothesis is that it suggests specific experiments not only in humans, but also in nonhuman animals. If religious experience is a uniquely human phenomenon, as is widely assumed, then we are limited to those experimental techniques that can be used with human subjects. If, however, the neurology of the sensation of God's presence is intimately related to the neurology of infantile crying and mother-infant bonding—behaviors that occur in nearly all mammals—then many new experimental possibilities appear, and a large body of existing research becomes relevant to the problem.

Part II is a journey through those parts of the brain where various aspects of the innate model of the mother appear to be implemented. The discussion comprises five chapters, each of which deals with one aspect of the innate model as follows:

- Crying
- Innate knowledge and feelings
- The appetitive longing for mother
- The sensed presence
- Short-circuit certainty

For each of these I describe evidence that associates the behavior or sensation with various brain regions. Where relevant I also discuss the role of these neural networks in some adult behaviors, such as maternal caregiving, adult sexual pair-bonding, or drug addiction. My goal here is to try to explain what these parts of the brain do and how they do it. Although it is mainly infantile experience that guides us to these places, the rationale for exploring them is that they are likely to be involved in spiritual emotion and experience.

The story that emerges is that there is probably not a specialized part of the brain dedicated exclusively to neonatal behavior. Instead it appears that the innate neural model of the mother is the seed around which crystallizes the *social brain*—those widely distributed neural networks that we use as adults to read the minds of others through facial expressions, body language, gestures, tone of voice, and other nonverbal cues and that generate socially appropriate emotions and behaviors in response.[1] The social brain includes areas in the prefrontal cortex that do not fully mature until adolescence and so have long been considered irrelevant to neonatal cognition and behavior. Recent evidence, however, paints a more subtle and complex picture in which even these late-developing regions embody innate knowledge that is accessible in infancy. This revised neurological view is consistent with behavioral and psychological evidence that the mother-infant bond is the template for adult social relationships.[2]

Throughout development, however, that neural seed probably remains intact and distinct, partly by virtue of the unusual sensory stimuli and conditions of infancy to which it is attuned, but also because of the way innate information is represented in the brain. It is this innate neonatal kernel of the social brain, activated in an abnormal adult context, that is the proposed neural substrate for the illusion of God's presence—the "trick of the brain" that Dan Barker found so strange and wonderful. When seen from this perspective, however, the illusion also appears remarkably similar to one that has long been familiar to neurologists: the phantom limb of the amputee, spawned by the expectation of the patient's brain that the missing limb should still be there. In chapter 7 I argue that the innate neural circuitry that expects the presence of mother can spawn a phantom divine presence in much the same way. Hence the title of the book: God is a phantom, conjured by a brain that expects a primordial savior to exist and compels that belief, both in infancy and, for many people, in adulthood.

Part III reexamines neurotheology from the new perspective developed in parts I and II, and it meanders into related research not specifically aimed at religious experience. Important themes include embodiment; the sense of agency; insights from neuropsychiatric illness; specialization of the left and right hemispheres; and the role of memory, expectation, and attention in conscious perception—both real and illusory. The book concludes with a discussion of predictions that may be empirically testable.

In my first book, I used the image of hiking to a high mountain summit and gazing at the landscape we had traversed as a metaphor for the new perspective on religion I had tried to present. In this book, the trail is steeper, the route is confusing in places, and the summit is higher. At times you may feel it is beyond your ability, but I urge you to persist. I warn you of the challenge ahead only to prepare and encourage you. You need not master every neuroscientific detail to see and understand the big picture. What you most need are the curiosity and courage to seek truth wherever that journey leads. Neuroscience is amazing, fascinating, and magnificent. On this trek, the summit experience is well worth the effort.

Part of what makes this book challenging is the complexity of the brain itself. There is no way to avoid this complexity and still do the subject justice. I shall therefore use conventional neuroanatomical terms for parts of the brain and their spatial locations. Some of these words have entered the popular lexicon, but most have not. I have tried to make it clear from the context that I am referring to a part of the brain (nucleus accumbens, orbitofrontal cortex, temporoparietal junction) or a spatial direction in the brain (rostral, caudal, medial, lateral, rostroventral) or a neurotransmitter or neuromodulator (dopamine, acetylcholine, endorphin) or an experimental method (fMRI, NIRS, EEG, MEG). It should be possible to follow the essential ideas without knowing exactly what all of these strange words mean, but if they make you uncomfortable, I recommend a careful reading of appendix A. It explains some basic neuroscience and includes links to helpful websites. Chief among these is Wikipedia, which often provides excellent overviews and illustrations when searched using the technical name of a part of the brain.

A few excerpts from this book appeared in condensed form in chapter 12 of *The Illusion of God's Presence*. I hope the repetition is more reinforcing than tedious. Also, feel free to skim or skip part I of this book if you already know that material.

I include cross-references to related ideas in different chapters of this book. Some readers of my first book objected to this, expressing frustration that I would tease them with hints of what was coming without simply making the point in one place. To reduce that frustration, I've tried to flesh out these connections with more complete descriptions, but there are good reasons for retaining the cross-references. Complex ideas that rise in conceptual layers are best explained sequentially, and that is how I write. But a good scientific theory must cohere throughout its structure. If an experimental result supports one aspect of the theory yet contradicts another, then something is wrong. Conversely, if a result supports not only the part it was intended to test, but also other aspects, then the whole edifice is strengthened. This is why I often point out links between seemingly disparate experiments or ideas scattered throughout the book. Theology, by contrast, is notorious for its lack of coherence.[3] For example, we are told that a loving, omnipotent, and omniscient God allows evil and suffering because he wants us to have free will, yet we are also told, in other contexts, that this omniscient God knows all that will happen in the future. This means that all future events are determined, even if known only to God, and therefore none of us—God included—can have free will.[4]

Finally, I write in a time when religious groupthink has seized a significant minority of Americans and, through their disproportionate influence, much of the American government. *Seizure* is the right concept, as a grand mal episode of hyperpartisanship leaves our nation aimlessly convulsing in the face of real and serious problems. Religious groupthink is not the only factor behind our political polarization, but it is one of the most difficult to fix. Elsewhere I have written about how science and mathematics can help us with the more easily remedied defects in our electoral system.[5] This book is not about politics and it won't save the world, but it may at least shed some light on one of the most vexing puzzles of human nature: our strange compulsion to believe in God.

Part I

BEHAVIORAL FOUNDATIONS

CHAPTER 1

Why Is God Two-Faced?

> [God] has a special place full of fire and smoke and burn-
> ing and torture and anguish where he will send you to live
> and suffer and burn and choke and scream and cry forever
> and ever till the end of time—but he loves you!—George
> Carlin[1]

George Carlin's joke is funny because it's true—not about *God's* nature, of
course, but about ours. Why do believers tell you their God will damn you
to hell, yet in the same breath insist that he loves you unconditionally? How
can Christians not see this blatant contradiction at the core of their theology?
How can a Buddhist monk, who espouses compassion and the universality of
human interconnectedness,[2] fan the flames of religious violence and ethnic
cleansing in Myanmar?[3] How can Muslims read that Allah is an all-forgiving,
all-merciful *dispenser of painful punishment* without batting an eye?[4] The two
faces of God show up in every religion, albeit with differing emphasis on his
cruelty or love. That most believers are blind to the contradiction strongly suggests that the dichotomy arises from distinct and powerful human intuitions,
not careful reasoning. That the intuitions are ubiquitous across human cultures suggests they are innate—as do twin studies, which show that religiousness is significantly heritable.[5] And if our tendency to believe in two-faced gods
is a product of our evolution, then those faces probably correspond to two
distinct selective pressures that shaped the evolution of human nature. What
are those selective pressures, and how have they made us susceptible to specific
kinds of religiousness?

Religion and Our Social Nature

Most of the scientific study of religion has emphasized the judgmental side of God, and the prevailing view is that the selective pressure behind this dimension of religion has to do with the great complexity of human social behavior relative to that of our more apelike ancestors. I refer to this selective pressure as the social root of religion.

Humans evolved in tribes that lived by hunting and gathering. This way of life requires cooperation—for example, in hunting large prey or fighting off competing tribes—and this presents an evolutionary puzzle. In shaping human nature, natural selection should favor a gene that produces behaviors that ultimately yield more copies in the population *of that gene*. This is what Richard Dawkins was referring to in the title of his famous book, *The Selfish Gene*.[6] That's the puzzle: how can "selfish" genes produce individuals that show unselfish behavior toward other individuals? As Dawkins explains, evolution has found several different solutions.

The answer is simple in social insects, where individuals in a colony are all closely related. A gene that makes an individual forgo reproduction to support the colony ensures the reproductive success of identical copies of itself in other individuals.[7] Yet this cannot explain the altruistic grooming behavior of many primates, which go to great effort to remove parasites from unrelated members of their social group. In this case, altruism is favored by natural selection if it is reciprocal and includes some mechanism for enforcing the reciprocity. For apes and monkeys, the enforcement mechanism is simple: you scratch my back, I'll scratch yours. The exchange happens right away.[8] But this is seldom the case in a tribe of cooperating humans, where many unrelated individuals must work together, often making unequal contributions for rewards that are not immediate.

This is called *delayed* reciprocal altruism, and making it work takes effort. Each individual of the cooperating group must know and recognize every other member of the group, observe and remember acts of cooperation and cheating, and somehow reward the former and punish the latter. This requires significant brain power for perception and memory, which explains why delayed reciprocal altruism is relatively rare in nature.[9] In humans it depends in part on moral intuitions that operate through powerful emotions: pride and pleasure when one cooperates, affection and gratitude when one is helped by others, guilt when one cheats, and outrage when others cheat.[10] Experiments in developmental psychology show that infants, in some cases as young as six months, can feel guilt, embarrassment, and empathy, strongly suggesting that the essential core of the moral intuition system is largely innate.[11] Reciprocal altruism in humans also depends on other mental systems, including good long-term memory for salient events, an astonishing capacity for recognizing human faces, and an uncanny ability to perceive the mood and intentions of others through subtle cues like tone of voice, body language, gestures, and facial expressions. Although the human brain is highly specialized for all of this, the larger the tribe, the harder it is to keep track of who cooperates and who cheats. As the size of the group increases, at some point delayed reciprocal altruism breaks down.

This is where the social root of religion comes in. If members of the group all believe in omniscient gods or spirits who know whether or not we cheat and who reward cooperation and punish cheating, then this relieves some of the computational burden that reciprocal altruism otherwise demands. It is easier to observe who in the group shows devotion and fealty to the gods than it is to keep a detailed account of individual acts of cooperation and cheating for each member. Religious devotion becomes a reliable proxy for proof of cooperation, but only if the religion demands costly, hard-to-fake sacrifices as proof of one's belief. The image of God as cruel, punishing, and tyrannical naturally emerges from such a mechanism—as does his lust for sacrifice.[12]

The intense selective pressure for hypersociality in humans has apparently made us prone to believe in this kind of god, but there is another wrinkle to the story. Cruel and punishing tyrants who demand costly sacrifice are not merely figments of our religious imaginations. *Human* tyrants dominate the chronicles of human history, much as they dominated their followers, and their personalities so closely match those of the cruel gods that they probably served as models for the most fearsome deities of ancient mythology. And like the gods of the social root, their personalities were also likely shaped by evolution.

Today we can see this connection most clearly in cults. By *cult* I mean a group of devoted followers led by a charismatic and exploitative leader, a social structure in which ties to friends and family outside the group are discouraged, control is enforced through fear of ostracism, loyalty of members must be repeatedly demonstrated through costly and continually escalating sacrifices, and defection from the group is extremely difficult or even life threatening. To most outsiders, cults seem so oppressive, humiliating, and dangerous that their existence is utterly bewildering. Why would anyone join a cult? My answer, of course, comes from the foregoing discussion of human sociobiology. Throughout our vast prehistory of cooperative hunting and gathering, evolution has shaped human nature to make confident, charismatic, domineering people appealing as leaders. More than that, it has sprinkled genes across our population in such a way that a few of us have a heritable tendency to be such leaders, while many more have a heritable tendency to fall under their spell.

Central to this idea is the observation that cult leaders typically have narcissistic personality disorder (NPD),[13] a psychopathology that comprises a strange mix of traits: a grandiose sense of importance and entitlement; boasting and pathological lying; impulsive self-indulgence; insatiable craving for attention and admiration; cruelty and lack of empathy; a lust for sacrifice by others as proof of their devotion; fits of rage when demands are not met; shallow and exploitative relationships; and secret feelings of insecurity, shame, and humiliation.[14]

Two studies strongly suggest that narcissism in humans is not only a product of evolution, but also, in some contexts, an adaptation. The first was a twin study that examined eighteen dimensions of personality disorder. Of these, narcissism was the most heritable, with 64 percent of the variance in this trait attributable to genetic factors.[15] The second study measured the prevalence of NPD in a sample of more than thirty-four thousand American adults. It found that 6.2 percent of the sample had NPD at some time in their adult lives, with a significantly higher rate for men (7.7 percent) than for women (4.8 percent).[16] This is about ten times higher than the lifetime prevalence of schizophrenia.[17] The authors noted that many NPD subjects have the symptoms only during adolescence or early adulthood, but the data suggest that at least half have a more severe and persistent form of the condition that lasts a lifetime. That NPD is both common and heritable suggests that it is adaptive. So does the success, both financial and reproductive, of many high-functioning narcissists who often excel in attention-grabbing careers like acting, preaching, politics, or business leadership. But if NPD contributes to reproductive success, then why are we not all narcissists? And if it is not an adaptation, then why does it persist at such high prevalence?

The answer to this puzzle may lie in a subtle facet of evolutionary theory called frequency-dependent selection, in which two distinct, interacting, and significantly heritable behavioral variants emerge in a population and persist at a stable ratio to one another.[18] The ratio is stable because a decrease in the number of individuals of one variant increases the fitness of that behavioral type and decreases the fitness of the other. The concept is most easily grasped through example,[19] and I describe several in chapter 12, where frequency-dependent selection proves helpful in understanding the differential specialization of the left and right halves of the brain.

But for understanding cults and narcissism, the most analogous and illustrative example is a familiar lament from office politics: "Too many chiefs and not enough Indians." Being a malignant narcissist pays off in evolutionary terms only if there are enough people around you who are seduced by your charisma, submit to your dominance, and make sacrifices to prove their loyalty. It doesn't work if everyone is a narcissist or even if most people are. It also doesn't work if we have all Indians and no chiefs. Evolution appears to be selecting for both, but only in the right proportion: few leaders and many followers. This means that the narcissistic personality is only half the story of the cult phenomenon.

The other half is a complementary personality type, the kind of person who is easily seduced into following a narcissist. From my reading of the literature on cults, there appear to be at least three salient traits that characterize this "follower type": a powerful need to belong; intense fear of ostracism; and a strong desire, or at least willingness, to sacrifice for a cause greater than oneself. If evolution has selected for distinct leader and follower types as the basis of human tribal behavior, then we should expect these follower traits, like the narcissistic ones, to be significantly heritable, and twin studies confirm this.[20] But if human cultishness is a product of evolution, and evolution cares only about reproductive success, then what accounts for such colossal reproductive failures as the mass suicides of Jonestown and Heaven's Gate? Something must be wrong with this picture.

What's wrong is that modern cults are not tribes of hunter-gatherers. They are an aberration, a strange and pathological amalgam of aboriginal and civilized life in which a self-serving, exploitative leader purges from the group any member who shows the slightest trace of assertiveness or disloyalty. Extant hunter-gatherers restrain dominant leaders and depose them if necessary,[21] just as coalitions of subordinate male chimps sometimes dethrone the alpha male.[22] In self-destructive cults, the leader's escalating lust for costly, hard-to-fake sacrifices runs freely to its inevitable conclusion.

In cult leaders we see a distillation of the cruel, authoritarian, and judgmental side of the two-faced god, the god of the social root. Surprisingly, however, we also find in cults a concentrated manifestation of god's unconditional love in equal proportion. "Love bombing" is the technique most often used to recruit new members: the initiate is overwhelmed with praise, flattery, kindness, and generosity.[23] And just like the two-faced god of religion, the cruel and fearsome cult leader is simultaneously seen as an unconditionally loving parent to his helpless, dependent followers. Cult leaders deliberately strip their followers of wealth and familial connections to make them helpless and dependent. The most extreme example was Jim Jones, who isolated his followers in a compound deep in a South American jungle. Yet even the overworked and fearful residents of Jonestown saw their leader as an unconditionally loving father, affectionately addressing him as "Dad."[24] What is the selective pressure behind *that* side of the two-faced god? A helpful clue can be found in what is probably the most atheistic demographic group: elite scientists.

Religion, Infancy, and the Illusion of God's Presence

Argument from authority and truth by revelation are anathema to scientists, who seek truth through reasoned arguments based on empirical evidence from nature. Not surprisingly, most scientists are repulsed by religion, especially the dogmatic and authoritarian aspects of its social dimension. The skepticism is greatest among elite scientists, like members of the National Academy of Sciences in the United States or Fellows of the Royal Society in the United Kingdom, of whom only 7 percent and 5 percent, respectively, believe in a personal God who answers prayer.[25] As Neil deGrasse Tyson has pointed out, however, the important point here is not that these percentages are so low. The astonishing thing is that they are not zero.[26] Why do roughly 6 percent of the most accomplished scientists in the world believe in a personal God who answers prayer? As Neil put it, "There's something else going on that nobody seems to be talking about," something mysterious and alluring about religion that can seduce even the most brilliant of scientific minds. He went on to argue that, if we really want to understand religion, these are the people we should be studying. Whatever it is that can make believers out of them must be the most concentrated essence of the thing we are trying to understand.

Luckily a few of these elite religious scientists have written about their beliefs, one of the most prolific being Alister Hardy, a marine biologist and Fellow of the Royal Society who died in 1985.[27] Hardy did not accept the dogma of any formal religion,[28] but instead believed in a benevolent God whose presence he had personally experienced. These mystical encounters began during his childhood, especially when he was exploring nature:

> From very early days I was a keen naturalist, and, when out on country walks by myself looking for beetles and butterflies, I would sometimes feel a presence which seemed partly outside myself and, curiously, partly within myself. My God was never "an old gentleman" out there, but nevertheless was like a person I could talk to and, in a loving prayer, could thank him for the glories of nature that he let me experience.[29]

Hardy was so moved by these experiences that he undertook a scientific study of them late in his career. Like any good naturalist exploring new territory, he began by collecting and categorizing specimens: accounts of mystical experiences he and his colleagues solicited from the general public through requests in the popular press. As of 1979 he and his collaborators had identified ninety-two distinct characteristics for classifying the reports.[30] For the first three thousand reports collected, the seven most common characteristics were:

1. Initiative from within self, response from beyond; prayers answered
2. Sense of security
3. Sense of joy
4. Sense of presence

5. Sense of certainty
6. Sense of purpose or new meaning to life as consequence of experience
7. Experience triggered by despair

I find this list revealing and important, but a feature of much lower resolution is also worth noting. The experiences seem to fall into two broad categories, one of which comes on gradually and persists, possibly for a lifetime. This may be an ongoing feeling of God's love and presence or just a vague sense that the universe is somehow mindful and benevolent. Hardy's personal account is one of these. The other broad category involves a sudden, dramatic, and transient experience of God's presence. Often these come in moments of personal crisis, despair, or helplessness. Item seven from Hardy's list captures this aspect, as does this anonymous report from his collection:

> Then, just as I was exhausted and despairing—I had the most wonderful sense of the presence of God. He was in a particular place in the room about five feet from me—I didn't look up, but kept my head in my hands and my eyes shut. It was a feeling of an all-embracing love which called forth every ounce of love I had in me. It was the tenderest love I have ever encountered and my sins were blotted out completely.[31]

Note the powerful emotions and the spatial specificity of the description: in a particular place, at a specific distance. This is the kind of mystical experience that may save a hopeless addict, comfort the bereaved, assuage crushing guilt, or repair what is sometimes called "spiritual brokenness."[32] Many people claim to have felt the presence of a loving God in this way, and an innate longing for and receptivity to God is sometimes described as a "God-shaped vacuum" in the human heart.*[33]

I do not question the reality of these feelings; I've had them myself.[34] Unlike Hardy, however, I just see them as a fascinating puzzle of human nature that cries out for a scientific explanation, not evidence of anything divine. I think the feeling of God's presence is an illusion that has a completely natural explanation, one that has to do with the peculiar constraints of human infancy, and I have formalized this idea as a hypothesis that draws heavily on my background in ethology. It not only accounts for six of the top seven items in Hardy's list,† but it also identifies the selective pressure behind the unconditionally loving side of the two-faced God.

---

* Francis Collins (2006) used the phrase, but it appears widely in Christian literature. It is commonly attributed to Pascal, who was deeply interested not only in God, but also in vacuums of the physical kind. He expresses the thought, if not the exact phrase, in *Pensées* (7.425).

† Item six in Hardy's list of characteristics is the one not specifically covered by my hypothesis.

My hypothesis has three parts:

1. A newborn infant has an innate neural model of its mother. This is not a detailed image of a specific person, but only a sense of the existence of some vague and amorphous primordial savior out there somewhere. More specifically, a newborn infant knows that its mother:

   - exists,
   - knows how to help the infant,
   - is able to help,
   - wants to help,
   - and will help when she hears crying.

   There is also some physiologically critical information about mother providing nourishment through nipples that can be recognized by scent and touch and taste; about her having a melodic voice of feminine pitch; and about her having a face, including some rough idea of what a face is and what it looks like.

2. The circuitry that implements this innate model of the mother persists into adulthood but normally lies dormant.

3. The existence of this circuitry in the adult can give rise to the illusory feeling of a loving presence, especially (though not necessarily) under conditions that mimic the helplessness of infancy. These feelings are often interpreted in religious terms because a crude mother-shaped vacuum is easily misperceived as a God-shaped vacuum.

This hypothesis is consistent with what we know about the newborns and hatchlings of many vertebrate species. Having some kind of innate knowledge is the rule, not the exception. A newborn wildebeest has an innate drive to stand and walk within minutes after birth, and it knows it must follow the herd in general and its mother in particular.[35] Sea turtle hatchlings know how and when to dig their way out of the nest, after which they orient to the brightest part of the horizon to find the ocean.[36] Another good example is the behavior of the herring gull chicks studied by ethologist Niko Tinbergen. They have an innate visual model of the parent's head and bill, including a prominent red spot near the tip. They peck at that spot to solicit feeding, and newly hatched chicks will do this without any prior experience. Tinbergen discovered that the chicks would also peck at artificial models of the bill, and if he exaggerated thinness, redness, and contrast near the tip, he could make a stimulus even more effective than the natural one. This is a classic example of what ethologists call a *supernormal stimulus*.[37]

In the same way, I suggest that God is a supernormal stimulus for the innate neural model of mother. When compared to the expectations of that model, God is a super-parent:

Mother	God
exists	feeling of certainty and sensed presence
knows how to help the infant	omniscient
is able to help	omnipotent
wants to help	perfectly loving
and will help when she hears crying	answers prayer

These exaggerated attributes of God excite that crude model of a mother even better than a real mother does, and when it is excited, the innate certainty of the mother's existence feeds the certainty of God's existence—often precipitating the illusion of his (or her) sensed presence. This, I suggest, is where the unconditionally loving side of the two-faced God comes from. Of course, the selective pressure behind this—which I call the neonatal root of religion—is directed not at an adult's connection with God, but at an infant's attachment to its mother. That it makes adults susceptible to belief in a loving God appears to be a side effect or by-product of its role in assuring the survival of helpless neonates.*[38] It gives rise, not only to the image of God as unconditionally loving, but also to a rich vein of infantilism† that permeates the doctrines, hymns, liturgies, prayers, art, myth, and sacred scriptures of the world's religions. Christians believe they must be born again—become infants again—to enter the kingdom of heaven,[39] and they raise both arms in fervent prayer, like toddlers wanting to be picked up by Mom.[40] Hindus offer similar prayers of supplication to the maternal goddess Shakti.[41] In a fascinating role reversal,[42] Sufi Muslims defang the god of the social root by venerating him as a helpless infant who must be nurtured in the heart of every believer.[43] Hindus do the same in their worship of baby Krishna,[44] as do Christians with their images and figurines of baby Jesus.[45]

---

* For a good explanation of the concept of an evolutionary by-product, see Gould and Lewontin (1979). A plausible case can be made that belief in the God of the social root is, or once was, adaptive as part of the mechanism of delayed reciprocal altruism and social cooperation in humans—though there is no general consensus on this point (Sosis 2009). If it turns out that belief in the God of the social root is adaptive, then we cannot easily dismiss the God of the neonatal root as a mere evolutionary by-product, because the two roots are deeply intertwined. If there was selective pressure for theistic belief as a mechanism for enhanced social cooperation, then anything else that contributes to belief in God, like the innate neural model of the mother, would come under indirect selective pressure as a consequence of that contribution.

† Unfortunately, the word *infantile* often has a pejorative tone. When I use the word to describe religious rituals, ideas, or imagery, I mean it only in a strictly neutral and literal sense: *of or having to do with infancy.*

Interestingly, the perceptive Alister Hardy recognized the infantile nature of his own mystical experiences. He suggested they were

> an "extra-sensory" contact with a Power which is greater than, and in part lies beyond, the individual self. Towards this, whatever it may be, we have a feeling, no doubt for good biological, or psychological, reasons (linked with the emotions of an early child-parent affection, but none the worse for that) of a personal relationship, and we can call it God.[46]

He argued that this connection with infancy "is not necessarily destructive of the idea of a theistic relation; on the contrary I believe it binds our natural theology more closely to the biological system and so gives it a more rational validity."[47]

Of course I disagree with him on that last point. By calling his project "natural theology," Hardy was echoing William Paley's pre-Darwinian desire to imbue theistic belief with an aura of scientific rationality,[*][48] despite the fact that Hardy fully accepted modern evolutionary theory and rejected the so-called argument from design.[49] It was the feeling of a personal relationship that Hardy was trying to explain, one that to him felt like the bond between parent and child. Yet rather than follow that completely naturalistic path, one already paved by biologists and psychologists, Hardy invoked extrasensory perception and the supernatural. The problem with this is that it is merely a god-of-the-gaps argument: we don't understand these feelings, therefore God did it. It explains nothing, makes no testable predictions, suggests no experiments, cannot be falsified, and is therefore scientifically worthless. How could such an accomplished biologist go so far off the rails of the scientific method?

I suggest that Hardy was led astray by a cognitively impenetrable illusion,[50] one so compelling that even a complete understanding of how it works is not enough to make it go away: the illusion of God's presence.[†][51] The feeling of certainty that accompanies the experience—along with the eerie, ineffable, and nonvisual sensation that another being is nearby—makes the illusion *feel like evidence* of something real. Apparently Hardy, like many other people, could not resist taking it that way.[‡][52] Because this intuitive sense of certainty can somehow

---

* For a lucid critique of Paley's argument written by one of Alister Hardy's students, I recommend Dawkins (1986).

† For a more detailed exposition of this argument with respect to Francis Collins, another elite religious scientist, see Wathey (2018).

‡ An alternative explanation commonly offered for the religiousness of elite scientists like Alister Hardy is that they must have had unusually strong religious indoctrination as children and so retained their faith into adulthood. The evidence, however, does not support this view. The study of Fellows of the Royal Society found that childhood religious upbringing was not significantly related to their religiousness as adults (Stirrat and Cornwell 2013).

avoid a scientist's normal path to truth—the long and laborious route through reason, evidence, and hypothesis testing—I call it short-circuit certainty.[53]

Attachment to Mother and God

Of course Hardy was not the first to notice the infantile nature of spiritual experience. Freud pointed it out more than a century ago,[54] though his explanation of religious infantilism in terms of wish fulfillment and the Oedipal complex was nearly as untestable as Hardy's, and—to the extent it could be tested—it mostly failed.[55] By contrast, Joseph Campbell, in his monumental study of religious mythology, *The Masks of God*, offered a truly scientific interpretation of religious infantilism that resonates with my own. He argued that there is a deep connection between the appeal of images of the gods, an infant's innate fascination with faces, and the innate animal behaviors studied by ethologists.*[56] But by far the most productive research into the infantile dimension of religiousness has grown out of Lee Kirkpatrick's attachment theory of religion.[57] At first glance, this theory might appear to be a competing alternative to my hypothesis of the innate model of mother as the basis for religious infantilism, but in fact the two approaches nicely complement and reinforce one another.

Kirkpatrick sought to explain the infantile aspect of religion using the attachment theory of child development pioneered by John Bowlby and Mary Ainsworth.[58] This theory had been extensively tested and refined, through decades of empirical research, by the time it attracted Kirkpatrick's attention. Central to Bowlby's theory is the idea that early attachment experience between a child and his primary attachment figure (typically the mother) shapes the child's internal working models of the self and of others. A securely attached child—one whose attachment figure is both sensitive and responsive to the child's needs—sees himself as worthy of love and affection and sees others as loving and trustworthy. By contrast, if the attachment figure is not consistently sensitive and/or responsive to the child, the child will develop one of several forms of insecure attachment. This may result in a model of the self as unworthy of love or a model of others as unreliable, untrustworthy, or threatening or some combination of problems in both of these internal models. Another central concept is that attachment behaviors are not themselves instincts, though the biological mechanism behind them is innate and highly adaptive, especially in the context of the hunter-gatherer existence of our distant ancestors. Although the motivation to form attachments in infancy is innate, the resulting pattern

---

* Although Campbell was not specifically trying to explain the illusion of a sensed presence, in a remarkable coincidence, he introduced his ethological interpretation with a description of the behavior of sea turtle hatchlings—just as I did before I had read Campbell.

of attachment behavior is adaptively shaped by the infant's environment during the first few years of life, and the most critical aspects of that environment are the responsiveness and sensitivity of the attachment figure. And although adults seldom need attachment in the way an infant does, adult romantic relationships, caregiving behavior, and social behavior in general are heavily influenced by internal models of self and other and so to that extent are also products of the attachment system.[59] For Lee Kirkpatrick, that last thought raised an interesting question. Is an adult believer's relationship to God an attachment relationship, in the sense that Bowlby and Ainsworth used the term?

Kirkpatrick makes a strong case that it is, citing many examples of infantilism in religion and emphasizing compelling analogies with attachment behavior in small children.[60] During the last few decades, he and his colleagues, along with Per Granqvist's group in Sweden, have tested the attachment theory of religion using subjects from diverse religious and cultural backgrounds. With some adjustment and refinement along the way, the theory has held up well. Kirkpatrick's extension of Bowlby's theory retains the original emphasis on internal working models of self and other, but with the "other" category enlarged to include God as an attachment figure. Unlike my proposed innate model of the mother, however, the internal models hypothesized in attachment theory are *learned* during a critical period in early childhood and thereafter are resistant, though not impervious, to change.[61]

One of the most consistent findings from the empirical studies of the attachment theory of religion concerns a subtle interaction between a subject's security of attachment and the religiousness of the subject's parents, as summarized in figure 1.1.[62]

	Religious parents	**Nonreligious** parents
Secure attachment to parents	*More religious*	*Less religious*
Insecure attachment to parents	*Less religious*	**More religious**

Attachment theory can't completely explain this group.

See themselves as unlovable

See others as untrustworthy

Prone to sudden religious conversion

Figure 1.1. Securely attached subjects have religiousness similar to that of their parents; the insecurely attached tend to have religiousness opposite to that of their parents. *After Fig. 5.4 of Kirkpatrick (2005) p 113 (annotations added).*

Subjects who remember a secure relationship with their parents (top row) tend to have religiosity similar to that of their parents. Secure attachment can therefore predict either high or low religiosity, depending on the beliefs of the parents. By contrast, subjects who remember insecure parental attachment (bottom row) tend to have religiosity opposite to that of their parents.[63]

Intriguingly, the same studies reveal a very different relationship between attachment and religiosity when they examine the rate of change of religious belief in the subjects. Those who report sudden religious conversion are more likely to recall insecure attachment with their parents, whereas those whose faith developed gradually tend to remember secure relationships with parents. These correlations hold in the opposite direction as well; that is, having an insecure attachment history significantly predicts sudden religious conversion rather than gradual religious change.[64] The quality of religious experience also figures into this picture in a significant way. The subjects who reported sudden religious conversions often had them during times of great emotional stress or crisis, and their perceived relationships with God were more emotional in nature than those whose religious beliefs developed gradually. Conversely, subjects who became religious without sudden conversion reported greater concordance between themselves and their parents with respect to religious affiliation, values, beliefs, and degree of religious involvement—all indicators of religiosity based on socialization to parental norms.[65]

The results just described support two distinct but complementary hypotheses in the attachment theory of religion. The correspondence hypothesis holds that secure attachment in childhood leads to positive internal working models of the self as worthy of love and of others as reliable, loving, and trustworthy. These persist into adulthood as templates for other social relationships and therefore serve as foundations for a potential relationship with a God perceived as close and loving. They also serve, during childhood, to facilitate socialization of the child to parental norms, including religious ones. The result is a tendency, in individuals with secure attachment histories and religious parents, for religiosity to match that of the parents and to develop gradually throughout childhood and adolescence (upper left cell of figure 1.1). In insecurely attached individuals whose parents are religious, however, the negative models of self or other, combined with poorer socialization to parental norms, usually lead to lower religiosity in adulthood (lower left cell).[66]

By contrast, the compensation hypothesis comes into play in the most interesting group of insecurely attached individuals: those who become highly religious later in life. These tend to have less religious parents, negative models of self and other, and relatively poor socialization to parental norms (lower right cell). The compensation hypothesis proposes that, in adulthood, especially during times of great emotional stress or crisis, these individuals are prone to emotionally based, sudden religious conversion, in which God is seen as an ideal substitute attachment figure. The new relationship with God compensates for the insecure models of self and other that ultimately derive from the lack of parental sensitivity and responsiveness in childhood.[67] The process is much like the compensatory behavior of most toddlers—including secure ones—who seek comfort from alternative caregivers, teddy bears, or blankets when the primary attachment figure is absent.[68]

There is, however, a serious problem here. Attachment theory cannot completely explain the religious conversion of the individuals just described (lower right cell). The compensation hypothesis holds that insecurely attached adults turn to God as a substitute attachment figure in times of crisis, thus precipitating sudden religious conversion. *But why would they do this?* These are people who see themselves as unworthy of love and others as unloving, unreliable, and untrustworthy. Yes, God is often advertised as unconditionally loving, but how and why would an insecurely attached adult, with mental models to the contrary burned deeply into his unconscious mind, believe the sales hype?

One possibility is that compensation is a higher-level, conscious, and deliberate process—one in which, through sheer effort of will, the rationalizing conscious mind forces belief in a loving God onto a reluctant unconscious mind. The problem with this idea is that it runs counter to nearly every description of the subjective experience of sudden religious conversion. Psychologists William James and Edwin Starbuck, for example, emphasized that it is the abandonment of willful effort, along with a highly emotional and intuitive experience of unconscious origin, that most distinguish sudden religious conversion from the more gradual form.[*69] This general pattern was also found in a more recent study that compared forty sudden religious converts to thirty age-matched religious controls who had not experienced sudden conversion.[70]

Well aware of the highly emotional nature of sudden religious conversion, Kirkpatrick likens it to the overwhelming experience of falling in love, with God taking on the role of an idealized romantic partner. He describes many similar comparisons that have appeared in the psychological and religious literature, which together hint at a deep connection between the human mechanisms for infantile attachment and adult sexual pair bonding.[71] I further explore this important connection in chapters 2 and 5.[72] Here I only point out that, by itself, this analogy to romantic love still does not explain

---

* See lectures 9 and 10 of James (1902) and chapter 8 of Starbuck (1911). Starbuck's table 8a separately categorizes the sudden conversions of 102 female and 72 male subjects into five bins of various relative contributions of conscious and unconscious processes, judged subjectively from the subjects' written descriptions of their conversion experiences. The data are skewed toward unconscious or "automatic" processes for both groups but more heavily so for the females.

why some insecurely attached adults surrender themselves to a loving God in times of crisis; they are just as insecure in their sexual relationships as in any others. In wrestling with this problem, however, Kirkpatrick makes the intriguing suggestion that "*something* happens to a sizeable minority of these individuals later in life that in effect activates an otherwise dormant love mechanism."[73]

This is where the idea of an innate neural model of mother as the basis of sudden religious experience meshes neatly with the attachment theory of religion. The innate model of the mother assumes the presence of a loving, sensitive, and responsive attachment figure. In effect, it biases every newborn toward secure attachment, regardless of the real mother's performance as caregiver. An infant who receives insensitive or unreliable care eventually learns an alternative model of the mother, one that normally inhibits and takes precedence over the innate one. Yet even in these insecurely attached individuals, the innate model persists as dormant neural circuitry that can be activated in adulthood by conditions that mimic the extreme stress and helplessness of early infancy. This, I suggest, is what happens in compensation, when a distressed adult with an insecure attachment history suddenly feels compelled to seek unconditional love and finds it in an ineffable sensed presence.[74]

The idea of the innate model of mother does more than merely fill a hole in the attachment theory of religion. It links the scientific study of religion to disciplines previously considered irrelevant, like the perceptual and cognitive abilities of human newborns and the biology of mother-infant bonding in nonhuman mammals. It shines the light of ethology on human religious experience, revealing deep connections to many complex and highly specialized animal behaviors that sprout from a foundation of innate knowledge and flourish through learning. In this light, the longing of a human infant for her mother is like the sea turtle hatchling's longing for the ocean. More to the point, the illusion of God's loving presence, felt by a woman in despair, is like the misleading allure of artificial lights on beachfront homes to an adult female sea turtle trying to find the ocean after laying her eggs.*[75] In both cases, a crude and innate model of reality is hyperactivated by a supernormal stimulus. From this perspective, religious experience appears not to be as uniquely human—or as supernatural—as once thought. For one of my astute readers, however, this ethological interpretation raised a fundamental concern about my hypothesis, and the question he asked merits special attention.†

---

* Disorientation due to artificial light kills far more sea turtle hatchlings than adults, but nesting adults are also sometimes lured to their deaths by beachfront lighting (Oberholtzer 2014).

† I thank Adam Lee for this question and many helpful suggestions.

Isn't This Illusion Maladaptive?

If a newborn has an innate neural model of the essence of mother, then why should the illusory sense of presence of this being—with the attendant feeling of comfort—be adaptive? Would it not make the infant stop crying for its *real* mother?

The innate model gives the infant a rudimentary understanding of what its mother is and does, along with the expectation that she exists, is nearby, and will respond to its cries. The expected sensation, however, is physical contact, movement, warmth, and a nipple in the mouth. The transition from longing to satisfaction happens only with those stimuli.

There may, however, be an important intermediate mental state between the stressful and painful state of helpless isolation and the blissful state of enfolding love at the mother's breast. Studies of pleasure-seeking behavior in many contexts reveal that it comprises two qualitatively different phases: appetitive (wanting the reward before it is attained) and consummatory (liking the reward itself).[76] When the stress of a human infant crying in isolation reaches some critical level, it may trigger a transient but illusory sense of maternal presence, which could switch the infant's mental state from social pain to appetitive pleasure seeking. This could have the paradoxical effect of simultaneously reducing the stress response while maintaining or even increasing the intensity of crying.

There is evidence for this idea in the behavioral and hormonal responses of infant monkeys that have been separated from their mothers. When an infant rhesus or squirrel monkey is totally isolated from its mother in an unfamiliar cage, it cries vigorously, and the concentration of the stress hormone cortisol increases in its bloodstream. If, however, the isolated infant can sense the mother's presence—by seeing her in a separate cage in the same room, by seeing her in another room through a window, or merely by detecting her voice and scent from the other side of an opaque barrier—then the infant's response is surprisingly different. Its stress response, as measured by cortisol concentration, is much less than in total isolation, but its rate of isolation crying is much greater.[77]

The scientists who did these experiments interpret infantile isolation crying not merely as a signal to the mother, but also as a coping mechanism for the infant. For example, when an infant squirrel monkey is totally isolated, its rate of crying rapidly increases and reaches a peak after about one hour. By that time the infant seems to begin losing hope that its mother will respond, and its rate of crying gradually declines over the next several hours. Meanwhile, however, its cortisol levels steadily rise throughout the first twenty-four hours of separation. The investigators suggest that the stress response is triggered mainly by uncertainty, that crying is an infant's way of exerting some control over its uncertain situation, and that the sensation of the mother's presence greatly reduces the infant's feeling of uncertainty, even if she remains out of reach.

Apparently something similar happens in a variety of rodents. Infantile isolation crying is accelerated by a brief sensation of the mother's presence—a response called maternal potentiation—though the effects of this on stress hormones have not been measured. I examine maternal potentiation in some detail in chapter 5.[*78]

These examples show that an innate expectation of the mother's presence could help an infant cope with separation and reconnect as soon as possible. But there is also a deeper sense in which the innate neural model of the mother is adaptive. It has to do with the way humans perceive and learn about the world around them. As with many other animals, much of human learning occurs through reinforcement. Sensations and behaviors that bring pleasure are rewarded and reinforced. Those that bring pain are remembered and avoided, if possible. This strategy cannot work, however, unless the brain has some innate knowledge of stimuli that are intrinsically rewarding or punishing—the so-called primary reinforcers. It is through association with these that new experiences acquire value, and this process is the essence of reinforcement learning. As I explain in chapter 4, an infant's innate model of its mother associates her presence with pleasure and her absence with pain, thus providing primary reinforcers essential to the learning of human social behavior and morality.

Similarly, innate knowledge may play an equally vital role in sensory perception. The human brain does not merely take in sensations and interpret them. It is biased to perceive some things better than others, and in familiar contexts it expects to see some things rather than others. Perception appears to be a fluid and iterative process in which the brain is continually generating multiple guesses about what is present in the environment. It estimates a likelihood of correctness for each of these competing interpretations, both from sensation and expectation. In this calculation, information flow is not only from the bottom up—from sensory areas to higher cognitive centers—but also from the top down, with higher centers biasing low-level sensation. Normally the system quickly settles into the most likely percept—the only one to reach consciousness—and we are unaware of the struggle among the alternatives or of the interplay between top-down and bottom-up information streams. Yet we know this is how perception works because cleverly designed ambiguous images, like the one in figure 1.2,[79] have revealed some of the action behind the scenes.

If you have never seen this image before, you will probably see only randomness in it, but that is an illusion. The image as whole does have meaning—vivid,

---

* Maternal potentiation behavior has been found in mice (Moles et al. 2004), rats (Shair 2007), guinea pigs (Hennessy et al. 2006), and prairie voles (Robison et al. 2016), though most of the research has been done using rats.

Figure 1.2. Is this a meaningful image or just random squiggles? *Copyright © 1957 by the Canadian Psychological Association Inc., reproduced with permission.*

striking, and surprising meaning that clearly conveys a specific emotion. Yet, even knowing this, you could probably stare at it for hours and never see it. Illusions like this happen when a neural model differs from reality. Often this happens because the brain takes shortcuts in the way it represents reality. These shortcuts embody useful assumptions that improve our perception and reproductive success under most conditions, but they can fail in unusual circumstances.[80]

To see the meaning in figure 1.2, try turning it upside-down.* It may also help to make the image smaller, so look at it from a few steps back. Even with these helpful suggestions, it may take some time for the perception to burst into your consciousness. When it does, it will come instantly and vividly. If you still cannot see the meaning, I will tell you what it is: the image is of a young woman's face in profile, partly in shadow, and looking to the right. Her mouth is open in an exuberant smile.

This is called a Mooney face, after psychologist Craig Mooney, who used images like these to study perception.[81] It is a grossly impoverished image, stripped of all color, shades of gray, and most spatial detail. By introducing it upside-down, I made reality differ from your neural model of a face, which evidently expects faces to be upright. This simple assumption by the brain greatly speeds our perception of faces, and, along with similar built-in expectations about the way faces are organized, it makes it possible for us to distinguish and recognize thousands of individual faces over the course of a lifetime. It also fills in the information missing from the stripped-down image of figure 1.2, making it possible to see the woman's face and her happy expression. This filling-in is a superb example of high-level information flowing from the top down during conscious perception. Although our neural model of a face is mostly learned, some of it, including the assumption that faces are always upright, is innate.[82]

For a newborn infant, innate sensory expectations of this kind confer the advantage of better perception of the environment—more sensitive, less prone

---

* If you're reading this on a screen that can't easily be turned upside-down, see figure 7.2 in chapter 7.

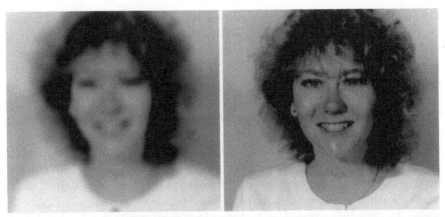

Figure 1.3. On the left, what a mother's face looks like to a newborn infant, with its still-developing visual system, from a distance of about thirty centimeters (twelve inches). On the right, how the same face appears to an adult at the same distance. *Copyright © 1998 by Alan M. Slater, reproduced with permission.*

to distraction, and more reliable—because they provide some rudimentary top-down processing capability for those stimuli that matter most. The most critical of these stimuli is shown in figure 1.3.[83]

The image on the left shows what a mother's face looks like to the visual system of a neonate, which is deficient in acuity,[84] contrast sensitivity,[85] color,[86] and depth perception,[87] relative to the visual system of an adult. The infant sees an impoverished image, rather like a Mooney face, so you can see how it would be helpful to a newborn to have some innate capacity for holistic face perception. I am suggesting they have more than that, especially in the cortex that will ultimately support complex social behavior in adulthood, but that in infancy supports attachment to Mom through an innate model of her as available, competent, and unconditionally loving. I revisit the subject of top-down perception in chapter 9, in the context of the illusion of a sensed presence in adults, and in chapter 10, where I try to make sense of hallucinations.

Finally, there is something unusual about the way that humans care for their infants that has relevance here. Human offspring are among the most costly to raise to self-sufficiency in the whole of the animal kingdom, yet our hominin ancestors produced them at double the reproductive rate of other great apes. This partly explains the rapid dispersal of our species during the Pleistocene, but it also means that ancient human mothers were incapable of raising their children alone.[88] Unlike our closest cousins among the great apes, human mothers do not hold their infants all the time. We can see this in hunter-gatherer cultures in which babies are nearly always held but not always by their mothers.

Grandmothers, fathers, aunts, and older siblings routinely help out when mothers need to search for food or juggle multiple children.[89] Sociobiologist Sarah Hrdy argues that this strategy of cooperative breeding has been at work in our hominin ancestors for hundreds of thousands of years, possibly all the way back to *Homo erectus*, and that it has deeply affected the mother-infant bond in humans.[90]

It has made a human mother's commitment to her infant more conditional than that of a chimp or bonobo mother, because a human mother must evaluate her prospects for support from others in providing for her baby. If those prospects look poor, and especially if her infant is small and sickly, she may do better by abandoning it, thus conserving resources for her other children while waiting for better times. Such behavior is almost never seen in other apes but is well documented in hunter-gatherers and other mammals that practice cooperative breeding. In humans the decision to abandon an infant usually occurs within a brief period after birth. Once lactation and suckling begin, hormonal and neural transformations in the mother's brain normally push her into total commitment.[91]

Presumably this state of affairs also exerts selective pressure on the physical, behavioral, and cognitive attributes of neonates. This may explain why human babies are born fatter than other primate newborns and why some young infants seem to cry excessively: both may be signals of the infant's health and vigor.[92] More to the point, however, it is one more reason—and an especially compelling one for the neonate—why human newborns are innately sensitive to social cues and endowed with knowledge of social primary reinforcers. In the aboriginal human condition, a newborn must make an emotional connection with its mother immediately after birth because failure at this increases the risk of abandonment.

Beyond those risky first days lie other special challenges that favor a sensory system biased with the expectation of a mother's presence. The demands on aboriginal mothers required that they sometimes put their infants down or, more likely, hand them over to a helper, but the mother remained the infant's primary source of nourishment and security. Unlike an infant chimp that is always in contact with its mother, a human infant must monitor its mother's whereabouts in a sea of other faces and voices. It may be comforted at a distance by subtle visual cues or soothing words that signal her presence, but it must be able to return a signal of distress when that presence vanishes or when hunger or thirst demand greater maternal attention.*[93] Hrdy argues that the effects of these selective pressures on the minds of infants instigated the rapid evolution of our

---

* This idea is fleshed out by Falk (2004) and Hrdy (2009, chapter 4), and it is not merely hypothetical. A one-month-old infant lying on a sofa spends significantly more time maintaining eye contact with its mother than when it is being held by the mother (Lavelli and Fogel 2002).

hypersocial human nature—especially our interest in the thoughts and feelings of others and our ability to read them from nonverbal cues.[94]

Similarly, the innate neural model of the mother is the developmental foundation of these abilities and, I suggest, of the spiritual longing in adulthood for that mystical *other*. But is there any empirical evidence for this innate model in newborn babies?

CHAPTER 2

Evidence for the Innate Neural Model of Mother

> The animal, directed by innate endowment, comes to terms with its natural environment not as a consequence of any long, slow learning through experience, through trial-and-error, but immediately and with the certainty of recognition.—Joseph Campbell[1]

> Neither in humans nor any other ape does the initial impetus to connect need to be learned. Rudimentary wiring for intersubjective engagement seems to be there.—Sarah Hrdy[2]

In recent decades a large body of research has discredited the view that the mind of a newborn is a blank slate.[3] Clever behavioral and neuropsychological tests have begun to reveal the innate foundation from which a human mind develops, but the field is new, controversies abound, and many questions remain unanswered. Here I briefly summarize the findings most relevant to the rest of this book.[4]

The Sense of Motion

At least four months before a fetus sees the light of day, it feels itself being moved. The vestibular system of the inner ear detects accelerations and the steady pull of gravity, and it is one of the first sensory systems to develop.[5] Vestibular signals appear to be necessary for the proper development of the circuitry that controls movement, not only before an infant can crawl, but even before it is born. Premature babies show significant delays in various aspects of neuromuscular development unless they are occasionally rocked in their incubators.[6]

But the vestibular system is far more important after birth, especially in its role in stabilizing the infant's visual world by cancelling out head movements with compensatory eye movements—an unconscious behavior called the vestibulo-ocular reflex. This is present in newborns and accurate enough at birth to compensate for the rapid head movements that occur during shifts of attention.[7] In adults, the vestibular system is also important in the perception of body movement and orientation with respect to gravity. This is essential for the sense of balance when standing and walking, especially in the dark, but it also has a less obvious but profoundly important role in the subjective connection between mind and body known as embodiment or body ownership, which I explore in chapter 6. Here I only point out that the vestibular system is sufficiently developed in neonates that it likely gives them some sense of body ownership and orientation, despite their inability to stand or walk.

Gentle, rhythmic vestibular stimulation is normally pleasant—hence the popularity of porch swings and rocking chairs—and nowhere is this more evident than in the response of crying infants to gentle rocking. A study based on reports from seven hundred breastfeeding mothers found that rocking or walking with a crying baby were highly effective calming techniques, surpassed only by holding and breastfeeding.[8] A smaller study that used more direct and quantitative measurements of crying behavior found that vestibular stimulation was significantly more effective at calming crying newborns than was mere physical contact.[9] But rocking the baby is more than just a convenient trick to ease the lives of parents. It is essential for the emotional well-being of the child. Romanian orphans raised in conditions of severe social deprivation rock themselves, as do infant macaque monkeys raised in isolation, apparently as a means of self-comforting.[10] By contrast, infant monkeys raised on artificial mothers that move rhythmically do not develop this pathological self-rocking.[11]

The significance of this may lie in the simple fact that most of the vestibular stimulation experienced by an infant signals the presence of another person—usually its mother, but always a larger, stronger, and more able person, one who can protect and nurture. The vestibular sense, and especially the association of gentle rocking with maternal presence, therefore probably constitutes a big piece of the innate model of the mother. If this is correct, then we should expect the onset of vestibular stimulation to increase the level of attention of a newborn, priming it to seek its mother through other senses, and this prediction has been confirmed. Newborns given vestibular stimulation are more attentive at tracking a moving target with their eyes than are unstimulated controls[12] and more likely to respond to the facial expressions of others.[13]

The Smell and Taste of Mother

Like all mammalian neonates, human infants are born with chemical senses and instinctive behaviors that direct them to the nourishment upon which their survival depends. If one breast of a mother is washed clean of its natural scent, and her newborn is placed prone on the center of her chest, the infant will preferentially crawl to and suckle from the unwashed breast.[14] Breast odor alone is sufficient for this discrimination: a newborn placed onto a warming table, with a breast pad just out of reach, is more likely to crawl to the pad if it bears the scent of its mother's breast than if it is clean.[15]

An important component of this olfactory signal comes from the secretions of glands in the skin surrounding the nipple, the areola. When this areolar scent is presented in isolation to a three-day-old sleeping infant, it inhales more deeply and responds with mouth movements related to suckling (pursing its lips or sticking out its tongue), even if it has been exclusively bottle-fed since birth. Infants respond in this way significantly more to the areolar scent than to various control odors, including water, cow's milk, formula milk, and human breast milk.[16] The odor of breast milk is not, however, a neutral stimulus. In a selective attention test, awake, two-week-old infants who have never experienced breastfeeding prefer the scent of breast milk to that of their formula.[17]

Human neonates are especially receptive to learning new odors during their first hour after birth, and this sensitivity greatly diminishes by twelve hours.[18] This postdelivery imprinting appears to be a consequence of the stress of uterine contractions, which puts the brain of a newborn into a state of heightened arousal and attentiveness.[19] Beyond that first hour, however, olfactory imprinting still occurs in human infants, but only in conjunction with reinforcing stimuli like breastfeeding, which is one of the most effective. Newborns exposed to the otherwise neutral odor of chamomile while nursing prefer chamomile to a novel control odor when tested two or three days later. They also show roughly equal preference for chamomile and the odor of breast milk when given a choice between them—unlike infants not previously exposed to chamomile, who prefer the scent of breast milk in this test.[20] The memory of an odor that has been associated with breastfeeding in this way turns out to be remarkably persistent. A long-term study of these infants found that the neural imprint of their neonatal experience with chamomile lasted at least until age twenty-one months, the oldest age tested. It was not, of course, a conscious memory, but instead an unconscious preference for objects scented with chamomile.[21] These experiments show that something about the experience of breastfeeding is innately rewarding; in the language of psychology, it is a primary reinforcer. Chamomile, by contrast, is a secondary reinforcer, because its perceived value as a reward in these subjects arises only through association with the unlearned primary reinforcer.

Rewarding Touch

Gentle touch is also innately rewarding for human newborns. In an elegant demonstration of this, one-day-old infants were given ten brief exposures to a citrus odor while receiving gentle massage. Control infants received only the odor, only the massage, or both stimuli but not at the same time. The next day the infants were tested with the citrus odor, and only the group that had previously experienced it while being massaged turned their heads toward the fragrance. The main point of the study was to demonstrate that human newborns can rapidly learn new odors through classical conditioning, but it also shows that gentle human touch is a primary reinforcer. Interestingly, although the person giving the massage stood at the head of the bassinet and tried to remain out of sight of the infant, several of the infants turned their heads as if trying to see who was stroking them.[22]

This makes sense, of course, if the feeling of being touched is one of several innate cues to the presence of another person—cues that excite and redirect the infant's attention to that person. Other studies are similarly suggestive. One that compared the efficacy of touch and rocking at calming crying infants found that each by itself was effective, though rocking more so.[23] In another, premature infants received fifteen minutes of gentle massage twice a day for fifteen days, which significantly reduced their stress hormone levels and increased their sleeping time relative to control infants.[24] A neuroimaging study of full-term infants a few days old found that touch elicits neural activity in broad areas of the cortex that extend well beyond the primary sensory cortex for skin sensations.[25] The excited areas include a region known to be important in adults for perceiving the presence, actions, and mental states of other people.* Although we do not yet know what this part of the brain can do in a newborn, the result is consistent with the idea that an infant in some sense understands that gentle touch comes from another person. And just as gentle touch appears to be an innate cue to the presence of mother, so does a special kind of gentle voice.

Motherese

Older children, adults, and especially mothers speak to infants in an unusual way—more slowly than with adults, using higher pitch and greater range, emphasizing vowels and significant words—all with prosody that borders on the musical. This is *motherese*, also more formally known as infant-directed speech.[26]

---

* I am referring here to the transition zone between the temporal and parietal lobes, the temporoparietal junction, which figures prominently in chapter 6.

It may have evolutionary roots that predate our species, as there are hints of it in the vocalizations of baboons and rhesus monkeys and in the infant-directed gestures of gorillas.[27] Motherese occurs in nearly all human cultures, and its prosodic features are similar even in greatly dissimilar languages.[28] There is even a purely gestural equivalent in the signing of deaf mothers to their infants.[29] Mothers with higher oxytocin levels tend to use more motherese, which suggests a role in emotional communication and bonding.[30] Motherese is more effective at getting an infant's attention than is adult-directed speech, and this preference is clearly evident in newborns.[31]

All of this strongly suggests that motherese is a highly adaptive signaling behavior supported by innate neural circuitry, both in its production by mothers and in its reception by infants. We cannot, however, rule out the possibility that prenatal learning contributes to a neonate's preference for motherese, because a fetus can hear and develop a preference for the sound of its mother's voice before birth.[32] Whether or not prenatal learning is involved, the brain of a newborn obviously has some kind of internal model of motherese, and several laboratories have tested infants with artificially simplified variants of it to try to identify its most salient attributes. The results vary with age over the first year, but one-month-olds prefer the real thing to several simplified versions.[33] This shows that the internal model is fairly complex and includes multiple acoustic features, including the high fundamental pitch and the natural pattern of low-frequency modulations that constitute prosody. It also raises the broader issue of the predisposition of the neonatal brain for language acquisition. Infants do not begin babbling until about five or six months of age, but studies of their responses to speechlike sounds show that their brains are highly attuned to human speech, even at birth.[34] I discuss this research in detail in chapter 7.

The Perception of Faces

Face-to-face interactions are so critical to mother-infant bonding that a newborn's innate model of its mother, if it is to be at all useful, should include at least some rudimentary expectation that its mother has a face, along with an innate preference for looking at faces. Many experiments during the last few decades have shown that human newborns prefer to look at images of faces or face-like drawings, rather than non-face objects or schematic faces with their features scrambled.[35] There remains some controversy, however, regarding exactly which aspects of a face attract the attention of a newborn, what we can infer from this about neonatal thoughts and feelings, and whether the brain treats faces as a special category of visual stimulus distinct from all others.

To appreciate what a newborn is likely to see in a face, we must first step back and look at face perception in adults. Face perception seems special because as adults we are so good at it. We are better at remembering and recognizing images of faces than images of non-face objects. Part of the reason for this is that the brain processes a face holistically, rather than as a collection of parts. We saw an example of this earlier in my discussion of the Mooney face (figure 1.2), which is strangely incomprehensible when upside-down. Similarly, in memory tests we are better at recognizing faces than houses, but that advantage disappears if the images are presented upside-down.[36] Adults are able to recognize as different two images of a face that differ only slightly in the separation of the eyes, but not when the images are upside-down.[37] A more striking example is the Thatcher illusion, so named because it was first inflicted on a photograph of British Prime Minister Margaret Thatcher. The eyes and smiling mouth of poor Margaret were turned upside-down, rendering her a hideous and malevolent ghoul. Amazingly, though, the ghoulishness of her altered countenance becomes imperceptible if the whole image is turned upside-down.[38] These are all examples of the face inversion effect, which shows that the brain is specialized for processing whole, upright faces.

Most experiments on the neonatal face preference examine the infant's spontaneous head turning and visual fixation on either of two simultaneously visible images, at least one of which is in some way face-like. As I mentioned earlier, newborns have relatively poor vision for fine spatial details, and this is evident in their innate preference for faces.[39] Early experiments led to the suggestion that the preference arises from a sensitivity of the neonatal visual system to a crude innate model of a face: an elliptical outline containing three dark blobs, two for the eyes and one for the mouth. This was thought to be merely an unconscious mechanism for directing an infant's attention to faces during the first two months of life, during which a different neural system begins to learn the detailed understanding of real faces that underlies conscious perception of them.[40] Although there is much evidence for the essence of this idea,[41] recent discoveries complicate the picture.

When tested with stimuli that probe more subtle aspects of face perception, newborns respond in ways that suggest a surprising level of competence at holistic face processing. For example, newborns can perceive the invariant identity of two views of the same person's face, one seen frontally and the other with the head turned 45 degrees to one side. This requires at least a rudimentary representation of rotational invariance.[42] Newborns prefer an upright Mooney face to a similar non-face image, and they spend more time looking at an upright Mooney face than the same image presented upside-down.[43] Newborns perceive as different a photograph of a real face and a "Thatcherized" version of the same image (eyes and mouth rotated in place 180 degrees), but not if the images are upside-down. This implies a capacity for the analysis of detailed facial features, but only in the context of a whole upright face.[44] Given a choice between two images of the same face, one that makes eye contact and the other with eyes averted, neonates prefer the one that makes eye contact. This reveals not only feature analysis, but also sensitivity to an important cue of social engagement. The preference vanishes if the images are upside-down.[45] In a more elaborate version of this experiment, newborns were shown a series of images of a schematic face that initially makes eye contact, then turns its eyes to one side. A new visual target then appears on one side of the face, either on the same side to which the eyes just turned or on the opposite side. When the eyes turn toward the place where the target is about to appear, infants shift their gaze to the target more quickly than when the eyes turn in the opposite direction. This is a rudimentary form of gaze following, an important aspect of human social interaction.[46] It suggests an understanding of the face as part of another person and as a cue to that person's intention. Taken together, these experiments argue for an innate model of a face far more elaborate than a mere ellipse containing three dark blobs.

A rare neurological case offers unique insight into the importance of innate knowledge in face perception. A sixteen-year-old boy who suffered damage to his face-perceptive cortex as a one-day-old infant has severe prosopagnosia—an inability to recognize faces that leaves relatively intact the visual recognition of non-face objects. This case demonstrates that the brain does not treat faces as just another class of object discovered through visual experience. If it did, the intact parts of this boy's brain that learned to recognize many non-face objects would have learned to recognize faces as well. That the damage occurred at birth shows that this special treatment of faces, as distinct from all other objects, must be genetically specified. This stunning lack of plasticity in the face-recognition system reveals it to be a brittle, inflexible, special-purpose cortical network dedicated to one function, despite the fact that that function—representing the identities of thousands of faces—requires a vast amount of learning throughout life. This is yet another example of the brain using an innately specified perceptual system as an obligatory foundation for subsequent learning.

Seeing Biological Motion

Most studies of face perception in neonates have used still images. More impressive responses might be evoked by using realistic faces with meaningful and changing facial expressions. One promising approach, often used with older infants, is the still face. In a typical experiment of this kind, a woman engages the infant's attention through motherese, making eye contact, raising her eyebrows, and smiling. Then, for a minute or so, she becomes still and silent, putting an unchanging, neutral expression on her face, after which she springs back to life. Infants a few months old find the still-face period disturbing. Typically they stop smiling, look away from the experimenter, and may even start crying. The technique is a sensitive assay for the degree of emotional connection of the infant with the experimenter.[47] A recent study tried this on newborns and infants aged one and a half and three months. The older infants responded in the typical way, but the newborns did not, mainly because they were less engaged than the older infants before and after the still-face period.[48] The experiment was intriguing, however, because the data hinted at a slight still-face effect in the neonates just shy of statistical significance. Other researchers have since looked for a still-face response in newborns using longer trial periods and with analyses of more aspects of the infants' responses. This approach yielded statistically significant effects of the still face on eye contact, distressed facial expression, crying, and sleeping episodes in the neonates, though, unlike older infants, they showed little response to the reengagement phase.[49] In both of these studies, the infants were reclining in a motionless baby seat throughout the experiment. More striking results might be obtained if the newborns were periodically given a brief vestibular stimulus to arouse their attention.[50]

Other studies have taken a completely different approach to exploring the perception of biologically relevant motion, one that does not involve faces at all. They use a visual stimulus developed by Swedish psychologist Gunnar Johansson for isolating the motion of an animal from its form.[51] This involves placing points of light on the main joints of the limbs and body and creating a video that shows only those points as the person or animal moves. Any single frame appears to be a meaningless assemblage of dots, but when viewed as a movie, a vivid perception of the whole moving animal is evoked.* This is called a point-light display, and it has been used for decades to study how the brain perceives form and behavioral intent from a minimal representation of biological motion.[52] Interestingly, the part of the brain that specifically responds to point-light displays is also intimately involved in the perception of speech and sign language, lip reading, the interpretation of facial expressions, and the feeling of a sensed presence—but that is a subject for chapter 6.†[53] Here our concern is with the behavioral responses of newborns to point-light displays of biological motion.

Newborn human infants have an innate preference for displays of biological motion, and the studies that first revealed it used a point-light movie of a walking hen rather than of a walking person. The infants preferred the biological motion display to any of three alternative patterns of movement: the same points moving randomly, the fixed points rotating about a vertical axis, or the

---

* For examples see www.youtube.com/watch?v=rEVB6kW9p6k (accessed 14 December 2021).

† This is the superior temporal sulcus (Allison et al. 2000), which partly overlaps an area implicated in the illusion of a sensed presence, the temporoparietal junction.

biological motion display shown upside-down. Using the habituation procedure, the authors found that the neonates could discriminate between the point-light display of the walking hen and the same points displayed with their starting positions randomized, though the infants had no preference given a choice between those two. The authors used a nonhuman point-light display in these experiments in part to ensure that the stimulus was different from anything the infants had seen prior to testing. They emphasized that this appears to be an innate system that biases newborns to pay attention to animal-like movement.*[54] Like other examples we have seen in this chapter, it is useful at birth but greatly refined through subsequent learning. By age six months, for example, infants can perceive the direction of walking implied by a point-light display of a person walking on a treadmill.[55]

Imitation and the Expectation of Mother

To this point I have been more concerned with what infants can perceive than with what they can do. Aside from the eye and head movements associated with selective attention, newborns seem unable to do much of anything besides crying, suckling, and moving their limbs in apparently meaningless ways, and even these modest achievements are often dismissed as mere thoughtless reflexes. Careful analysis, however, reveals unexpected subtleties of control and purpose that belie the supposedly reflexive nature of neonatal behavior.[56] Unlike a true reflex, the so-called rooting reflex—the tendency of a newborn to turn its mouth toward whatever touches its cheek—is highly context dependent. The behavior obviously facilitates suckling, but it tends to happen only when the infant is hungry and only if the touch comes from another person, not from the infant's own hand. This simple finding is deeply significant, because it shows that a newborn can distinguish between herself and another person.[57]

Sucking must be more than a reflex, because neonates continually adjust their sucking pressure in response to the rate of milk flow.[58] Obviously sucking is one of the most completely developed of neonatal behaviors, but a surprising experiment has revealed just how much an infant can control it and how much of a mind must be at the controls. Given headphones and an electronically monitored pacifier that controls whether a newborn hears a recording of his mother's voice or that of an unfamiliar woman, the infant quickly learns to adjust the rate or timing of his sucking to obtain the playback of Mom reading a story in motherese.[59]

---

* To compensate for the visual limitations of human neonates, the points were enlarged and displayed as black dots on a white background, and the movies were displayed at reduced speed; see Bardi et al. (2011) and Simion et al. (2008).

The most impressive of neonatal behaviors, however, is their ability to imitate the facial expressions and manual gestures of adults. Neonatal imitation is not obvious. To see it in a convincing way requires careful procedure and proper controls. In a typical experiment of this kind, an infant a few hours or days old is tested when in a calm and alert state. The experimenter gets close to the infant and engages her attention. Then in alternating blocks of, say, twenty seconds, the experimenter puts on either a neutral facial expression or one of several distinct facial gestures, like sticking out the tongue, pursing the lips, or opening the mouth. The faces of the experimenter and the infant are separately recorded by two video cameras for later analysis. The recording of the infant's face is scored in each twenty-second period for resemblance to the expressions used in the study, but the judges are unaware of which expression the infant was watching during each period. Statistical tests are then done to ensure consistency of scoring among several independent judges and to see if the infant's expressions match those of the experimenter to a statistically significant degree.[60]

The results show clear evidence of imitation, typically of simple gestures like tongue protrusion and mouth opening, but one especially interesting study found that neonates also imitate facial expressions of happiness, sadness, and surprise.[61] This not only strengthens the case that basic human facial expressions are innate,[62] but also suggests that a need for emotional connection motivates neonatal imitation. In this study the experimenter held the newborn and got her attention by auditory and vestibular stimulation (making tongue clicks and doing knee bends) before producing each facial expression. As I noted earlier, vestibular stimulation increases the attentiveness of neonates during a visual tracking task,[63] but its efficacy at priming a newborn to imitate suggests that it enhances not merely attention, but also a specific expectation of engagement with another person. Another result from this study was that infants did most of their mimicry during the middle third of a series of trials of the same expression. The decrease in imitation toward the end of the series was probably a consequence of habituation, but the delayed appearance of imitation at the beginning suggests an effortful process of neural computation and error correction. This interpretation is supported by other studies that found gradual improvement in the fidelity of neonatal imitation over repeated trials.[64]

Neonatal imitation seems to be most robust for facial gestures and expressions, but other imitative behaviors have been seen. In one study the experimenter simply held up her closed hand to the infant and extended the index finger. The infants were significantly more likely to extend their index fingers following this display than during a baseline period before the experiment. Interestingly, their spontaneous hand movements during the baseline period occurred with roughly equal frequency on the left and right hands, but the finger movements during the imitative period were predominantly on the left hand, regardless of which hand the experimenter used. This may be a consequence of right-hemispheric dominance in the neonatal brain, a subject I take up in chapter 12.[65]

The most amazing thing about neonatal imitation is what it says about the neonatal mind. A newborn knows, through her proprioceptive sense, what her facial movements feel like, but she has never seen her own face. She knows what the experimenter's face looks like, but not what he feels. The fact of neonatal imitation implies that a newborn somehow makes the connection between what a gesture looks like on another person and what it feels like to herself and that the neural hardware for doing this is at least partly innate. The most promising explanation for this is that actions are represented in the brain in a way that is independent of sensory modality.[66] Neurons that participate in such a representation would therefore respond not only to the proprioceptive feeling of sticking out your own tongue, but also to the experience of seeing someone else sticking out *his* tongue. Neurons that fit this description—aptly called mirror neurons— have been discovered in monkeys and adult humans.[67] If this interpretation of neonatal imitation proves correct, then human infants are born with mirror neurons. The newborn imitates by remembering the pattern of neural activity evoked by the sight of the experimenter's protruding tongue and tries a behavior that produces a similar pattern of activity. In part this would be a trial-and-error search, which would explain why neonatal imitative behaviors are tentative and incomplete at first but then quickly improve.

The idea that a newborn infant could do this was a radical departure from the conventional wisdom on neonatal cognition when it was first suggested in the 1970s. Many in the field simply did not believe the early findings, and skepticism grew when a few early attempts at replication failed. Since then, however, neonatal imitation has been replicated in many independent laboratories using a wide variety of protocols and rigorous controls.[68] In one study newborns were even tested immediately upon delivery and imitated the first face they saw.[69] Neonatal imitation has also been found in chimps and monkeys, which bolsters the case in humans and suggests a pre-hominin origin of the trait.[70] This growing evidence and the other findings on neonatal cognition I have described in this chapter bolster the case for the reality of neonatal imitation, though some studies still fail to replicate the phenomenon and controversy persists, largely over methodology.[71] In fairness to the contrarians, research of this kind is difficult, the infants must be studied when in an attentive state, and individual human newborns vary significantly in their propensity to imitate.[72]

The meaning of that individual variability is unknown for human infants, but a study of neonatal imitation in rhesus monkeys found a correlation between the propensity to imitate and subsequent motor development. Monkeys that were good imitators during their first week of life did significantly better than non-imitators at reaching and grasping tasks a few weeks later. The authors emphasize that such tasks depend on a cortical network that integrates vision, proprioception, and motor commands with behavioral intent—the same network that contains mirror neurons. They interpret the correlation of neonatal imitation with subsequent proficiency at reaching and grasping as evidence of earlier maturation of this network in imitators than in non-imitators and argue that neonatal imitation depends on this same cortical network.[73]

Another revealing example is the imitative aspect of language acquisition. Infants only three to five months old can imitate vowel sounds,[74] and newborns produce lip movements that suggest they are trying to imitate phonemes, despite their inability to vocalize.[75] When four-month-olds are simultaneously shown two video clips of faces pronouncing two different vowels while the sound of one of them is played, they prefer to watch the face that matches the vowel heard.[76] Like other findings in neonatal imitation, these results suggest a neural representation of speech that is independent of modality and that connects related sensations of vision, hearing, and proprioception. The rudiments of these representations appear to be innate and constitute the foundation upon which more refined neural models of the native language develop during the first year.[77] Most of the imitation involved in language acquisition does not, however, happen in the piecemeal, turn-taking manner of, say, imitating a surprised facial expression. Instead, infants build upon, refine, and elaborate their neural representations of the speech they hear during their first months, as just described. Then, at about age six months, they spontaneously begin to babble. Although babbling initially seems random, it appears to be an infant's attempt to imitate her internal neural model of speech.[78]

Some especially intriguing evidence for the link between social engagement and neonatal imitation comes from a study in which changes in heart rate were monitored as neonates imitated adult tongue protrusions. The experimenter repeatedly stuck out her tongue until she was certain the infant had seen the gesture, then waited for imitative responses. After several cycles of imitation, the experimenter then allowed more time between her tongue gestures, from two to five minutes, to look for attempts by the infant to initiate more cycles of imitation. Of the forty-five newborns tested, thirty-five imitated and seventeen appeared to initiate cycles of imitation during the long delays. Although it is possible that the putative provocations by the infants were merely delayed imitative responses, the heart rate data suggest otherwise. During imitative responses, the infant's heart rate accelerated, indicating an increase in general arousal. When the infant stuck out his tongue after a long delay, his heart rate decelerated.[79] In a wide variety of experimental settings, heart rate deceleration accompanies orientation, increased selective attention, and a state of expectation.*[80] The authors

---

* In chapter 10 I explore the importance of selective attention and expectation in the illusion of a sensed presence.

argue that the newborns were sticking out their tongues in expectation of a response from the adult and that the imitative exchanges seen in the experiment constitute a form of protoconversation.[81]

Taken together, the studies of infantile imitation reveal it to be a continual process of refining highly abstract and modality-independent cognitive representations that are present in rudimentary form at birth. The things represented include movements, gestures, emotional expressions, and the building blocks of language. Central to this process is interaction with another person, which appears also to be the ultimate goal of neonatal imitation. Far from being merely an unconscious reflexive reaction to an arousing stimulus, neonatal imitation embodies an infant's first efforts to understand the actions and intentions of another person—most often and most importantly those of her mother.

So far in this chapter I have described direct evidence, in newborns and young infants, of the innate neural model of mother. But there is also indirect evidence of it in the behavior of adults, one of the most striking examples of which apparently arises not only from our evolutionary history, but also from the peculiar way evolution works.

Tinkering and the Religious Obsession with Sex

If your eye were the product of an intelligent designer, its retina would not have been installed backward—with the light-sensitive cells facing the inside of your head, rather than the outside world they are trying to see—and you would not have a blind spot where your optic nerve exits your eyeball.[82] Kludges like the vertebrate retina illustrate a fundamental principle of biology: evolution works only by making small changes to what already exists, without any grand plan, design, or forethought. Evolution is a tinkerer, not an engineer.[83]

Hack jobs of this kind often show up in the brain and, therefore, in human nature. John Bowlby[84] and Lee Kirkpatrick[85] have independently suggested that infantile attachment to mother, maternal caring for infants, and adult sexual pair-bonding in humans all share a common underlying neural mechanism, one that manifests as behavioral cross talk among these three critical systems. If an innate neural model of mother contributes to religiousness as I have suggested, then this cross talk can explain several vexing puzzles of religion, including the greater religiousness of women relative to men and the greater religiousness of physicians relative to nonphysician biologists. Here, however, I explore a different puzzle of this kind, arguably the most bizarre, and one that we will encounter again in subsequent chapters.[86]

Why does God want us to mutilate our genitals, especially those of our children? Why does the creator of this infinitely vast universe care so much about what we do with our sexual organs as adults? Why is lust considered sinful? Why was Jesus born of a virgin, and, according to Catholics, his mother also? Why did the early Shakers not only practice celibacy, but also dance naked?[87] Why do religious fundamentalists oppose not only abortion, but also the most effective means of preventing it, contraception? Why do televangelists rant about the evils of adultery, pornography, masturbation, fornication, and homosexuality, only to be caught with their pants down doing exactly those things? *Why is religion so obsessed with sex?*

One answer concerns the control and sexual subjugation of women common to many religions. This is an important but narrow aspect of the question, one with a distinct biological basis more closely tied to the social root of religion.[88] Here I tackle the broader question of why religion is obsessed with all dimensions of human sexuality.*[89]

My answer involves the evolutionary hack that conflates infantile, maternal, and sexual love. I have argued that an innate neural model of the mother primes a newborn to expect an unconditionally loving other. This model shares the same mechanism of affiliation and emotional commitment—what we commonly call love—that in adults is central to maternal caregiving and sexual pair-bonding. It also serves as a template for other social relationships and, as I explain in chapter 4, it is a foundation for a special kind of learning essential to the formation of those relationships. These entanglements not only ensure the persistence of the innate model into adulthood, but also cause interesting problems once adulthood arrives.

One of these is vulnerability to belief in deities that maximally excite the innate model under the right conditions, especially those of infantile helplessness and desperation. As the core of a believer's religious love of God, the innate model colors religious experience with infantile emotions. Yet this is happening in the brain of an adult, in whom the circuitry of affiliation and commitment has been redirected toward sexual pair-bonding. The more intensely pious the believer becomes, the more his infantile experience of God's presence takes on an incongruously sexual quality, as in this famous description by Saint Teresa of Avila of her mystical ecstasy at the hands of an angelic visitor:

> In his hands I saw a great golden spear, and at the iron tip there appeared to be a point of fire. This he plunged into my heart several times so that it penetrated my entrails. When he pulled it out, I felt that he took them with it, and left me utterly consumed by the great love of God. The pain was so severe that it made me utter several

---

* The toxic effects of religion on sexuality have ramifications beyond the scope of this chapter. See Ray (2012).

moans. The sweetness caused by this intense pain is so extreme that one cannot possibly wish it to cease, nor is one's soul then content with anything but God. This is not a physical, but a spiritual pain— though the body has some share in it—even a considerable share.[90]

Perhaps this is to be expected, considering that novitiates are told that their celibate life to come is the price paid for the privilege of being wedded to Christ. Yet Catholicism teaches that the desires of the flesh are sin, fornication is filthy and evil, adultery is of Satan, and so on. The perceived sinfulness of sex is the main reason for the Immaculate Conception and the celibacy of priests and nuns. The quote from Teresa of Avila is unusual in that it *embraces* the emotional spillover into sexual territory. For the most part, religion condemns it, at least as a matter of doctrine. *But why?*

The big problem here is that, despite this intrusion of the sexual, religious emotion remains primarily infantile, with God normally in the role of a greatly exalted parent. Sexual feelings for this parental figure therefore activate the incest avoidance system, an innate mechanism that generates feelings of disgust and revulsion at the prospect of sex with close kin.*[91] Because of evolution's careless shortcuts, the intensely pious tend to be flooded with love, lust, and revulsion— conflicting emotions of unconscious origin, generated by innate mechanisms selected for neonatal survival, adult sexual pair-bonding, and incest avoidance. The predictable result of this conflict is an obsession with controlling or prohibiting sexual behavior, coupled with an almost irresistible drive to violate those prohibitions. How these people deal with this conflict—as individuals and as religious institutions—determines the variously entertaining or tragic manifestations of sexual obsession in religion.

If these ideas are right, then we should expect sexual misbehavior among the clergy to be significantly greater than in comparable secular professions, and several studies bear this out. Two surveys of predominantly male American Protestant ministers found that more than 12 percent of respondents admitted to having sexual intercourse with a member of their congregation other than their spouse.[92] This is roughly double the incidence of sexual intercourse with clients committed by male clinical psychologists†[93] and physicians.‡[94] The studies

---

* The incest avoidance system is itself an evolutionary hack, apparently built in part upon the neural mechanism of disgust, a system that evolved mainly for the avoidance of contaminated food (Lieberman and Hatfield 2006).

† Of the 347 male psychotherapists anonymously surveyed by Holroyd and Brodsky (1977), 5.5 percent admitted to having intercourse with clients. Among 310 female psychotherapists, the incidence rate was only 0.6 percent.

‡ Kardener et al. (1973) found the incidence of physician-patient intercourse to be either 5 percent or 7.2 percent in a sample of 460 physicians, depending on how the question was asked. A more recent study of 1,891 physicians found a 9 percent rate of incidence of "sexual contact" by subjects with their patients, but it did not restrict any question specifically to sexual intercourse (Gartrell et al. 1992).

also hint at a more general problem of sexual conflict, guilt, and obsession. One of them found that 37 percent of Protestant ministers admit to the vague sin of "sexual behavior inappropriate for a minister."[95] Responses to other questions in the study suggest that at least some of these transgressions involve sexual fantasy, viewing pornography, and masturbating in private—acts of negligible consequence or moral weight in the broader society.[96]

The pastors, however, see it differently. Their religiously warped view of human sexuality is especially evident in *The War Within*, an account of one Protestant pastor's tormented struggle with lust, published anonymously in a journal for Protestant clergy.[97] Although he was married and regularly had sex with his wife, the author tells a sad tale of his obsession with various aspects of sexuality off-limits to good Christians. Like Eve tasting the forbidden fruit, he saw himself perpetually tempted by Satan to indulge his sexual curiosity—visiting seedy strip joints, peep shows, and X-rated theaters; exploring pornographic magazines and videos; and fantasizing about women other than his wife. Although he never committed adultery, he felt that he had, and for years he prayed in vain for deliverance. He mentions only in passing a "repressed childhood" that may have contributed to his obsession with lust, but beyond that glimmer of insight he appears utterly blinded by his religious worldview to the essence of his problem: lust is healthy and normal, but a religious obsession with it is not. *The War Within* is only one piece of anecdotal evidence, but the editors of the journal note that it prompted more letters from readers, by far, than any other article before or since. Some of these condemned it for its explicit lewdness, most praised it for bravely addressing a common problem among the clergy, but all implicitly support it as evidence of a struggle with love, lust, and revulsion among the most intensely religious.

That struggle takes a darker turn among Catholic clergy, who are disproportionately likely to sexually abuse vulnerable children and adolescents. Catholic apologists deny this, often citing a study that found that only 4 percent of Catholic priests sexually abuse minors.[98] They argue that this is the same as among the general male population, implicitly suggesting that there is no special relationship between Catholicism and the sexual abuse of children.[99] There are, however, several problems with this rationalization. One is that other studies find higher rates of child sexual abuse in the Catholic clergy, more than 5 percent,[100] and even that is likely to be an underestimate, given the reluctance of victims to come forward and the intense effort by the Catholic hierarchy to protect offending priests and cover up the scandal.[101] Another problem is the specious comparison to the general male population. The Catholic Church claims to be a high moral authority—God's representative on Earth—and its clergy are supposedly transformed by God into a special state of sanctity. If this were true, we should expect exemplary behavior from Catholic priests, especially with respect to the sexual abuse of innocent children, one of the most odious crimes imaginable. If the Catholic Church really were what it claims to be, the incidence of child sexual abuse by priests should be zero.

The appropriate comparison, of course, is to non-Catholic Christian denominations. Although Protestant clergy outnumber Catholic priests and brothers in America by six to one, Catholic priests predominate in samples of clergy sexual offenders.[102] Similar results come from studies that compared rates of child sexual abuse by Anglican and Catholic clergy in Australia, where the rate of such abuse was about seven times higher among Catholic Church personnel than among Anglican.[*][103] There is also a striking qualitative difference that suggests possible explanations for the numerical disparity. In contrast to the victims of child sexual abuse in the general population, most of whom are prepubertal girls, the victims of predatory Catholic priests are mainly postpubertal adolescent boys.[104]

The most obvious culprit here is the Catholic obsession with the sinfulness of sex for pleasure, especially homosexual sex, along with a peculiar sexual proscription for clergy: celibacy. Intentionally or not, the celibate priesthood specifically attracts applicants who are at best confused or conflicted about their sexuality and at worst deliberately seek positions conducive to sexual exploitation of the vulnerable. Although Catholicism condemns homosexuality, priests appear to be disproportionately gay. The prevalence of homosexuality among priests has not yet been properly measured via random sampling, but when 101 American gay priests were asked to estimate this prevalence, most of them guessed 40 percent or more[†][105]—a rate far higher than in the general population.[106] A different study of a thousand priests found that 20 percent of them were in homosexual relationships and another 20 percent in stable relationships with women. Most of these were verified by interviewing the sexual partners.[‡][107] Aside from being an ineffective sham, the doctrine of celibacy for the priesthood apparently attracts young Catholic homosexual men, either as a means of suppressing a sexual orientation they perceive as evil or—in certain Catholic subcultures—as a safe venue

---

* Parkinson (2014, 5–7) cites counts of victims of clergy *convicted* of child sexual abuse in Victoria and separate counts of *complaints* of child sexual abuse by clergy in Melbourne. By both methods, the number of victims of such abuse by Catholic clergy was roughly tenfold higher than by Anglican clergy. After adjusting for the greater size of the Catholic Church in the studied areas, the disparity in prevalence is reduced only slightly, to about sevenfold.

† More than 90 percent of this sample of gay priests estimated that 40 percent or more of priests are gay; their mean estimate was 48.5 percent (Wolf 1989, 59–63). By itself this result is difficult to interpret, but in conjunction with the study by Sipe (1990), it seems safe to conclude that Catholic priests are disproportionately gay.

‡ This study (Sipe 1990) was not a survey based on random sampling, but an ethnographic study of about five hundred priests who sought psychotherapy and another five hundred whom the author met through professional meetings.

for hooking up with other gays.*[108] But even with all of its faults, Catholicism turns out not to be the most sexually obsessed belief system.

The religious obsession with sex is most extreme in cults, though the manifestations differ greatly among them. The full spectrum was evident in three nineteenth-century cults in the state of New York: the Shakers, who ostensibly practiced absolute celibacy and segregation of the sexes for all members; the Mormons, who indulged the lust of their most favored males with polygyny; and the Oneida Community, who banned monogamy, encouraged sexual promiscuity, and pursued eugenics among the cream of their crop.[109] More recent examples are no less bizarre. Rather like the Oneida Community, the UFO cult known as the Raëlian Movement discourages monogamy and advocates free love.[110] Heaven's Gate, another UFO cult, practiced celibacy like the Shakers but outdid them by encouraging castration among its males. At the direction of its sexually conflicted leader, the cult came to a sudden and tragic end by mass suicide in 1997.[111] Perhaps the most despicable example, however, is the cult known as the Children of God, which for many years systematically subjected children and teens to sexual abuse and prostitution.[112]

The great magnitude of sexual obsession, abuse, and humiliation in cults likely arises from the extreme degree of religious infantilism fostered among their followers, running amok because of the evolutionary hack at the core of these emotions. But the extraordinary power and control wielded by cult leaders are also essential elements, and these derive largely from the other side of the two-faced God: the social root of religion I discussed in the previous chapter.

Ultimately the best way to test such ideas is to trace the circuitry in the brain that gives rise to these behaviors. If the religious obsession with sex derives from the incongruous intersection in the pious of infantile longing for mother and adult sexual pair-bonding, then the neural substrates of these behaviors should significantly overlap in the brain. With specific hypotheses of this kind to guide us, it makes sense to seek an understanding of religion in the brain. In the circuitry of embodiment, for example, we will discover a deep connection between the illusion of God's presence and an equally compelling illusion, the phantom limb. We might even find some surprising connections between the diametrically opposite natures of the two-faced God. Our journey begins in the next chapter.

---

* Ironically, Rome itself is one of these subcultures; see Thomas (2014).

Part II

CIRCUITRY OF THE SENSED PRESENCE

CHAPTER 3

Crying and the Neural Alarm System

> Do not try to approach God with your thinking mind. It may only stimulate your intellectual ideas, activities and beliefs. Try to approach God with your crying heart. It will awaken your soulful, spiritual consciousness.—Sri Chinmoy[*][1]

The kind of spiritual experience that culminates in the sensation of a divine presence is often precipitated by an intense emotional crisis, a desperate and helpless longing, or spiritual brokenness.[2] In chapter 1, I argued that the primordial instance of this feeling is the desperate helplessness of early infancy and that an innate neural model of the mother in infancy is the basis for the sensation of God's presence in later life. If this is so, then the path to a neurobiological understanding of spiritual experience begins with the neurobiology of infantile crying.

The Mammalian Way of Life

To my knowledge, as of late 2021, there have been no neuroimaging studies of the brains of crying human infants.[†][3] Such studies will eventually be done, but in the meantime we can infer a great deal from studies of crying in other

---

[*] Chinmoy (1970). This admonition to replace reason with infantile emotion comes, not surprisingly, from a cult leader; see Tamm (2009) for that story.

[†] For technical reasons, neuroimaging of crying infants would be extremely difficult with fMRI (Laureys and Goldman 2004), though feasible, at much lower spatial resolution, with NIRS imaging. NIRS was used in two neuroimaging studies of adult subjects crying in response to emotionally evocative movies; crying was associated with increased neural activity in the medial (Sato-Suzuki et al. 2007) and ventromedial (Kamiya et al. 2008) prefrontal cortex. The anterior cingulate cortex, a major focus of this chapter, is part of the medial prefrontal cortex.

mammals. Infantile crying is nearly universal among the mammals, and there is evidence that the evolutionary origin of innate vocalizations—not only mammalian crying and laughing, but also territorial and mating calls in a wide variety of vertebrates—occurred early in vertebrate evolution, long before the mammals appeared.[4]

Nearly all mammalian infants emit some kind of cry when separated from their mothers, though the acoustic qualities differ greatly among species. In some species, such as rats, mice, and guinea pigs, crying occurs only during infancy.[5] In others, like dogs, humans, and squirrel monkeys, it persists into adulthood, typically as an isolation call given as a signal to reestablish contact when an individual is separated from its social group.[6] The crying behavior of humans is unusual in that we are the only species that sheds tears while crying,[7] probably as a hard-to-fake social signal.[8] The crying of adult dogs is also unusual, in that it is often elicited by isolation from human companions, and the presence of a human can be more comforting to an isolated puppy than that of its own mother."[9] Crying behavior, whether in separated infants or socially isolated adults, is not merely an acoustic signal. It is typically accompanied by a stress response that includes increases in heart rate, blood pressure, respiration, and the level of stress hormones in the bloodstream.[10]

Experiments that electrically stimulate or locally destroy parts of the brain have shown that crying is generated in the hindbrain and midbrain in a wide variety of mammals.[11] This is consistent with a few anecdotal reports that crying is present, if somewhat abnormal, in human infants in which the forebrain is either incompletely developed or entirely missing.[12] Similar evidence comes from experiments in which rat pups are chilled to the point of being comatose: they produce ultrasonic cries as they begin to warm but well before they regain consciousness.[13] For many years this brainstem model of cry production prevailed, and the forebrain structures that subserve human consciousness were thought to play little or no role in the crying of human infants. The absence of any role for conscious cognition in human infantile crying would be fairly strong evidence against the idea that a human newborn has an innate cognitive model of its mother.

---

* Pettijohn et al. (1977). Leuba (1925, p 280) quoted the following passage from Darwin to suggest an analogy between the sensed presence illusion and the comfort a dog obtains from the mere presence of a human companion: "It is curious to speculate on the feelings of a dog who will rest peacefully for hours in a room with his master or any of the family, without the least notice being taken of him; but, if left for a short time by himself, barks or howls dismally." Darwin (1874, p 103). Four pages earlier in his book, Darwin cites several authorities who suggest that a dog looks upon his human master as a god. Hardy (1966, pp 173–75) makes a similar suggestion.

Social Isolation and Other Stressors

But this view has changed during the last few decades with the discovery that at least one forebrain structure, the anterior cingulate cortex, is critically important for normal mammalian infantile crying and adult isolation calls. The first clues came from studies in which direct electrical stimulation in this part of the brain elicited cries or isolation calls, along with various other kinds of vocalization, in several species of monkey.[14] But one of the most intriguing discoveries came from a study in which the anterior cingulate cortex was destroyed in squirrel monkeys.

An adult squirrel monkey makes isolation calls in two distinct contexts: either when separated from its social group or in reply to the isolation call of another monkey that is separated from the group. Destruction of the anterior cingulate cortex in adult squirrel monkeys eliminates the production of isolations calls in the first context but not the second.[15] This shows that the basic neural circuitry for producing the cry sound is still intact in these animals, but the circuitry that perceives and reacts to the animal's own social isolation is not. This suggests a role for the anterior cingulate cortex not merely in crying behavior, but specifically in social cognition.

Many other experiments support this view and reveal the anterior cingulate cortex to be a mysterious and multifaceted brain region, one that becomes active in such a wide variety of stimulus conditions and behavioral responses that it defies simplistic interpretation. Here is a sample of the many sensations or tasks that in some way engage this part of the brain, taken mostly from neuroimaging studies of human subjects:

- *Maternal response to infant stimuli.* The sound of an infant's cry or the sight of its face activates the anterior cingulate cortex in human mothers,[16] and damage to this structure severely degrades maternal behavior in rats, mice, and hamsters.[17]
- *Conflicting stimuli and error detection.* The classic example of this is the Stroop test, in which subjects are briefly shown a word on a computer screen and must quickly say the color of the letters on the screen.[18] When there is a conflict between the meaning of the word and the color in which it is displayed—for example, the word "red" displayed in blue letters—the anterior cingulate cortex is activated, especially so when the subject gives an erroneous response.[19]
- *Effortful activity and cognitive load.* The anterior cingulate cortex is activated during effortful tasks, whether the effort is physical or cognitive in nature. In memory tasks its activity increases with the number of items that must be remembered—the cognitive load—but whether the task is a memory test or

an isometric exercise, the anterior cingulate cortex generates an autonomic response that affects heart rate and blood pressure.[20]

- *Vocalization.* In addition to isolation cries, the anterior cingulate cortex appears to be involved in a wide variety of vocalizations in mammals,[21] and in humans it has a similar role in nonverbal vocalization and in the emotional prosody of speech. For several weeks after injury, some human patients become completely inactive and mute after suffering extensive damage to the anterior cingulate cortex bilaterally—that is, in both the left and right hemispheres of the brain. Normally their ability to speak gradually returns, but their speech lacks emotional tone.[22] Patients with smaller lesions that involve only the rostroventral part of the anterior cingulate cortex lack such drastic symptoms, but they are impaired in the perception of emotionality in another person's speech.[23]

- *Physical pain.* The anterior cingulate cortex is one of several parts of the brain that are consistently activated in response to painful stimuli. Part of it, mainly the dorsal sector, is activated no matter where on the body the stimulus is applied, which suggests a role not so much in sensory perception as in cognitive, emotional, and behavioral response to pain.[24]

- *Social pain.* Rejection, exclusion, and ostracism activate the anterior cingulate cortex, much as does physical pain.[25]

- *Empathetic pain.* Seeing another person receive a painful stimulus activates regions in the anterior cingulate cortex that tend to overlap, but do not exactly match, those activated in response to directly experienced physical pain.[26]

- *Bladder control.* The anterior cingulate cortex is activated in neuroimaging studies that involve filling of the bladder, urine retention, and voiding. Its role has been interpreted as being related to the emotional aspects of bladder control.[27]

- *Depression and sadness.* The rostroventral part of the anterior cingulate cortex has significantly less gray matter volume in patients with heritable major depressive disorder, and this part of the brain is abnormally active in these patients during depressive episodes.[28] A similar pattern of activity occurs in normal subjects when they become sad while contemplating an unhappy memory.[29] The target of most modern antidepressant drugs is the serotonin transporter, a membrane protein that reduces the concentration of serotonin in synapses. The highest density of serotonin transporter molecules in the entire human brain is in this same part of the anterior cingulate cortex,[30] and direct electrical stimulation of this region is now being used to treat the most intractable cases of depression.[31]

- *Social behavior.* Damage to the anterior cingulate cortex causes monkeys to spend less time in close proximity to one another[32] and in human patients causes deficits in sociability as evaluated by close family members.[33] In subjects with autism spectrum disorder, the anterior cingulate cortex is less active than in control subjects during tests of social cognition, such as perception of emotion in facial expressions or taking the mental perspective of another person.[34]

Obviously the anterior cingulate cortex is much more than just a "crying center" in the brain. The bewildering variety of sensations, emotions, and behaviors that in some way involve this neural structure has generated much controversy among those scientists who struggle to make sense of it.

One insight that brings some order to the chaos is that the anterior cingulate cortex is a heterogeneous area, parts of which contribute in different degrees to the responses seen in these various experiments. The dorsal part tends to be selectively activated by cognitive tasks, like the Stroop test for colored letters. The rostroventral part tends to be activated in more emotional tasks, like the perception of emotion in tone of voice.[35] The validity of this division is supported by other differences between these two subregions, like the details of their cellular organization and connectivity with the rest of the brain,[36] but the cognitive-versus-emotional interpretation of these subregions may be oversimplified. Painful stimuli, which evoke powerful aversive emotions, typically activate the supposedly cognitive part, and highly emotional isolation cries are evoked by stimulation of, and prevented by damage to, this same region.

The Urgent Response

A complete understanding of the anterior cingulate cortex is not yet within reach, so I shall merely emphasize two recent interpretations that seem to capture some of the essence of what this part of the brain is doing.

One is the idea—emphasized by neurophysiologist John Allman and others—that the anterior cingulate cortex is the place in the brain where emotion and cognition meet, specifically for the purpose of choosing the behavioral and/or physiological response that will maximize reward or minimize punishment.*[37] Support for this idea comes not only from the convergence of emotion and cognition that is evident in the experiments just described, but also from the connections of the anterior cingulate cortex to other parts of the brain. It receives an extremely dense innervation from the dopamine-secreting neurons of the midbrain that send signals of reward expectation to large areas of the brain.[38] The dorsal part of the anterior cingulate sends output fibers to areas of the motor

---

* Recent insights from computational theory cast these ideas into a more detailed framework in which the anterior cingulate cortex controls a hierarchical system for choosing between alternative behavioral plans based on anticipated rewards and perceived errors in the currently active plan (Holroyd and Verguts 2021).

cortex that generate and control complex conscious behaviors.[39] Its rostroventral part sends output fibers to the hypothalamus,[40] a collection of neurons that initiate essential innate behaviors like feeding, drinking, mating, and maternal caregiving—in part by sending neural commands to brain stem circuits that generate these behaviors, but also by stimulating the release of pituitary hormones that prime various organs of the body for digestion, water retention or excretion, sexual arousal, lactation, or the fight-or-flight response to an external threat.[41] In addition to these hypothalamic mechanisms, the anterior cingulate cortex also directly innervates brainstem centers that generate innate behaviors, like those in the midbrain that produce the crying response to infantile separation or adult social isolation.[42]

My other favorite interpretation of the anterior cingulate cortex essentially echoes the one just described, but it uses an especially compelling metaphor. This is the idea of psychologists Naomi Eisenberger and Matthew Lieberman that this structure is a neural alarm system that detects unusual, critical, or threatening conditions and initiates the appropriate and urgent response to those conditions.[43] They emphasize the remarkable similarity of response in the anterior cingulate cortex to both physical and social pain, and argue that this similarity can be traced back to the evolutionary origin of the mammalian way of life and the mother-infant bond.[44] An infant that depends on its mother for nourishment must have a way of detecting its separation from her and responding with a distress signal. A complementary system that responds to this signal must also be present in the mother. In both cases, the motivation to respond to separation must be powerful and urgent. In developing this cognitive and emotional neural alarm system, evolution evidently co-opted parts of a more ancient system for responding to, and learning to avoid, another kind of urgent distress: physical pain.

The neurochemistry of mammalian mother-infant separation supports this view. In many mammalian species, drugs that block opiate receptors intensify infantile separation cries and adult isolation calls, and opiates like morphine— in doses low enough not to cause sedation—reduce or eliminate such cries.[45] Opiates also affect maternal behavior, but in more complex and variable ways. In rats and dogs, opiate blockers increase a mother's motivation to retrieve her pups but interfere with her competence at this behavior.[46] In rhesus monkeys, opiate blockade decreases a mother's protectiveness and grooming of her infants without affecting other maternal behaviors.[47] Interestingly, the density of opiate receptors in the anterior cingulate cortex is among the highest in the entire brain,[48] and a decrease in binding of endogenous opiates to these receptors correlates with the subjective experience of sadness in humans.[49] As we shall soon see, however, opiates are not the whole story in the neurochemistry of mother-infant bonding.

Our pursuit of infantile crying in the brain led us to the anterior cingulate cortex, a complex forebrain circuit at the junction of emotion, conscious perception, social cognition, behavior, and emotional expression. Its role in the perception of and response to social isolation makes it a likely component of the innate model of the mother, but there is much more to that model, and more of the brain must be involved. How is a mother's presence perceived? Where and how is innate information stored? What is the source of the complex perceptions on which the anterior cingulate cortex operates? If we follow its neural connections to nearby regions of the prefrontal cortex, we find that it is part of two elaborate and tantalizing neural networks, which together hold some of the answers to these questions.

CHAPTER 4

Innate Knowledge and the Circuitry of Gut Feelings

> My father used to quote an unanswerable argument, by
> which an old lady, a Mrs. Barlow, who suspected him of
> unorthodoxy, hoped to convert him:—"Doctor, I know
> that sugar is sweet in my mouth, and I know that my
> Redeemer liveth."—Charles Darwin[1]

In chapter 2 I described behavioral evidence for innate knowledge in human infants: the scent of a mother's breast, a taste for her milk, an ear for motherese, a need for warmth and touching, a preference for face-like visual stimuli, an ability to perceive and imitate some essential facial expressions, and possibly also an innate association of vestibular rocking sensations with the close physical presence of mother. Is there any neurobiological evidence for such innate knowledge?

The Forbidden Tree

Deep within the human brain, buried between the two frontal lobes and just above our eye sockets, lies much of the circuitry of reward and punishment, of approach and withdrawal, of seeking and aversion, of pleasure and pain, of love and fear, of right and wrong. In it lies much of what we call human nature. It is the place where learned and innate knowledge converge as we make decisions. It may be the closest thing we have yet found to the mythical tree of the knowledge of good and evil. Above all it seems ideally suited as a repository of an infant's innate knowledge of its mother, as a connection between that knowledge and the feelings essential to the mother-infant bond, and as a foundation for a lifetime of learning to function as a social animal.

Our window into this mysterious place was opened, in part, by a series of difficult and tedious experiments that traced the connections of many small regions within the prefrontal cortex of monkeys. These studies revealed two distinct but related networks of highly interconnected cortical areas, one of which—the medial network—lies mainly on the medial surface of each frontal lobe and includes the anterior cingulate cortex. The other—called the orbital network—lies on the ventral surface of the frontal lobe, just above the eye socket or orbit, and comprises most of the region known as the orbitofrontal cortex.[2] These experiments revealed not only the web of connections among these areas, but also their long-range connections to other parts of the brain—high-level areas involved in vision, hearing, touch, taste, smell, memory, emotional state, and behavior.[3] Although experiments of this kind cannot be done in humans, the various subregions that constitute these networks have distinctive features— in their neurochemistry and cellular organization—that can be seen in human brains and thus allow a tentative mapping of the equivalent human networks.[4] These anatomical studies identified the medial and orbital networks as related entities and hinted at their importance by virtue of their extrinsic connections, but their function could only be discovered by other methods that eavesdrop on their constituent neurons.

Reinforcement Learning

Some of the most revealing experiments of this kind were done in the laboratory of neurophysiologist Edmund Rolls.[5] He and his collaborators studied the electrical activity of individual neurons in the orbitofrontal cortex of monkeys as they responded to stimuli and learned new tasks. As might be expected from the great diversity of sensory input to the orbitofrontal cortex, many of its neurons respond to various tastes or smells, to the feel or texture of different kinds of food in the mouth, to the visual appearance of food items, or to various combinations of these and other stimuli. These neurons are not, however, merely registering sensations. Each response appears to indicate the value of the stimulus as a reward or punishment. This is most clearly evident in the response to a primary reinforcer: a stimulus that has an innate association with pain or pleasure, like the appealing taste of sugar. As the monkey drinks a sweet fruit juice until it is satisfied, the response of such a neuron gradually decreases along with the monkey's desire for more juice. Other rewarding primary reinforcers include the feel of a fatty food in the mouth, the savory taste of protein, and the gentle touch of velvet on the hand.

But the main function of orbitofrontal neurons appears to be to associate primary reinforcers with sensations that are not innately rewarding or punishing, thus attaching value to those sensations and making them secondary reinforcers. A good example is the learned association of a fruity odor with a rewarding sweet taste, thus yielding the sensation we commonly call flavor. Associations of this kind are essential in the life of a monkey in the wild as it learns to recognize ripe and tasty fruits by their scent and color.

In a laboratory setting, however, orbitofrontal neurons show their versatility by easily learning associations to a wide variety of unnatural secondary reinforcers. An orbitofrontal neuron that responds to a sweet taste but not to the appearance of a blue triangle on a computer screen can quickly learn to respond to that visual stimulus if it is paired with a sweet-tasting reward. Often the neuron learns the association in one or a few trials, after which the visual stimulus alone excites the neuron. But if the visual stimulus is then repeatedly presented without the reward, the neuron quickly unlearns the association or replaces it with a new one if a different visual stimulus is paired with the rewarding taste. Throughout all of these learning experiences, however, the neuron's response to the primary reinforcer persists.[6]

Rolls and his colleague Alessandro Treves argue that the elements of innate knowledge—primary reinforcers and their various associations with reward or punishment—are represented in the orbitofrontal cortex as nonmodifiable synapses onto neurons of the kind just described. Secondary reinforcers are represented as modifiable synapses that change in strength according to the degree of synchronization of their activity with that of the synapses of the primary reinforcer.[7]

Social Primary Reinforcers

The plasticity just described is what makes the orbitofrontal cortex useful and important in the normal behavior of an adult monkey or human. But that marvelous learning circuitry is useless without the foundation of innate knowledge of those essential, fundamental stimuli—the primary reinforcers—that are intrinsically rewarding or punishing, attractive or aversive, good or bad, because a long history of evolutionary experience has discovered them to be so. Some of this knowledge is exactly what an infant monkey or human must know at birth, and it is not surprising that many rewarding primary reinforcers are the sensations of nursing at the breast: gentle touch, attractive scents, sweet tastes, and the feel of fat in the mouth. If, however, there is an innate model of the mother of the kind I have suggested, there should more. There should be innate knowledge specifically about another being.

There is. Some neurons in the orbitofrontal cortex respond when the monkey sees a face. This is not surprising, because the orbitofrontal cortex receives input from visual cortical areas in the temporal lobe that are specialized for the perception of faces. The face-sensitive neurons in the orbitofrontal cortex are, however, different in several important ways from those in the temporal lobe. They are more selective: unlike temporal face neurons, they tend to respond poorly if at all to a two-dimensional image of a face but instead respond vigorously only to a real three-dimensional face. Many of them appear to encode information about which specific individual is present because they respond differently to the presentation of different faces familiar to the monkey. Perhaps most importantly, many face-sensitive neurons in the orbitofrontal cortex respond specifically to socially significant facial expressions, like a smile or an angry face. Rolls suggests that some of this information is innate—that the essential facial expressions may be primary reinforcers that are intrinsically attractive or aversive—whereas other aspects of facial information in these neurons, like the identity of a face and the association of that individual with reward or punishment, are learned.[8]

Other neuroscientists have found in the orbitofrontal cortex evidence for innate knowledge of faces that concerns the parental side of the parent-infant bond. Using MEG imaging, which has excellent resolution in the time domain, they found in young adult human subjects an extremely rapid response to images of unfamiliar infant faces in the medial orbitofrontal cortex. By contrast, there was no such response in the orbitofrontal cortex to images of adult faces adjusted to the same emotional valence, attractiveness, and other visual properties, though both kinds of faces equally excited the fusiform face area of the temporal lobe at a longer latency. The authors argue that the extremely rapid and specific response to infant faces is direct evidence of a long-hypothesized innate releasing mechanism for parental care, implemented in the orbitofrontal cortex.[9]

Nonvisual information reaching the orbitofrontal cortex can also signal the presence of another being, and some of these complex stimuli—like the sensation of being gently touched—appear to be primary reinforcers.[10] Others include the innate vocalizations monkeys produce for social communication. The orbitofrontal cortex receives input from an area in the temporal lobe specialized for the perception of these sounds, all of which signal the presence of another monkey and convey something about that monkey's emotional state or intentions.[11] Another important signal of the presence of another being—especially for an infant monkey or human—is the vestibular sensation of being held, carried, or rocked. Responses to vestibular stimulation have not been described in orbitofrontal neurons, but this may be merely a consequence of the great difficulty of producing such sensations in a natural way without ruining the experiment. Rocking an animal during a neurophysiology session can disturb the delicate placement of the microelectrode that is listening in on the neuron, and high-resolution neuroimaging in humans requires immobilization of the head. For this reason, most attempts at mapping vestibular areas in the cortex have used unnatural stimuli that do not require physical movement—like direct electrical stimulation of the vestibular nerve or abrupt temperature changes of the inner ear on one side[12]—stimuli that may have little effect in a cortical area highly tuned for complex natural sensations. There is, however, indirect evidence that the orbitofrontal cortex could process vestibular information: it receives input fibers from the insula, an adjoining region of cortex that is in part specialized for internal feelings from the body, including vestibular sensations.[13] Although technically challenging, the search for vestibular responses in the orbitofrontal cortex could be a productive direction for future research.

Whatever innate knowledge it contains, the orbitofrontal cortex cannot be the only part of the brain that stores such knowledge. Complex percepts such as facial expressions and meaningful vocalizations are the result of elaborate computation at many levels of the nervous system. The lower-level parts of the brain that inform the orbitofrontal cortex must also store elements of this innate knowledge. For example, the visual face-processing area in the fusiform gyrus of the temporal lobe knows how eyes, mouths, and noses must be arranged to make a whole face, because its neurons respond best to images of whole, upright faces. By contrast, neurons in the face area of the superior temporal sulcus know about facial expressions and movements, and those in the occipital face area know about parts of faces, but neither of these areas has neurons selective for whole faces.[14] Throughout a person's life, the neurons in these visual areas become more finely tuned and specialized for the perception and recognition of the faces and facial expressions of real people, and most of the information they store comes from experience. But if a facial expression is to have an innate value as a source of pleasure or pain, then even at birth there must be some innate circuitry for the perception of that expression throughout these different levels of the visual system or, at the very least, an innate developmental program for its subsequent emergence. Components of the innate model of the mother must therefore be broadly distributed in the brain, down to and including subcortical circuits, though the conscious expectation or perception of this other being may be localized to a few high-level areas like the orbital and medial networks and other areas with which they interact.*[15]

---

* The innate ability of human newborns to perceive faces may depend in part on information in the superior colliculus, a visual center in the midbrain (Johnson 2005). Interestingly, neural transplantation experiments in newly hatched domestic chicks and Japanese quail show that their innate auditory preferences for the maternal calls of their species can be interchanged by transplanting neural tissue of the developing midbrain (Long et al. 2001). I return to this subject in chapter 7.

Aversion, Attraction, and the Unique Other

One of those other areas is the amygdala, a structure in the rostral part of the medial temporal lobe that is heavily interconnected with the medial and orbital prefrontal networks. For many years it was considered a fear center in the brain because it is essential for a kind of learning that involves threatening conditions. In a typical experiment of this kind, a rat learns to associate a previously neutral stimulus, like an audible tone, with a painful or aversive stimulus, like an electric shock. Once this association is made, the tone alone evokes the stress response and avoidance behavior normally elicited by the shock, and neurons in the rat's amygdala also respond to the tone. Most importantly, a rat will fail to learn the association if its amygdalae have been destroyed.[16] Several neuroimaging studies are consistent with this interpretation of the amygdala because threatening stimuli and images of fearful faces excite the amygdala in human subjects.[17]

Recent research, however, suggests that the amygdala has a broader role, similar in some ways to that of the orbital network. In monkeys, some individual neurons of the amygdala respond to rewarding stimuli, others to punishment, and still others to both kinds of emotional valence. Like the orbital network, the amygdala appears to be involved in learning both positive and negative associations between neutral sensations and emotionally salient experiences. The two structures cooperate in some kinds of learning tasks and act antagonistically in others.[18] Severing the connection between them can be as disruptive to learning as is complete destruction of either of the two.[19] The amygdala also stores innate knowledge about primary reinforcers and their rewarding or punishing value. A dramatic example is a rhesus monkey's innate fear of snakes,[20] which is eliminated by bilateral destruction of the amygdalae.[21]

Perhaps the most important similarity between the amygdala and the orbital network, however, is their role in the perception of and emotional response to another being. In rhesus monkeys, many neurons of the amygdala are selective for the identity and/or emotional expression of faces. Among those neurons responsive to facial expressions, most are excited by threatening faces and inhibited by appeasing ones.[22]

The amygdala is part of the temporal lobe, much of which is concerned with seeing, hearing, and identifying objects, other animals, and—in the left temporal lobe of humans—words. The most rostral part of it—the cortex of the temporal pole that surrounds the amygdala—contains neurons of stunning conceptual complexity and selectivity.*[23] A rare study of the activity of single neurons in the human temporal pole and amygdala found neurons that were

---

* Like the amygdala, the surrounding cortex of the temporal pole is directly connected with the medial and orbital prefrontal networks in macaque monkeys (Kondo et al. 2003).

selective for the concept of a familiar individual person. A neuron of this kind is selectively excited by images of that person's face, either directly viewed, or in various degrees of profile, or by a cartoon caricature of that person, or even by the sight of his written name.[24] A single neuron of this degree of conceptual sophistication and specificity is obviously the product of much learning and experience. But if the neonatal brain has any innate concept, however rudimentary, of the existence of a specific other being, the temporal pole is likely to contribute to that representation. The role of the amygdala in this, however, may not be simple.

The complexities of the contributions of the amygdala to mother-infant bonding are evident in a study in which infant rhesus monkeys were given bilateral lesions of the amygdalae at age two weeks.[25] When separated from their mothers at age six months, they did not make isolation cries—unlike control animals that had received either sham operations or lesions elsewhere in the medial temporal lobe—and they showed no preference for reunification with their own mother over another adult female when offered the choice. These results suggest a role for the amygdala in the normal infant's perception of the threatening condition of isolation from mother and in learning to recognize and distinguish her as a unique individual whose proximity is greatly rewarding. There was, however, one other significant difference in the behavior of the amygdala-lesioned infants that complicates this picture: throughout their infancy, they spent more time in physical contact with their mothers than did the control infants.

This seemingly contradictory result makes some sense, however, in light of other experiments that examine the role of the amygdala in establishing interpersonal distance limits in humans. In most social situations, humans have a sense of personal space that causes them to keep some minimal distance from others, and the intrusion of another person into that space is perceived as unpleasant or threatening. One way to measure this distance limit is to ask an experimental subject to stand still while another person walks toward the subject. The subject must tell the approaching person to stop at the distance they find most comfortable. A study of this kind found that a woman who has bilateral amygdala damage has an interpersonal distance limit about half that of a group of normal control subjects.*[26] Using fMRI scans of another group of normal subjects, the same study found that violation of personal space increases neural activity in the amygdala. Interpersonal distance limits appear to be the result of a balance between neural processes of attraction and repulsion, and the amygdala is essential for the repulsive process. Perhaps the infant monkeys with lesioned

---

* Her preferred interpersonal distance averaged thirty-four centimeters versus sixty-four centimeters in the normal controls, though she told the experimenters that she was comfortable at any distance. On some trials she did not give the stop signal until the approaching person touched her. See Kennedy et al. (2009).

amygdalae spent more time in close contact with their mothers because of the consequent shift in this balance toward attraction.

Other experiments support this interpretation. An adult rhesus monkey with bilateral amygdala lesions is less socially inhibited than normal, especially when interacting with unfamiliar monkeys. It will elicit more grooming behavior than normal from an intact monkey when the two are interacting as an isolated pair,[27] but in larger groups its abnormal sociability tends to elicit stress responses and aggression from other monkeys, resulting in its social isolation.[28] In rat pups there is a specific neurochemical mechanism whereby the amygdala's bias toward social wariness and aversive learning is temporarily suppressed during a critical period in early infancy. A rat mother often steps on her pups or knocks them around in the nest—aversive experiences that might otherwise interfere with mother-infant bonding. During this critical period, however, activity in the amygdalae of the rat pups is inhibited by increased levels of endogenous opiates and by a decrease in the stress hormone corticosterone. This inhibition not only suppresses the learning of aversive associations, but actually causes the pups to be attracted to stimuli that they would normally avoid. For example, if a rat pup of this critical age is trained to associate the scent of lemon with an electric shock, what it learns is a paradoxical preference for the lemon scent, not the aversion that an older rat would learn.[29] Interestingly, this paradoxical learning normally occurs only if the rat pup is in the mother's presence, the sensation of which evidently triggers the release of endogenous opiates and the decrease in corticosterone in the amygdala.[30] The peptide hormone oxytocin appears to play a similar if less extreme role in the amygdalae of adult humans: when taken as a nasal spray, it enters the brain, significantly reduces activity in the amygdalae, and increases the subject's feelings of trust toward other people.[31]

The Visceral Core of Our Moral Nature

Obviously there is much we do not yet understand about the parts of the brain that store innate knowledge in humans, and especially about how that knowledge contributes to neonatal cognition, but a critical insight has recently emerged from the behavior of human patients who have suffered damage to these circuits. In response to sensations of great emotional significance and urgency, like physical pain, hunger, cold, or isolation from its mother, an infant's amygdala and the medial prefrontal network trigger autonomic responses in the body through their connections to the hypothalamus and brain stem. These responses—changes in heart rate, blood pressure, respiration, and other visceral effects—not only prepare the body to cope with the problem, but also generate additional sensations from the body that reach the orbital network. This visceral sensory feedback is likely to be important in neonatal cognition because it appears to be essential for the adult social behavior that will ultimately depend on these same neural structures.[32]

The most striking evidence of this is seen in patients with damage to the medial and orbital parts of the prefrontal cortex. Such lesions cause profound impairments in planning, social interaction, and other decision-making tasks that require proper attribution of emotional consequences to the alternatives. The result is often a drastic change in personality, in which a previously hard-working, prudent, and responsible adult becomes impulsive or indecisive, aimless, rude, profane, and prone to socially inappropriate behavior. Typically such patients are inconsiderate of the feelings of others, take foolish risks, fail to learn from mistakes, and repeatedly make self-destructive choices.[33]

One of the best-documented cases of this kind is patient EVR, who suffered damage to the medial and orbital prefrontal cortex during the removal of a tumor. Prior to his surgery, EVR was a successful accountant at a small company, was married, had two children, and was a respected member of his community. After the surgery he began making poor decisions in almost every aspect of his life. He lost his job, went bankrupt, his marriage failed, and he was forced to move in with his parents. He went on to marry a prostitute then divorced her after two years. By almost every conventional test, his intelligence and memory were intact. When tested on rules of social conventions and morality, his answers demonstrated that he still knew these rules and understood them at an intellectual level.*[34] What he lacked was a visceral response to emotionally charged situations involving moral judgment—what we commonly call gut feelings. Without such feelings to guide his choices, his life fell apart.[35]

Similar effects occur in the early stages of frontotemporal dementia, a devastating degenerative brain disease that often begins with atrophy of the medial and orbitofrontal cortex but that ultimately destroys much more of the frontal and temporal lobes.[36] A tragic case of this kind was described in a PBS news story about Keith Jordan—a well-educated, successful, and loving father whose behavior became increasingly strange.[37] He would repeatedly tell childish or inappropriate jokes and showed no consideration for the feelings of others. His physicians tried various medications and psychotherapy without effect. On one memorable camping trip, he was so indifferent to the safety of his own six-year-old daughter that he pulled her into a deep and swiftly flowing river and let her go; she would have drowned had her mother not rescued her. Jordan

---

* Patients like EVR have an abnormally high tendency to choose utilitarian options when confronted with moral dilemmas that pit an intellectually defensible option (the greatest good for the greatest number of people) against an emotionally aversive one (smother the crying baby to protect a group of civilians hiding from hostile soldiers). See Koenigs et al. (2007).

had become so different from the person he once was—so alien, so bizarre, so immoral—that his desperate family went so far as to try an exorcism.

The ritual had no effect, and eventually Jordan's condition was properly diagnosed, but it was no accident that his suffering family came to see his dementia through the lens of God, demons, and religion. Morality is, after all, much more than a set of arbitrary rules. It comprises rules about how we treat other people, rules that are attended and enforced by powerful emotions, the source of which is an evolutionary imperative for cooperation within human social groups, for pair-bonding of human mates, and for the bond between parent and infant that is essential to human survival. Like our innate fear of snakes and love of sugar, our elementary prosocial behaviors—like protecting and nurturing our children—arise from potent primary reinforcers.[38] It should come as no surprise that all of these behaviors depend on common neural circuits and that elementary moral behavior requires the conjunction of the innate knowledge in these circuits with the gut feelings they generate. If these parts of the brain—the medial and orbital networks—generate an infant's innate image of a loving and nurturing other being, and if this becomes the basis for an adult's illusion of God's presence, then their destruction in Keith Jordan's brain produced for his family an illusion of opposite quality: the presence of Satan.

The most extreme symptoms of this kind occur in patients whose injury to the medial and orbital prefrontal cortex occurs in infancy.[39] Unlike adult-onset cases, these patients never learn any concept of right and wrong. They are delinquent as children and criminal as adults. They lie, steal, have careless sex, and become dangerously negligent parents. They show no guilt or remorse. Their lack of empathy causes so much suffering for the people close to them that their condition is known as acquired sociopathy.

Cases of this kind underscore the critical importance of the innate information about relations with other humans that is stored in this part of the brain. Without this knowledge of social primary reinforcers and their innate associations with punishment or reward, other parts of the brain that normally learn elaborate social norms evidently cannot do so. This is in stark contrast to the neural plasticity that follows damage to the much larger cortical regions that support language. Even if the entire language-dominant cortical hemisphere is surgically removed in early childhood, the cortex in the remaining hemisphere eventually reacquires language through learning.[40] The profound moral deficits of patients with acquired sociopathy suggest that the circuitry of our innate moral intuition is an indispensable foundation for normal adult morality.

Some results of neuroimaging studies are consistent with this interpretation. The medial prefrontal network is specifically activated in tasks that involve taking the perspective of another person, especially when the situation involves a moral dimension.[41] The ventral part of the medial network is engaged when subjects take the perspective of other people they consider similar to themselves, whereas the dorsal part is engaged when they imagine the mental states of dissimilar others.[42] When subjects experience the gut feeling of regret after a losing decision in a gambling task, the orbital network is activated.[43] The orbital network is more highly activated when mothers view photographs or video clips of their own infants than when they see unfamiliar infants or adults, and the amount of activation is correlated with their subjective ratings of positive mood at seeing their infants.[44] In mothers of toddlers, the orbital network is more active when they watch the feeding behavior of their own children than when they watch unfamiliar ones, and higher activation of the orbitofrontal cortex is associated with less parenting stress.[45] Similar responses have been found in the orbitofrontal cortex of fathers in response to their infants.[46]

The Neural Response in Infants

Of course, the neuroimaging studies of greatest relevance here would be ones that examine the brains of infants as they perceive and interact with their mothers. As I explain in appendix A, research of this kind is technically difficult, but some recent studies using the NIRS technique indicate that portions of the frontal lobe in infants, probably the orbitofrontal cortex and perhaps also part of the frontal pole, are activated when an infant perceives its mother. In one study, one-year-old infants watched brief video clips that showed their mothers either smiling or with neutral facial expressions. The response in the infants' orbitofrontal region was significantly greater to the smiling than to the neutral facial expressions.[47]

Responses in the orbitofrontal cortex of newborns have been studied using stimuli in other sensory modalities, also with the NIRS method. One of these examined the response to motherese. In this experiment, two audio recordings were made of each infant's mother reading part of a children's story. For the first reading she was told to read as if she were reading to her infant, and for the second she was told to read as if for an adult listener. Although the sounds were played back to the infants while they were sleeping,*[48] their orbitofrontal cortex responded, and motherese elicited a significantly greater response than did adult-directed reading.[49] Two other studies probed the orbitofrontal cortex of newborns using salient odors. One found a significantly greater response to maternal breast milk than to formula milk.[50] The other found a response to the scent of colostrum that diminished over the first few days of life.[51] In the near

---

* The sensory cortex of newborns is more responsive to stimuli during sleep than is the adult cortex (Lauronen et al. 2012, Martynova et al. 2003).

future we can expect more experiments of this kind, along with the development of better neuroimaging techniques for use with human infants.[52]

Meanwhile we can turn to studies of the infant brain in other species. Like human infants, rat pups are born helpless and completely dependent on maternal care. Much like human babies, they cry when separated from their mother, but their rate of crying greatly accelerates after a brief sensation of her presence—an unusual behavior that might be a useful animal model for the illusory sensed presence in humans. The search for a part of the brain that is specifically associated with this behavior leads us a little deeper into the brain, to a place that is, not surprisingly, heavily interconnected with the amygdala and the prefrontal networks we have just examined. It is a place where learning, motivation, love, and craving converge.

CHAPTER 5

Imprinting, Maternal Presence, and Addictive Love

Inside faith is longing; this is the secret of its energy. This longing is the desire to return to God, to creation itself, to the forgotten state of union we knew before we knew ourselves as individuals.—D. Patrick Miller[1]

I think religion is a drug. It's addictive. Can God deliver a religion addict?—Marjoe Gortner, former evangelist and child preacher[2]

If there is any recurring theme in the story of innate knowledge in the human brain, it is the seemingly paradoxical observation that the parts of the brain that store such information are primarily dedicated to learning. The paradox is merely apparent, not real. Innate knowledge is the essential skeleton on which a human mind is built through a lifetime of learning. Our evolutionary history informs our neonatal brains about the rewarding value of sweet tastes, but it cannot know whether we will eventually find kiwi fruit or peaches in our foraging. We learn from experience how to recognize these in their ripe condition, but our preference for them over less nourishing alternatives is inherited. Similarly, the innate neural model of the mother I described in chapter 1 is the skeleton of an infant's first and most important interpersonal relationship, but it is only that. The fleshing out of that bond begins immediately after birth, as the infant learns the details of its mother's face, touch, scent, and voice.

Crying Rats, Trembling Hands, and Babbling Chicks

This dependence of learning on a foundation of innate knowledge occurs in many other species. A good example is an intriguing aspect of mammalian mother-infant bonding that is most often studied in rats. When a rat pup is separated from its mother, it emits ultrasonic cries to summon her help and stops crying when reunited with her. If the pup has had about a week to learn some details of its mother's scent and the feel of her body—a restricted form of learning called imprinting—then it reacts in an unexpected way to a brief contact with her during its isolation. If the isolated pup is exposed to the mother only for a minute or so, the pup resumes its crying after her removal, but at a much higher rate, often double or even triple its initial rate of crying. This behavior, called maternal potentiation, appears to be the pup's enthusiastic effort to help the mother find it, spurred on by its recent sensation of her presence. The behavior makes sense in the wild because that recent sensation means the mother is probably nearby. In the laboratory, however, the sensation of the mother's presence is illusory during the pup's exaggerated crying because the pup is in a separate plastic chamber, and the mother is nowhere nearby.[3]

Variants of this behavior have been described in other mammals,[4] and it may have relevance to the illusory sensed presence in humans. If we could find a part of the rat brain that is in some way specifically involved in maternal potentiation, then we might have an important clue as to where and how to look in the human brain for correlates of the sensed-presence illusion. Such a place has been found, and what we know about it from other lines of research has much to say about the nature of the religious impulse—to the extent that the longing for God and an infant's longing for its mother spring from the same neural circuit. Most of all it promises to illuminate that deeply mysterious, powerful, and tenacious grip that religious emotion has on the human mind.

The discovery was made by Harry Shair and his colleagues, who have searched not only for a location in the brain, but also for a neurochemical sensitivity that is specifically related to maternal potentiation. They found that this behavior is specifically blocked by a drug, quinpirole, that mimics the action of the neurotransmitter dopamine at a specific family of dopamine receptors denoted D2. The effect was specific because the drug did not affect the rate of ultrasonic cries of the isolated pup prior to the brief exposure to the mother. Its only effect was the elimination of the increase in crying rate after that exposure.

The initial discovery was made by systemic injection of quinpirole into the rat pups and so revealed nothing about what part of the brain was involved.[5] More recent experiments, however, have used microinjections of the drug into tiny regions of the brain to find its critical site of action.[6] That critical site turns out to be the nucleus accumbens, which is in the ventral part of an elaborate group of neural circuits in the forebrain known collectively as the basal ganglia. This neural structure lies deep in the forebrain of all jawed vertebrates, which suggests a role in something basic and essential to the vertebrate way of life.[7] Its evolutionary origin may lie even deeper in the tree of life, as something like it has been found in the brains of arthropods—that vast invertebrate phylum that includes insects, arachnids, and crustaceans.[8] What is the normal function of the basal ganglia, and what exactly is the role of the part of it called nucleus accumbens?

As it is for many parts of the brain this far removed both from sense organs and muscles, our understanding of the basal ganglia is incomplete, tentative, and a bit vague in places. It comes from studies of the neurochemistry, anatomical connections, and electrical activity of these circuits in experimental animals; from the symptoms of human patients with pathologies of the basal ganglia; from studies of highly specialized animal behaviors that involve this part of the brain; from the neurology of social bonding; and from the darker side of human compulsions. I can best convey some of that understanding by describing a few specific examples, from which I draw some tentative conclusions about the role of the nucleus accumbens in the human longing for food, water, mother, mate, and God.

Most areas of the cerebral cortex have connections with some part of the basal ganglia. These connections form complex feedback loops that send signals from the cortex to basal ganglia and thence to several other subcortical structures that ultimately send signals back to the basal ganglia and cortex.[9] These subcortical structures include neurons that produce the neurotransmitter dopamine, and a disease that affects these neurons provided early clues about the role of the basal ganglia.

Parkinson's disease involves the gradual death of the dopamine-producing neurons that innervate the dorsal part of the basal ganglia. This region participates in feedback loops involving the motor cortex, and deficits in movement are the most prominent early symptoms of the disease. These include tremor, rigidity, abnormal posture, difficulty with fine motor control of the hands during complex sequences of movement, problems with initiating movement, and slow, poorly controlled walking.[10] The difficulty in initiating movement suggests a deficit not only in the control of behavior, but also in the motivation to act. The symptoms become debilitating as the disease progresses, but they can be dramatically relieved, for a while, by drugs that increase the level of dopamine in the brain.[11] The implications are that the basal ganglia have some role in motivation and in behaviors that involve complex sequences of movements and that the neurotransmitter dopamine is critically important in these processes.

A similar but more detailed picture of what the basal ganglia do emerges from the study of their involvement in birdsong, another complex learned behavior that develops on a foundation of innate predisposition. The basal ganglia circuits in birds have a different anatomical organization than in mammals, and the part of the circuit dedicated to singing is anatomically distinct. These differences, along with the relative simplicity of recording and measuring singing behavior, have made this an excellent experimental system for exploring this part of the brain.[12]

In many species only the male sings, and the song is used both to establish territories and to attract mates. A young songbird learns the characteristic song of its species from its father during a critical learning period—another example of imprinting. During this time, it is only memorizing the song it hears, not trying to produce a song of its own.[*][13] A bird that is raised in isolation and therefore deprived of a tutor's song ultimately produces an abnormal song that retains only some of the features of its species-typical song. The degree of similarity to the normal song varies with the species, but these experiments clearly reveal a crude innate knowledge or template of the song that must be fine-tuned through learning.[14] A specialized cortical area called LMAN is essential for this learning and may contain some of the innate aspects of the song template—rather like the orbital and medial prefrontal networks in mammals.[†][15]

Having memorized the tutor's song, the bird then learns to sing through a process reminiscent of human infantile babbling. Its first attempts are meaningless, disorganized outbursts of syllables in abnormally great variety and in no particular order. Over time, syllables not present in the memorized template are eliminated, whereas those similar to ones in the template are retained, refined, and gradually rearranged in sequence so as to match the correct song.

The basal ganglia circuit is essential for this second phase of learning to sing. If it is damaged before the bird starts trying to sing, its song will never advance beyond the early babbling stage. The circuit is not, however, necessary for the production of the song once the bird has learned to sing.[16] If it is damaged at that point, the bird continues to sing normally, though with a loss of some of

---

* This description applies to seasonal songbirds. For more rapidly developing species, like the zebra finch, there is some overlap between the periods of memorizing the tutor's song and learning to sing (Doupe et al. 2005).

† LMAN stands for lateral magnocellular nucleus of the anterior nidopallium. Birds do not have cortex, as the term is strictly defined for the mammalian brain, but both LMAN and mammalian cortex develop from the same part of the brain in the embryo: the pallium (roof) of the forebrain. The learning of the tutor's song depends on a specific kind of synaptic plasticity that occurs in LMAN only during a brief period in the early life of the bird, coincident with the critical period for this imprinting. Destruction of LMAN has essentially no effect in an adult bird that has already learned to sing, but if LMAN is lesioned in a young bird, its song crystallizes prematurely to an abnormally simple song, even less like the template than the song of an intact bird raised in isolation. This evidence strongly suggests a role for LMAN in the learning and representation of the song template, but other parts of the brain are also likely to be involved. See Adret (2004) and Doupe et al. (2005).

the subtle variability in singing that would otherwise occur in some natural contexts.[17] These and similar experiments show that young birds learn to sing by trying, initially with little success. Yet it is essential that they hear their own bad singing, which they compare to the memorized template. Partial successes must be recognized as such and reinforced. The basal ganglia appear to be involved in many aspects of this process, but especially in the reinforcement of matching sequences of syllables, the purging of erroneous ones, and the adjustment of pitch within a syllable.[18] This reinforcement learning requires a form of synaptic plasticity that depends on a dopamine signal at the D1 family of receptors, and the critical period for this second phase of song learning begins at about the time that dopamine-containing nerve fibers grow into the basal ganglia.[19]

Another important insight from studies of birdsong is that the basal ganglia are involved both in sensation and behavior. Neurons in the basal ganglia respond to the sound of the bird's own song played back to it, to the tutor's song, or to fragments of song containing a few syllables in proper sequence.[20] They are also active when the bird is singing, even if the bird is deafened, which means that the activity in this case is related to the behavior, not the sound of the song.[21] Even so, the basal ganglia are not necessary for the behavior, because their destruction in an adult does not eliminate singing.[22] It does, however, eliminate some of the variability in the song, and it prevents the deterioration in the quality of the song that occurs in deafened adults that have intact basal ganglia.[23] These experiments show that the basal ganglia are continually making fine adjustments to the song, even in adult birds. When an adult is deafened, this adjustment continues without the guidance of normal sensory feedback, and therefore the song gradually deteriorates. A similar process may contribute to the symptoms of Parkinson's disease, considering that surgical destruction of part of the basal ganglia can sometimes reduce the symptoms in the most intractable cases.[24]

Parkinson's disease involves a part of the basal ganglia that connects with the motor cortex. Learning to sing involves a part of the bird's basal ganglia that connects with a song-related cortical structure. But it was the longing of an infant rat for its mother that first brought us to a part of the mammalian basal ganglia called nucleus accumbens. With what part of the cortex is it connected, and what does it do?

Nucleus accumbens connects with the cortical structures we explored in the previous chapter, the orbital and medial prefrontal networks and the amygdala.[25] In these we found innate knowledge of attractive and aversive stimuli essential for survival—including social stimuli like facial expressions and vocalizations—along with powerful innate associations of these with reward or punishment and learned associations with other stimuli. Presumably the nucleus accumbens does for these aspects of emotion and cognition something analogous to what other basal ganglia circuits do for the motor cortex in humans and for the song-related cortex in birds. It is likely to involve both sensation and behavior—specifically the learning of complex or sequential behaviors in a process that requires reinforcement and sensory feedback. From what we know about the orbital and medial networks, those behaviors probably include such things as feeding, drinking, and basic social interactions like mating, maternal caregiving, and suckling. Neural activity in the nucleus accumbens probably also affects the motivation for these behaviors, and we already know from Shair's experiments with rat pups that dopamine is somehow involved.[26]

Prairie Voles and Social Bonding

My next example takes us back to the nucleus accumbens and confirms some of these expectations, though in the brain of a different rodent, the prairie vole. These inconspicuous animals live in burrows in the grasslands of central North America, but they are rising stars in the firmament of neuroscience because of their unusual sexual and social behavior. In the wild, prairie voles live in extended family groups and form monogamous pair bonds when they mate.[27] The strength of this bond can easily be measured in the laboratory: given a choice between two chambers, one containing its mate and the other an unfamiliar opposite-sex vole, a pair-bonded prairie vole spends much more time with its mate.[28] The act of mating creates this preference in both sexes, but more profound behavioral changes occur in the male. It becomes aggressive toward male competitors and toward females other than its mate, and it helps its mate care for the pups.*[29]

What most excites neuroscientists about the prairie vole, however, is its status as a natural experiment: several closely related species, like the meadow and montane voles, live more solitary lives and mate promiscuously, and only females care for the pups.[30] These profound differences in social bonding are evident even in the behavior of the pups. The highly social prairie vole pups cry vigorously for their mothers when separated, but montane vole pups emit few if any isolation cries—a rare exception among mammalian infants.[31] Comparison of the brains of monogamous and social versus promiscuous and asocial voles has begun to reveal the neurological basis for these behavioral differences, and some of the most interesting clues have turned up in the nucleus accumbens.

The most striking differences in the brains of these species involve oxytocin and its molecular cousin vasopressin. In mammals these hormones act in various

---

* Wang et al. (1997). There is evidence that female prairie voles also become aggressive toward strangers after pair-bonding (Getz et al. 1981) and that some pair-bonded voles occasionally cheat on their partners (Ophir et al. 2008).

parts of the body to regulate such basic functions as childbirth, milk production, blood pressure, water retention, and body temperature.[32] They are also released by neurons in some parts of the brain, where they act as neuromodulators.[33] The density and location of the nerve cells and fibers that contain oxytocin and vasopressin are similar in prairie voles and in their promiscuous cousins, but the distributions of the receptors for these molecules are strikingly different. Oxytocin receptors are abundant in the nucleus accumbens of the monogamous prairie voles, and vasopressin receptors adorn an adjacent group of neurons—the ventral pallidum—that receives output signals from the nucleus accumbens. By comparison, the corresponding parts of the brain in promiscuous voles are nearly devoid of these receptors, which suggests that pair-bonding depends on the action of oxytocin in the nucleus accumbens and vasopressin in the neighboring ventral pallidum.[*34]

Other experiments confirm that suggestion. Direct injection of a vasopressin receptor blocker into the ventral pallidum of a male prairie vole prevents the formation of a partner preference when it mates.[35] The same deficit occurs in a female prairie vole when its oxytocin receptors are blocked in the nucleus accumbens, and direct injection of oxytocin into the nucleus accumbens of a female induces a partner preference in the absence of mating.[36] These and similar experiments identify the nucleus accumbens as a central player in sexual pair-bonding, but there is more to the neurochemistry of this behavior than oxytocin and vasopressin.[†37]

The neurotransmitter dopamine also plays a big role in this story. The concentration of dopamine increases in the nucleus accumbens when a prairie vole mates, and this too is an essential neurochemical signal for pair-bonding.[38] Injections of drugs that mimic or block the effects of dopamine at the D2 family of dopamine receptors show that activation of these receptors in the nucleus accumbens is both necessary and sufficient for the formation of a partner preference in both sexes of prairie voles.[39] Activation of the D1 family of receptors, however, has more complex effects. Drugs that activate D1 receptors prevent the formation of a partner preference during mating, but, weeks later, D1 receptors contribute to a different aspect of pair-bonding behavior. After a male prairie vole has mated and lived with its partner for a week or two, it aggressively repels other females—even sexually receptive ones—and this behavior depends on an increase in the density of D1 receptors in its nucleus accumbens.[40]

---

* The nucleus accumbens and ventral pallidum are not the only places in the brain where there are differences in the densities of these receptors between monogamous and promiscuous voles, but they, along with the medial prefrontal cortex and lateral septum, are the places most critical for the formation of a partner preference. See Young et al. (2008) and Insel (2010).

† There is also more to the behavior than the formation of a partner preference. The development of paternal care of the young and aggression toward non-partner adults are other important aspects of the pair-bond, and these are also regulated by the neurochemical systems discussed here. See Young et al. (2011) for details.

The prairie vole story has implications for the role of the nucleus accumbens in maternal potentiation in rat pups, but it also addresses broader issues. That the nucleus accumbens and its affiliated circuitry are centrally involved in both adult sexual pair-bonding and maternal potentiation behavior supports the idea, which I discussed in chapter 2, that evolution co-opted some of the circuitry of mother-infant bonding in the subsequent development of monogamy in those mammalian species that have it. This commonality between sexual and mother-infant bonding also extends to neurochemical mechanisms—notably the oxytocin/vasopressin and dopamine systems—but with some significant differences in how these systems come into play.

In the pair-bonding of adult prairie voles, there are distinct and specialized roles for vasopressin in males and oxytocin in females. In rat pups, oxytocin decreases the rate of isolation crying[41] and is necessary for learning to associate an odor stimulus with the presence of the mother,[42] but its effects on these behaviors are the same in both sexes.

The role of dopamine is even more complex. In popular literature it is often called a pleasure molecule or the neurotransmitter of reward. There is good evidence for that interpretation, but it greatly oversimplifies the multifaceted action of dopamine in the brain.*[43] We have already encountered the two families of dopamine receptors denoted D1 and D2. The activation of D1 receptors usually increases the excitability of neurons, whereas activation of D2 receptors has an inhibitory effect.[44] Both of them act through a complex chain of biochemical reactions within the neuron but with essentially opposite influence on those reactions.[45] In neurons of the basal ganglia, including the nucleus accumbens, the two receptor types are segregated such that each neuron tends to have either mostly D1 or mostly D2 receptors, but most neurons have at least some of both.[46] Dopamine therefore appears to be simultaneously pushing and pulling on its target neurons in the nucleus accumbens.

The significance of this is not fully understood, but it suggests that these neurons are finely tuned and highly sensitive to changes in dopamine concentration. This may reflect another important dimension of dopamine signaling: its rate of release.[47] Some of the neurons that supply dopamine to the nucleus accumbens release it in a slow but steady trickle. Others release it only rarely but in extremely short bursts of high concentration. Those quick, transient high doses of dopamine are often released in synchrony with salient events in the

---

* Here I emphasize the roles of dopamine and the nucleus accumbens in appetitive reward seeking, as this appears to be the essence of what they contribute to mother-infant bonding. Like the cortical structures with which it connects, however, the nucleus accumbens is concerned with both rewarding and aversive stimuli. The same is true of the neurons of the ventral midbrain that supply dopamine to the nucleus accumbens: some signal unexpected reward, others unexpected punishment. For more thorough treatments of these aspects of the nucleus accumbens and the dopamine system, see Brischoux et al. (2009), Faure et al. (2010), Salamone et al. (2005), and Schultz (2010).

environment (especially ones that are unexpectedly rewarding) and reinforce the animal's associated behavior.[48] By contrast, the much lower but relatively steady concentration of dopamine determines the animal's general level of motivation. The higher this background concentration of dopamine, the more the animal is biased toward behavioral strategies that require more effort but promise greater reward.[49] The opposite extreme can be seen in mice genetically engineered to be unable to synthesize dopamine. They are so devoid of motivation that they do not eat and will starve to death unless given injections of a dopamine precursor.[50]

These two aspects of adaptive behavior—learning and motivation—are often evident in the distinct behavioral roles of the two dopamine receptor families. One recurring theme is that activation of one receptor type is involved in the learning of a behavior, after which activation of the other type motivates expression of the behavior. In many forms of reinforcement learning, like a rat learning to press a lever for a food reward or a songbird learning to sing, activation of D1 receptors is necessary during the learning phase.[51] There are, however, important variations and exceptions. Recall the sexual pair-bonding of adult prairie voles, in which the activation of D2 receptors is essential for the learning of a partner preference, whereas the D1 receptors become important in the expression of a subsequent aspect of pair-bonding—aggression toward other voles.[52]

In the maternal potentiation behavior of rat pups, the accelerated crying that follows a brief sensation of the mother's presence is specifically eliminated by the activation of D2 receptors in the nucleus accumbens by the drug quinpirole, which mimics the effects of dopamine at these receptors. This is an effect not on learning, but on the expression of a behavior. Maternal potentiation and its blockade by a dopamine-like drug both appear to be changes in the motivational state of the rat pup, precipitated by changes in that delicate balance between the pushing and pulling by dopamine. Upon sensing the mother's presence and then her disappearance, the pup evidently switches to a strategy involving greater effort (more vigorous crying) but also potentially greater reward (more rapid reunion with the mother).*[53]

This behavioral state may be qualitatively different from the pain and stress of isolation crying. Although the pup is crying much more vigorously during maternal potentiation, it may be in a pleasurable state of appetitive reward seeking, triggered by the sensation of its mother's presence. This interpretation is consistent with experiments involving infant monkeys that I described in chapter 1: when an isolated infant monkey can sense the presence of its mother yet

---

* It is not yet known whether this change in strategy is the result of a change in general dopamine level in the nucleus accumbens or is triggered by a high-concentration burst of dopamine at specific synapses. The quieting of the crying pup when it is reunited with the mother is also highly sensitive to dopaminergic signaling in the nucleus accumbens, and this sensitivity applies only to contact with the mother, not to littermates or nest material. See Shair et al. (2009).

still not reach her, its crying accelerates while its stress level (indicated by cortisol concentration) declines.[54] This view also comports with the dramatic change in subjective emotional state often described by humans in their accounts of the sensed presence during mystical experience: from painful loneliness and desperation to the perception of a loving and comforting being—enfolding, embracing, or just out of reach.[55]

A similar emotional state may attend the mother's part of the mother-infant bond. The response of a mother rat to her infant's cries also appears to be appetitive reward seeking because it too is driven by dopamine signals in the nucleus accumbens. Rat mothers lavish attention on their pups by licking and grooming them, but they are not all equally attentive. Individual differences in this behavior are abolished by drugs that eliminate differences in the dopamine signal in the nucleus accumbens.[56] The concentration of dopamine increases in the mother's nucleus accumbens while she is licking her pups, and an oxytocin signal from the hypothalamus evidently triggers this increase.[57] An even more critical maternal behavior—the retrieval of pups separated from the nest—is disrupted by the injection into the mother's nucleus accumbens of a drug that blocks activation of the D1 family of dopamine receptors.[58] But probably the most dramatic demonstration of this compelling appetitive drive is the fact that, in the first postnatal week, a rat mother prefers her pups to cocaine. Her maternal drive gradually declines with time, however, and by about postnatal day ten, cocaine wins the contest.[59]

Spiritual Methadone

This brings us to another form of compulsive and appetitive reward-seeking behavior: drug addiction. Nearly all addictive drugs exert their tenacious grip, either directly or indirectly, by increasing the concentration of dopamine in the nucleus accumbens.[60] They hijack the circuitry that normally compels an animal to seek rewards of such great importance that reproductive success is at stake—such rewards as attend eating, drinking, mating, caring for young, or seeking mother. With addiction, those healthy compulsions are short-circuited and redirected toward getting more of the drug. The relapse of a former addict who has just been handed a crack pipe may therefore have much in common with the enthusiastic crying of a rat pup that has just felt the presence of its mother.[61]

This deep connection between addiction and the mother-infant bond lends further support to the connection between that bond and the illusion of God's presence. If that illusion has its basis in an innate model of the mother, then it too represents a hijacking of this powerful motivational circuitry, and this could explain the extraordinary tenacity of religious belief. It might also explain the efficacy of twelve-step programs for recovering addicts. Much of their success stems from social interaction with supportive peers, which is yet another example of a rewarding behavior that increases dopamine in the nucleus accumbens,[62] but their emphasis on a mystical higher power may be equally important. In turning to their higher power, recovering addicts may be replacing one addiction with another—embracing a spiritual methadone that gradually weakens the tenacious grip of the drug it replaces. This interpretation predicts that the nucleus accumbens should be specifically activated during the kind of intense religious experience that evokes the feeling of God's presence. As I explain in chapter 8, at least one fMRI study of extremely devout subjects has confirmed this.

We originally ventured into the nucleus accumbens in a quest for the circuitry of the sensed presence. What we found seems mainly concerned with only one dimension of that illusion: the intense appetitive longing for that other being, the compulsive drive to reach it, the tenacious grip it exerts on the mind. The rewarding, attractive, and loving aspects of that sensed presence—as well as the painful or desperate feelings that elicit the illusion—fall squarely within the province of the nucleus accumbens, the amygdala, and the medial and orbital prefrontal networks. But the illusion is typically more than just these emotions. It also includes a sense of agency, of another conscious being with intentions distinct from the self and with spatial attributes—either at a specific and nearby place or completely enveloping the person who feels it. There is another part of the brain that deals with just this kind of thing, and it has connections with these prefrontal networks.[63] If we follow those neural fibers to their source, they lead us to an intriguing place farther back in the brain, where the temporal and parietal lobes meet.

CHAPTER 6

Self, Other, and the Illusory Sensed Presence

> Among the vague and curious anomalies of corporeal awareness (or body-image) occurring with lesions of the subordinate parietal lobe, an idea of a "presence" may intrude.—neurologist MacDonald Critchley[1]

How do you distinguish between yourself and another human? How do you perceive the intentions of others who are affectionate, angry, disgusted, sad, threatening, or indifferent? Most of us do these things so effortlessly and naturally that it seems pointless even to ask such questions. Yet behind them lies a computational problem of staggering complexity, one that has vexed cognitive scientists for decades. For many years psychologists assumed that our competence at these tasks is entirely a product of learning during childhood, and Jean Piaget went so far as to argue that a young infant cannot distinguish between itself and its mother.[*2]

Recent studies, however, contradict this view. In chapter 2 I described evidence that newborn humans can imitate several basic facial expressions, that they can distinguish a point-light display of biological motion from a display of the same points moving randomly, and that they prefer to look at the biological point-light display. These and similar experiments show that the human brain is wired at birth with some incomplete yet functional circuitry for distinguishing between self and other and between inanimate objects and living beings that act with intention.

---

* Or, for that matter, between itself and anything else. See Piaget and Inhelder (1969) pp 22–24.

Embodiment

That human newborns have some ability to imitate others suggests a strategy the brain may be using to solve these problems: simulation.[3] With this strategy, a part of the brain that generates a behavior—a facial expression, say—is also part of the circuitry that interprets that expression when it appears on the face of another person. The brain internally simulates the generation of the facial expression and, in so doing, internally generates the mood or intentional aspect associated with that expression.[4] In effect, the brain is automatically and unconsciously reasoning, "If I pulled up the corners of my mouth in that way, it would be because I felt happy. Therefore this person in front of me is happy." For this strategy to work, however, the brain must have not only that association between mood and expression, but also an association between the generation of that expression and its visual appearance on another face. The imitative abilities of newborns demonstrate that their brains have the latter association, at least for a few elementary expressions.

For the broader problem of reading intention from the bodily action of others, the brain must also have a deep sense of connection between its own intentional state and the position and movement of the body that contains it. Neurologists call that deep sense of connection embodiment, but, as psychologist Susan Blackmore explains, the concept comprises several distinct aspects of the relation of self to body:[5]

self-location (the position where I seem to be)
body ownership (the knowledge that this body is mine)
sense of agency (the feeling that I am in control of the body's actions)
first-person perspective (the position from which I am seeing the world)

Normally these coincide: I feel myself to be where my body is, I control it, and I experience reality through its sense organs.

Embodiment is such a fundamental aspect of human experience that it is difficult for most of us even to grasp the need for such a concept or to imagine that this could somehow be a problem the brain needs to solve. For neurologists, however, the need is obvious because they have seen in their patients what happens when the circuitry of embodiment goes awry. They have also found clues about where in the brain that circuitry lies.

Beneath a place on the head slightly behind and above the ear lies the transition zone between the temporal and parietal lobes—the temporoparietal junction.*[6]

---

* There are no precise anatomical boundaries for this cortical region. As I use the term in this book, the temporoparietal junction is a cortical area roughly centered on the posterior end of the lateral sulcus,

When this part of the cortex is damaged, especially on the right side,* human patients behave most strangely. If their brain damage does not extend into the motor cortex, they retain the ability to move their left arm and hand, but they may lose their sense of ownership of that arm. Its movements seem purposeless to them, and they may insist that another person is controlling their left arm.[7] This is the aptly named "alien hand syndrome" that was immortalized by Peter Sellers in his portrayal of Dr. Strangelove.[†8]

More extensive damage in this region causes the even more bizarre condition known as hemispatial neglect. Such patients neglect and often deny the existence of the left half of their body. They are not blind on the left side, but they have lost on that side that deep sense of connection between body and mind. They may struggle to shove their left limbs out of bed, convinced they are part of a stranger who crept into bed with them. They may shave or put makeup on only the right side of their face. When they dress, they may put clothes on only their right side. The neglect often extends to nearby extrapersonal space. For example, they may eat only the food on the right side of their plate. When asked to draw a clock face, they may draw a complete circle, but then cram all the numbers from one to twelve into the right half of the circle. In extreme cases they are unable to perceive that anything is wrong with their bodies, even if they have suffered a severe stroke and are paralyzed on the left side. When told that they cannot move their left arm or leg, they simply deny this or confabulate.[9]

As if that were not strange enough, some pathologies of the temporoparietal junction go even further, precipitating a total breakdown of embodiment: the illusion of an out-of-body experience.[10] Accounts of this fascinating phenomenon have been offered as evidence for mind-brain dualism and the existence of an afterlife, but properly controlled tests of that hypothesis consistently fail.[11] The illusion of total disembodiment appears to be an entirely natural consequence of neurological disturbances,[12] and we encounter it again in my discussion of schizophrenia in chapter 11. Here I am more concerned with what we

---

including the posterior part of the superior temporal sulcus and gyrus, along with the inferior part of the posterior parietal lobe. There are differences in specialization among the various parts of this cortical region, the details of which I mostly ignore in this chapter. For an interesting study of those details, see Bahnemann et al. (2010).

* The asymmetry appears to be a consequence of different specializations in the temporoparietal junction on the left and right sides of the brain. On both sides it deals with embodiment, agency, and orientation of the body in its surrounding space, but these functions are mainly a specialty of the right temporoparietal junction—so much so that it represents both the left and right sides of body and space. By contrast, the left temporoparietal junction is in part specialized for language comprehension; its more limited representation of body and space cover only the opposite (right) side.

† The kind of alien hand behavior that results from parietal lobe damage tends to be more random and aimless than that of Dr. Strangelove's malevolent gloved hand (Kubrick et al. 1964), but damage to the anterior corpus callosum, caudal anterior cingulate cortex, and/or other medial prefrontal regions can produce such symptoms—even the grasping at the throat (Banks et al. 1989).

have learned about embodiment and the temporoparietal junction in healthy human subjects and in monkeys.

It might seem helpful to examine the temporoparietal junction in monkeys, considering the deep insights into neuronal function we have already seen from studies of their orbitofrontal cortex and amygdala. Such experiments have been done, but the neurons of the temporoparietal junction in rhesus monkeys are mainly concerned with eye movements, visual orientation to interesting objects in nearby space, and visually guided reaching and grasping with the hand.* In humans, tasks of this kind activate the superior parietal lobe, an area just above the temporoparietal junction, and lesions there do not cause the profound hemispatial neglect syndrome that occurs with temporoparietal damage. In monkeys, no cortical damage in or around the temporoparietal junction has ever caused a neglect syndrome like the one in humans.[13]

These observations suggest that the temporoparietal junction of humans is a recent evolutionary invention and has no obvious homologue in monkeys. This is consistent with the high degree of specialization of this area on the two sides of the human brain, with embodiment on both sides but mainly on the right, and some aspects of language comprehension and arithmetic mainly on the left. Even in a parietal region that humans and monkeys have in common— the superior parietal lobe—humans have some interesting specializations. As in monkeys, this area is activated by tasks that require judging physical distances between objects, but in humans it is also activated during the evaluation of social distances—the degree of similarity or compatibility of personality as judged by a first impression of other people's faces.[14] As we shall see, and despite its obvious role in embodiment of the self, the nearby temporoparietal junction has an even greater interest in other people—one that may derive from its more general role in selective attention.[†15]

Reading Minds

Junction turns out to be an appropriate moniker, for reasons that go beyond the anatomical. Many different kinds of sensory information converge in the temporoparietal junction, including vision, hearing, touch, sensations of limb position, and the vestibular sense of bodily acceleration and orientation with respect to gravity.[16] These are exactly the sensations one would expect in a brain region that handles embodiment of the self, but the visual sensations in particular also suggest an interest in the bodies of others. Much of this visual input is already highly

---

* This description applies mainly to the inferior parietal lobe.
† In chapter 10 I revisit the role of the temporoparietal junction in directing one's attention.

processed, arriving through a part of the temporal lobe that is specialized for the detection of socially relevant stimuli like hand and limb gestures, lip movements, facial expressions, and eye contact.[17]

Given this proclivity for the sight of natural biological motion, it is not surprising that the temporoparietal junction is also specifically activated by biological motion in its most minimal and essential form: point-light displays of humans doing familiar things like walking, throwing, or kicking. It is much less activated by the same displays presented upside-down, and less still by displays of the same points moving exactly as in the upright biological motion case but with their starting positions randomly scattered. It is even activated when subjects are instructed to imagine a display of biological motion, though not as strongly as when they see the real thing. The response to biological motion tends to be greater in the right temporoparietal junction than in the left, and this bias occurs no matter where in the visual field the stimulus is presented.[18] These experiments show that this part of the brain is specifically involved in the subjective conscious perception of biological motion, but not that it is necessary for that perception. That proof requires a demonstration that interfering with the function of the temporoparietal junction interferes with the perception of biological motion.

This can be done using a method called repetitive transcranial magnetic stimulation (rTMS), which I discuss in some detail in appendix A. In these experiments, magnetic fields applied to the head induce electrical currents in a small region of cortex below the magnetic coil, and these currents temporarily scramble the neural computation in that part of the brain. When the activity of the right temporoparietal junction is disrupted in this way during the presentation of point-light displays, the subject's ability to perceive the biological motion is temporarily degraded. This effect is specific to the temporoparietal junction because application of the rTMS to a different part of the brain that also responds to visual motion has no effect on the subject's perception of biological motion.[19]

Biological motion can be ambiguous. For example, the sight of a moving hand could refer to one's own hand or to that of another person. Some visual areas of the brain make no distinction between these two cases, but the temporoparietal junction does. It correlates visual stimuli with signals it receives from the motor system so that it responds less, for example, to the sight of opening and closing one's own hand than to the identical sight presented on a video screen with a slight delay, which is perceived as the action of another person. This ability makes sense for a part of the brain that handles embodiment, but it is also useful as part of a mechanism for distinguishing self from other.[20]

The temporoparietal junction shows up in a great many neuroimaging studies that involve some kind of discrimination between self and other, even when the subject's sense of agency comes from the experimenter's instructions rather than from motor feedback. For example, in one such study the subject operated a joystick to move a small circular dot through a simple maze on a computer screen. On some trials he was told that his joystick controlled the dot, but on others he was told that the experimenter controlled the movement. In all trials, however, he was told to move the joystick as if he had control, so that his sensations of moving his hand and fingers were the same in all cases. The temporoparietal junction was active only when another person had control.[21]

A similar study found an important role for the temporoparietal junction in various aspects of imitation. The subject and experimenter manipulated identical sets of simple objects on identical tables. The subject could feel only his own tabletop and objects but could see the experimenter's work area on a video screen. The temporoparietal junction was activated, mainly on the left, when the subject imitated the other. It was active predominantly on the right when the subject saw his own actions being imitated by the other.[22]

A more direct test of the discrimination of self from other used a series of digitally altered images of each subject's face, morphed to various degrees into the face of a familiar friend of the same sex. The subject had to rate each image, presented in random order, as either "self" or "other." This task activates a network that includes the right temporoparietal junction.[23] In a follow-up study, rTMS applied there significantly diminished the subjects' performance at discriminating self from other, which shows that the right temporoparietal junction is necessary for this task.*[24]

Clearly this part of the brain is involved in perceiving the presence of another person, but it also does something much more subtle: it infers the intentions or mental states of other people from their actions. This is especially evident in situations that demand some extra effort to make sense of another person's mental state. In one such neuroimaging experiment, the subject sees an actress at a table with two cups on it. The actress turns her head and eyes to one cup, smiles with raised eyebrows and says "Ahhh!" She then turns to face the subject and picks up one of the cups. If she picks up the one toward which she just expressed a positive emotional reaction, her action is congruent with the intention implied by her expression, and the right temporoparietal junction is activated. If she picks up the other cup, however, the temporoparietal junction is even more highly activated, as if it were working extra hard to make sense of the incongruous behavior of this strange person.†[25]

Although it was the sight of biological motion that triggered the inference of intention in the "two cups" experiment, verbal stimuli are equally effective. This can be seen in neuroimaging studies in which the subject reads a simple

---

* The same rTMS treatment applied to the left temporoparietal junction had no effect.

† Similar results were obtained when the actress displayed emotional disgust toward one of the cups, in which case picking up that cup was the incongruous action.

story that requires the attribution of intention to the story's protagonist. For example:

> A boy is making a paper-mache project for his art class. He spends hours ripping newspaper into even strips. Then he goes out to buy flour. His mother comes home and throws all the newspaper strips away.[26]

The last sentence compels the inference of the mental state of the mother, specifically her false belief that her child just made a pointless mess and failed to clean up after himself. Reading this story activates the temporoparietal junction, typically on both sides of the brain. The right-side bias, common in the perception of faces and biological motion, is absent here, probably because the task involves language comprehension.[27]

Moral Judgment

Just as rTMS was used to show that the temporoparietal junction is necessary for the perception of biological motion and the discrimination of self from other, analogous experiments show that it is also necessary for moral mind reading. These experiments used stories like the following:

> Grace and her friend are taking a tour of a chemical plant. When Grace goes over to the coffee machine to pour some coffee, Grace's friend asks for some sugar in hers. The white powder by the coffee is just regular sugar. Because the substance is in a container marked "toxic," Grace thinks that it is toxic. Grace puts the substance in her friend's coffee. Her friend drinks the coffee and is fine.

After reading this story, the subject is asked to rate the morality of Grace's act of putting the white powder into the coffee. Although no harm was done—because the container was mislabeled—it is clear from the story that Grace believed the powder to be toxic and therefore must have intended to kill her friend. Subjects normally rate this act a highly forbidden moral transgression, comparable to any attempted murder. If, however, a brief pulse of rTMS is applied to the right temporoparietal junction exactly when the subject is asked to make the moral judgment, then the normal working of that neural circuitry is momentarily disrupted, and the subject considers Grace's behavior significantly more permissible. Since moral culpability depends on the perpetrator's intent, this experiment shows that a normally functioning temporoparietal junction is necessary for inferring the intentions and beliefs of other people.[28]

The Sensed Presence

We have found in the temporoparietal junction many computational features we would expect in a part of the brain responsible for the spatial and intentional aspects of the sensed presence: a confluence of many sensory modalities, embodiment, the perception of biological motion, the distinction between self and other, the inference of the mental states of others from their actions, and even moral judgment. The problems it solves seem deeply intertwined, as are the sensations it uses to solve them. The vestibular sense, for example, is essential for the problem of embodiment because it provides precise and immediate feedback on the effects of one's own movement. Vestibular disturbances appear to be central to many pathologies of embodiment,[29] and artificial vestibular stimulation can reduce the symptoms of hemispatial neglect.[30] For an infant who is being rocked in its mother's arms, however, the vestibular sense is even more important as a powerful cue to the presence of another person.

For the purposes of this chapter, two big questions remain about this part of the brain:

1. Is the temporoparietal junction specifically active when a newborn perceives the presence of another person?
2. Is it involved in the illusory perception of a sensed presence in adults?

Much more research is needed, but for both questions we have some intriguing evidence consistent with "yes."

Although the first question has not yet been explicitly tested in neonates, it has been tested in slightly older infants. In one study, four-month-old infants were shown computer animations of a person's face with the head turned slightly to one side and the eyes initially turned away. The eyes then turn to make eye contact with the infant, and one second later the face breaks into a subtle smile with eyebrows raised. This stimulus elicits activity on both sides of the brain in or near the temporoparietal junction, but mainly on the right side. The response is much less if the eyes never turn to make eye contact prior to the eyebrow movement and smile.*[31] A similar study by another lab found that when nine-month-old infants watch a person shift their gaze in an unexpected direction—away from a novel object, rather than toward it—their neural response is

---

* This study used NIRS in some trials and ERP in others. Localization to the temporoparietal junction was established by NIRS, but with much less precision than in fMRI. The ERP method is better at temporal resolution and could distinguish between responses to the change in gaze versus the subsequent eyebrow movement and smile. Both techniques revealed greater responses on the right side, though the difference was less in the ERP data. Responses were also detected in the right prefrontal cortex, consistent with activity in the rostral orbitofrontal cortex or frontal pole.

greater than when the gaze shift is in the expected direction.*[32] Similar responses were obtained in adult subjects, and the results suggest neural activity specifically directed at inferring the mental state or intention of the person in the video display—much like the enhanced response from the temporoparietal junction following incongruous behavior in the "two cups" experiment.

Another study used a more elaborate form of socially relevant biological motion. Five-month-old infants were shown video clips, either of a woman playing peekaboo, using highly expressive gestures of the face and hands, or of some nonbiological motion like a rotating mechanical toy. In a separate behavioral test, the infants demonstrated a clear preference for looking at the social stimulus versus the nonsocial one.†[33] The greatest neural response to the social stimulus relative to the nonsocial one was again in the posterior temporal area on both sides of the brain, consistent with activity in the temporoparietal junction. Similar results have been obtained in several other studies of infants younger than a year old, using frontal or profile faces,[34] upright versus inverted faces,[35] normal versus scrambled faces,[36] and the face of the infant's mother versus that of a stranger.[37]

As for the second question, the sensed presence illusion in adults is more difficult to study because no one has yet found a simple and reliable way to elicit the experience on demand in normal subjects. Even so, there are several lines of evidence linking experiences of this kind with the temporoparietal junction. Neurologists have long recognized the sensed presence illusion as a phenomenon that tends to occur with parietal lobe damage or that may accompany epileptic seizures that originate there or in the temporal lobe. Neurologist MacDonald Critchley summarized the early literature and presented some colorful examples.[38] In several cases the patients not only sensed a presence separate from the self, but also felt distortions or breakdown of their own embodiment, like a grossly swollen head or hand, a change in overall body size, or the delusion of being only a severed head. He noted that the feeling of a presence may accompany the anomalies of body image that occur with lesions of the temporoparietal junction and described one such patient who felt that the left half of his body did not belong to him.‡ While walking, this patient would sometimes get the notion that he was being followed, behind and to his left, by another person. When he had this sensation, he lost the feeling of strangeness that affected the left side of his body. A more recent survey documented thirty-one patients with sensed-presence experiences of neurological origin and found a slight preponderance

---

* This was an ERP study, so the anatomical source of the neural response could not be precisely localized. The data are consistent, however, with activity in the region of the temporoparietal junction.

† This study used the NIRS method. Some activity was also detected in the ventral prefrontal area, mainly on the right side.

‡ Critchley (1955, p 159) uses the term "subordinate parietal lobe"—more commonly called the inferior parietal lobule—at least some of which is generally considered part of the temporoparietal junction.

of lesions on the left side in those patients whose damage was restricted to one side."[39]

Two recent case studies describe epileptic patients who had never reported any sensations of a presence in relation to their seizures, but who had that experience in a compelling way when their left temporoparietal junction was electrically stimulated during a diagnostic procedure. When the stimulating current was turned on, one of these patients described an irresistible urge to look to the right because she felt another person was standing next to her on that side. She could not explain how she knew this, only that she considered this person a threat. Because her seizures began with the turning of her eyes and head to the right, her neurologists speculated that this illusion might occur at the onset of every seizure, but she could not remember it afterward because of seizure-induced amnesia.[40]

The other patient felt the presence of a person behind her when the stimulus was turned on. When she was lying on her back during the stimulus, the illusory person seemed to be lying beneath her. When she sat up in bed and clasped her arms around her raised knees, she felt the other person's arms around her upper body—a sensation she found disturbing. When asked to do some simple tasks with her hands during the stimulus, she said the other person was trying to interfere with her actions. The sensed presence so faithfully mimicked this patient's behavior that it was obvious to the neurologists that she was misperceiving her own body as an alien presence—a disruption of embodiment and failure of the self-versus-other distinction. Although the patient was aware of this mimicry, she never interpreted the presence as her own body, evidently because her mind could not penetrate the illusion that another person was there.[41]

Although much evidence implicates the temporoparietal junction in the spatial and intentional aspects of the sensed presence, I cannot leave the subject without mentioning four other parts of the brain that might also contribute. One is the medial prefrontal network, which I have already discussed in the context of moral judgment and the inference of the mental states of other people. The medial network and the temporoparietal junction appear to cooperate in the sense of agency and the distinction between self and other.†[42] The second is the precuneus, an area on the medial surface of the parietal lobe that is connected with the temporoparietal junction, anterior cingulate cortex, and several other parietal and prefrontal regions. Neuroimaging studies implicate the precuneus in self-related thought, autobiographical memory retrieval, the sense of agency, and consciousness.[43] The third is the extrastriate body area, a region in the transition zone between the occipital and temporal cortex that is specialized for the visual

---

* I discuss this study in more detail in chapter 12.
† This may explain why alien hand syndrome can also occur with lesions in the medial prefrontal area, as I noted above.

perception of static and moving images of parts of the human body other than faces. In the absence of visual input, it is specifically activated when subjects merely *imagine* moving their own body.[44] The fourth area of relevance is the insula, an area of cortex that lies deeply buried in the cleft between the temporal and frontal lobes. Like the temporoparietal junction, it receives sensory input of many kinds, including vestibular signals and sensations from internal organs like the gut and heart.[45] It also appears to be involved in body ownership and the sense of agency,[46] the sense of taste,[47] the conscious cravings of addiction,[48] and the subjective experience of pain.[49] At its anterior end it is heavily interconnected with other regions we have already explored: the medial and orbital prefrontal networks, and especially the anterior cingulate cortex.[50] Epileptic seizures in this part of the insula are sometimes accompanied by intense ecstasy,[51] and one such case involved the illusion of a sensed presence imbued with feelings of comfort and support.[52]

Probably all of the brain regions we have thus far examined in this book cooperate during an infant's perception of its mother and during an adult's perception of a mystical presence. There is, however, one more attribute of such experiences we still must explore.

CHAPTER 7

Certainty, Neonatal Cortex, and the Phantom God

> If you have intuitions at all, they come from a deeper level of your nature than the loquacious level which rationalism inhabits. Your whole subconscious life, your impulses, your faiths, your needs, your divinations, have prepared the premises, of which your consciousness now feels the weight of the result; and something in you absolutely *knows* that that result must be truer than any logic-chopping rationalistic talk, however clever, that may contradict it.—William James[1]

> More appealing than knowledge itself is the feeling of knowing.—Daniel J. Boorstin[2]

If you are honest with yourself, you can probably remember being absolutely certain about something, only to discover later that you were mistaken. You may also have experienced déjà vu, that unsettling feeling that the present moment—the setting, the people with you, the conversation you are having—all seem so familiar that you are certain you have previously lived through this episode, despite your objective knowledge to the contrary. Some epileptics experience déjà vu as part of their seizures, and they may experience it repeatedly—at the flick of a switch—when their temporal lobes are electrically stimulated in preparation for neurosurgery.[3] A few patients with temporal lobe damage spend most of their waking hours in a state of recurring déjà vu.[4] With the help of clever and subtle suggestion, even completely healthy subjects can be induced to incorporate and believe false memories.[5] These observations show that memory, and our certainty about the truth of specific memories, are related but separable things.*[6]

---

* Neurologist Robert Burton not only makes this point, but takes the somewhat radical position that certainty is a feeling that always originates through unconscious mechanisms, rather than as a consequence of reason and observation.

89

In this regard, the feeling of knowing often described by mystics as a central aspect of God's presence—what I called short-circuit certainty in chapter 1—is similar to déjà vu but not identical. Whereas déjà vu applies to verifiably real sights, sounds, and people and involves anomalous recollection of a false memory, mystical certainty applies to something or someone unseen. It is not so much a memory or sensory perception as an anomalous sense of agency, a feeling that there exists this other being with emotional, intentional, and spatial attributes—exactly those things that led us to the temporoparietal junction in the previous chapter. This unusual feeling of certainty seems to be inextricably bound with these aspects of the experience, which suggests that the temporoparietal junction may be part of its neural basis. Unfortunately, experiences of this kind cannot be reproducibly elicited on demand, so getting an experimental handle on short-circuit certainty is not easy. Like the drunk who lost his keys across the street but is looking for them under the lamppost, we may have no choice but to search where the light is better.

Searching for Lost Keys

One such lamppost is a study of the distinct neural signatures of true and illusory memories.[7] The stimuli for this experiment were lists of words taken from a database of common words organized by category—things like musical instruments, parts of the face, or farm animals. Within each category, the words were listed from most to least popular, where popularity is a measure of how often people think of that word when asked to name words from the category. In the first part of the experiment, each subject was shown a series of lists of four words, one list at a time. For each trial, the video screen displayed the category name and four words from the category, but the two most popular words for that category were deliberately omitted. In some trials, one of the four words shown did not belong in the category. Each list was displayed on the screen for four seconds, after which the subject had to push one of two keys to indicate whether or not the list contained an extraneous word. This was the only task the subjects were asked to do; they were not told to try to remember the words.

In the second phase of the experiment, a single word appeared on the screen in each trial, and the subject had to indicate both his memory of having seen the word during the first phase and his level of confidence—high or low—in that recollection. During this second phase, some of the words presented were the popular ones that had been deliberately omitted during the first phase. Because these words have such strong associations with the category, subjects tended to develop false memories that they had seen these words during the first phase when primed with the category in the second phase. For example, subjects who had seen a list of farm animals in the first phase often had high-confidence false memories of having seen the word "cow" because it is the most popular word associated with "farm animals."*[8]

The neural activity of the subjects was recorded during the testing phase using fMRI, and the signals were averaged within each response type—high-confidence true memory, low-confidence true memory, high-confidence false memory, and low-confidence false memory. The closest analog to short-circuit certainty in this experiment is high-confidence false memory. Relative to high-confidence true memories, high-confidence false ones activated two regions in or near the temporoparietal junction, along with parts of the medial prefrontal network—including some of the anterior cingulate cortex—and two other areas in the rostral and lateral prefrontal lobe.[9] The responses were on both sides of the brain, but mainly on the left, which may reflect the linguistic nature of the task. A similar pattern of activity was seen in the response to high-confidence false memory versus low-confidence false memory.[10] By contrast, high-confidence true memories activated memory-related structures in the medial temporal lobe.[11]

This study comes fairly close to teasing out the essence of short-circuit certainty mainly because of the way it cleverly isolates the experience of high-confidence false memory. The results are intriguing because they implicate the temporoparietal junction in this feeling of knowing, despite the fact that the task has nothing to do with other functions found there, like biological motion perception, embodiment, sense of agency, attribution of intention, or moral judgment. Also implicated is the medial prefrontal network, the part of the brain where we began our quest for the circuitry of the innate model of the mother, and one that appears to be involved in essentially all aspects of that model.

Another neuroimaging study of relevance here examined the brains of fourteen adult subjects as they evaluated the truth of various statements.[12] The statements were of two categories, either testable (matters of fact that could, in principle at least, be accepted or rejected on the basis of empirical evidence) or non-testable (ideas based on emotion, intuition, or opinion). For example, one of the testable statements used was "Karen is a more common first name than Nancy." An untestable one was "Life is like a journey." What is special about this particular study, for our purposes, is that the subjects had to answer in a way that measured their degree of certainty about the answer. For each item, the screen would show the words "I believe" followed by the statement. The subject had to indicate her level of agreement by pressing one of four buttons labeled "definitely," "possibly," "possibly not," or "not at all."

---

* This method for inducing high-confidence false memories is a variant of the widely used Deese-Roediger-McDermott (DRM) paradigm, and the false memories themselves are sometimes called the DRM illusion. For reviews of this literature see Gallo (2010) and Straube (2012).

One interesting result was that the subjects were much faster in responding to non-testable statements than to testable ones—an implicit endorsement of Daniel Kahneman's aptly titled book, *Thinking Fast and Slow*.[13] The two types of statement activated distinct and largely non-overlapping areas in the brain, but those are not the results we most care about in this discussion.[*][14]

What we most want to know is the pattern of neural activation when subjects express high certainty about their responses without regard to response category (belief or disbelief) or testability of the statement. That pattern included the left medial prefrontal cortex, the middle temporal gyrus near the temporoparietal junction in both hemispheres, and the left posterior cingulate cortex. There was no area specifically activated by high uncertainty. The only significant interaction between testability of the statement and certainty of the subjects was that the insula was more active during certainty about responses to non-testable statements, but it was less active during certainty about testable statements.[15]

Despite the many differences in experimental design, both this study and the one about illusory memories suggest that the medial prefrontal cortex and the temporoparietal junction are somehow involved when a person has a strong but unjustified feeling of certainty. Some caution is warranted, however, because the conditions of these experiments were far removed both from neonatal experience and from the illusion of a sensed presence. We may be searching for our keys under the wrong lamppost, so we should not yet feel certain that we understand the feeling of certainty. And specifically with respect to neonatal experience, there is yet another reason for caution, one that merits a significant digression.

Does a Newborn Use Its Cortex?

In the studies just described, the parts of the brain implicated in high-confidence false memories or certainty about belief statements were almost entirely cortical, and in previous chapters I presented evidence for cortical involvement in most other aspects of the putative innate model of the mother. For many years, however, the conventional wisdom of the neurology of infancy has been that the cerebral cortex does little or nothing in newborns.[16] Most of the cortex of neonates is anatomically and physiologically immature, and subcortical circuits are

---

* For the curious, activation was greater for testable than non-testable statements in the dorsolateral prefrontal cortex of both hemispheres and in the posterior cingulate cortex, also bilaterally. Activation was greater for non-testable than testable statements in the bilateral inferior frontal gyrus, left superior temporal gyrus, and the anterior part of dorsolateral prefrontal cortex bilaterally (Howlett and Paulus 2015, p 6). The activity in the inferior frontal gyrus in the latter contrast (non-testable > testable) is interesting in light of a study (which I discuss near the end of chapter 12) that found a correlation between skepticism and activity in the right inferior frontal gyrus (Lindeman et al. 2013).

thought to mediate much of their innate perception and behavior, as they do in a wide variety of nonhuman vertebrates. In several species of amphibians, birds, and mammals, innate visual pattern recognition is mainly a function of the midbrain, not the forebrain (which, in mammals, includes the cortex).[17] The same is likely true for innately recognized sounds, like the calls of mother birds for their chicks. Perhaps the most elegant demonstration of this was a series of neural transplantation experiments in which various parts of the brains of embryonic domestic chicks and Japanese quail were surgically exchanged. Exchange of the medial part of the midbrain was both necessary and sufficient to make the hatchlings prefer the maternal call of the other species.[18]

In human infants, however, the case for subcortical dominance is probably strongest with respect to vision. Newborns with damaged or even absent visual cortex have fairly normal visual orienting behavior and eye movements, though their visual evoked potentials and visual pattern preferences are abnormal.[19] The implication is that visual orientation and eye movements in neonates are largely handled by the visual part of the midbrain, not the visual cortex. The neonatal preference for faces could also be a function of the visual midbrain, given that this pathway contributes to rapid responses to emotional facial expressions in adults.[20] The response to rotating visual patterns, as measured by event-related potentials, also appears to be dependent on the midbrain pathway in infants aged two months or younger.[21] Conversely, an event-related potential in the parietal cortex that characteristically precedes voluntary eye movements does not appear in infants until about age twelve months.[22] Similarly, a physiological correlate of the perception of the illusory square in figure 7.1—a rhythmic pattern in the cortical EEG called the gamma oscillation—first appears at about age eight months.[23] It has even been suggested that neonatal imitation could be handled by subcortical circuits.[24]

There is special significance to the absence in newborns of a gamma oscillation in response to the Kanizsa square illusion. The gamma oscillation, a neural resonance typically around thirty to eighty cycles per second, is thought to represent a synchronization of firing of cortical neurons that is essential for perceptual binding: the holistic perception of an object from multiple separate pieces of sensory input.[25] In the case of the Kanizsa illusion, the holistic perception of the square emerges from the binding of the four Pac-Man figures. A more striking example is the perceptual challenge shown in figure 7.2.[26] This is the Mooney face* I presented upside-down in chapter 1 (figure 1.2), here shown upright. In adult subjects, the moment of conscious perception of the face coincides not only with a significant increase in the magnitude of the gamma oscillation, but

---

* The Mooney face gets its name from psychologist Craig Mooney, who introduced it as a means of studying how the brain fills in missing information during perception, a process he called perceptual closure.

Figure 7.1. The Kanizsa square illusion. The four "Pac-Man" figures suggest the presence of an occluding square, the corners of which partially obscure four complete circles.

also with widespread synchrony of its phase. This means that neuronal firing becomes synchronized in a group of neurons widely separated in the cortex—the likely neural correlate of perceptual binding.[27] The gamma oscillation is also critical in selective attention,[28] embodiment,[29] and in the recognition of previously seen words in memory tests of the kind I described in the previous section.[30] It develops gradually throughout childhood and adolescence, in parallel with increasing cognitive ability.[31] Perceptual binding represents an important aspect of conscious awareness that depends on cortical function, and if infants lack it prior to the age of eight months, then they probably also lack the requisite neural sophistication for an innate model of the mother prior to that age.

Recent studies have found, however, that the neonatal cortex is highly active under a wide variety of conditions, that newborns are capable of perceptual binding, and that their cortex can indeed generate the gamma oscillation characteristic of that capacity—albeit in a way far more limited than in older children and adults. This is most easily demonstrated in the auditory system, which in neonates may be more mature than the visual system because of prenatal exposure to somewhat muffled but otherwise natural sounds.

Experiments of this kind typically use the oddball paradigm, in which the subject hears a long series of brief sounds, each separated from the next by a brief silence, while responses in the brain are monitored via EEG electrodes on the scalp.*[32] Most of the sounds are identical, but—infrequently and at random times—an oddball sound occurs that differs from the normal ones in duration, pitch, volume, or some other more complex way. In adult subjects, the oddball stimulus produces a characteristically different response in the brain that shows up at the electrodes as a wave of negative polarity called the mismatch negativity. This originates mainly in the higher-order auditory cortex of the temporal lobe, though the frontal cortex is also involved. It is as if the brain has learned to expect the normal sound, and the violation of this expectation excites the cortex in a stereotyped way, regardless of what it is about the oddball that makes it different. All that is necessary is that the difference be of a kind and magnitude sufficient to make the oddball distinguishable from the normal sound.[33] This simple experiment suggests

Figure 7.2. A Mooney face. It may take you a while to see the face in this image. In that instant of recognition, neurons in your visual cortex resonate more strongly and with greater synchrony in the range of frequency called the gamma band. *Copyright © 1957 by the Canadian Psychological Association Inc., reproduced with permission.*

that learning and expectation are somehow essential to the way sensory perception works in the brain—a profound insight to which we return in chapter 10. Here we are more concerned with what this approach can tell us about cortical competence in newborns.

Before we turn to those experiments, however, we need to look more closely at the mismatch negativity in adults. An important and surprising aspect of the mismatch negativity is that it does not require the conscious awareness or attention of the subject. Adults who are told to ignore the sounds and concentrate on an unrelated visual task nevertheless have almost the same mismatch negativity response as subjects who listen carefully.[34] The mismatch negativity in adults does not even require full consciousness, as it can be elicited during REM sleep, though not in the deeper (non-REM) stages of sleep,[35] and not

---

* Other monitoring techniques that have been used in the oddball paradigm include MEG (Imada et al. 2006), NIRS (Mahmoudzadeh et al. 2013), and fMRI (Mangalathu-Arumana et al. 2012).

during propofol-induced general anesthesia.[36] The neural computation behind the mismatch negativity appears to be dedicated to building a meaningful context for each incoming sound from the sequence of sounds that preceded it. It is an automatic or pre-attentive process that does as much as possible, without conscious effort, to assemble the most recently heard sounds into holistic patterns that can then grab one's attention for conscious interpretation.[37] The mismatch negativity therefore represents a kind of auditory perceptual binding through time, roughly analogous to what happens with the Pac-Men in the Kanizsa square illusion, and it should come as no surprise that it is accompanied by a surge in the gamma oscillation.[38] This may also explain why it is abolished by propofol anesthesia, which disrupts long-range cortical synchrony like the gamma oscillation.[39]

That the oddball paradigm can be used with sleeping subjects is a boon to research with newborns because bodily movement creates noise and artifacts in the EEG signal. Sleeping newborns are ideal subjects, not only because they seldom move, but also because their brains are much more responsive to sensory input during all stages of sleep than are the brains of adults.[40] Newborns do indeed show a neural response to the oddball sound, though at greater latency than the mismatch negativity in adults, and with greater variability in size, shape, and polarity across different recording sites on the scalp.[*41] Interestingly, however, they also show a prominent increase in gamma synchrony following the oddball sound, and this proves to be a much more robust and consistent aspect of the response than the mismatch potential itself. These experiments show that the auditory cortex of newborns is capable of pre-attentive perceptual binding, using the same mechanism of cortical synchrony in the gamma frequency band observed in adults.[42]

There is, however, much more to the story than that. We can also use the oddball paradigm to explore the limits of processing power in the auditory cortex of newborn infants. Instead of using some simple acoustic difference to distinguish the oddball from the normal sounds—like frequency or loudness—we can try something more subtle and revealing, like a combination of the two, a repeating sequence of sounds, or fragments of human speech that differ in some interesting way. Such experiments show that the neonatal auditory cortex is surprisingly sophisticated in its capacity for perceptual binding, pattern recognition, and phoneme discrimination.

For example, one study used two values each of both frequency and loudness to generate four unique sounds: high/loud, high/quiet, low/loud, and low/quiet. Two of them, high/loud and low/quiet, were used as the normal sounds,

---

* Isler et al. (2012). The neonatal response to the oddball stimulus is more often positive than negative in many studies. This may represent a qualitatively different attention-shifting response superimposed on, and sometimes masking, a mismatch negativity; see Kushnerenko et al. (2002).

each presented on 45 percent of trials, so that altogether 90 percent of trials were normals. The other two, high/quiet and low/loud, were oddballs, each occurring on 5 percent of trials. The idea here is that neither frequency nor loudness alone is a useful cue for distinguishing normal from oddball. Only the conjunction of the two can make that distinction, and that requires the perceptual binding of frequency and loudness. The auditory cortex of sleeping newborns has no problem with this.[43]

Another study began as a simple oddball experiment in which the oddball sound differed only in pitch and was presented at random on 20 percent of trials. As expected, the newborns had both a mismatch potential and a gamma response to the oddballs. Then, only the order of presentation of the sounds was changed, from random to a repeating pattern: four normals followed by an oddball. Although the sounds and their overall frequency of occurrence were identical in the two situations, when they were grouped as a repeating pattern, the mismatch response to the oddball vanished. The neonatal auditory cortex recognized the pattern as a single auditory object—yet another form of holistic perceptual binding.[44]

Perhaps the most dramatic example is a study that used the oddball paradigm to test phoneme discrimination. It found that the brain of a sleeping newborn can tell the difference between the similar phonemes /pa/ and /ta/ spoken by the same person, while at the same time recognizing a series of either of these as being all the same syllable, even when each is spoken by a different person. The amazing thing about this is that the acoustic differences between utterances of the same phoneme by different people can be greater than those between two different phonemes spoken by the same person. This means that the auditory cortex of a newborn is capable of discriminating acoustic differences that are relevant to language comprehension, while ignoring others that are not.[45]

There are, however, limits to what neonatal auditory cortex can do. One study used a mixture of harmonic tones in the oddball paradigm, with the fundamental frequency deleted. Adults subjectively perceive the pitch of such a stimulus as the fundamental frequency, despite its absence, because the harmonics are sufficient to suggest its presence. The auditory cortex fills in the missing information, much as the visual cortex does with the Mooney face in figure 7.2. The auditory cortex of a newborn, however, evidently cannot do this; a neural response to an oddball difference in illusory pitch of this kind first appears at about age four months.[46]

Another study suggests that there are greater limitations in the speech-related cortex of the frontal lobe in newborns, relative to that of the temporal lobe. In older children and adults, both areas cooperate in the perception and production of speech, though Broca's area in the frontal lobe is more important in speech production, whereas the temporal speech areas are more concerned with perception and interpretation. In a study of phoneme discrimination, a response to the oddball was evident in the temporal but not frontal cortex in newborns, whereas both areas responded in six-month-old infants—the age at which babbling begins. For technical reasons this study examined only the cortex of the left hemisphere, which is the language-dominant side in most adults.[47]

The frontal lobe, especially its more rostral area known as prefrontal cortex, has long been considered the most slowly maturing cortical area, mainly for anatomical reasons. Its neural fibers do not become fully insulated with myelin until early adulthood.[48] It is an associative area known to be involved in higher-order cognitive functions like short-term working memory, the focusing of attention, the retrieval of long-term memories, complex planning and decision-making, and the generation of volitional behavior, including speech and other social behavior. Neonates lack adultlike competence at these cognitive skills, so perhaps it should not be surprising that the prefrontal cortex of newborns is unresponsive in a phoneme discrimination task. But the medial and orbital networks I discussed at length in chapter 4 are also prefrontal areas. The role I have suggested for them in the innate model of the mother seems implausible if the prefrontal cortex in general is nonfunctional in neonates.

Recent experiments using more sensitive techniques, however, are beginning to reveal a surprising degree of activity and responsiveness in the neonatal cortex, including prefrontal areas. One of these used an oddball paradigm with the NIRS functional imaging method to map cortical responsiveness to speech sounds in preterm newborns of about thirty weeks gestational age. At this stage of development, many of the neurons of auditory cortex have not yet migrated to their final positions among the cortical layers, and many synaptic connections are yet to be made. Amazingly, and in contrast to the study of Broca's area I just described, cortex of the temporal, parietal, and frontal lobes of these premature infants responded to oddball phonemes on both sides of the brain, but predominantly on the right side. An exception was the temporoparietal junction, which was more responsive on the left. When the oddball stimulus was not a different phoneme but a different voice (female versus male), the response was much weaker and exclusively in the right hemisphere. This shows that, almost from the moment it first becomes responsive to sounds of any kind, the prenatal cortex, including the prefrontal cortex, is capable of making subtle phonemic discriminations essential to language comprehension. Although language is learned, that learning arises on a foundation of the innately language-expectant cortex.[49]

Another series of NIRS imaging studies used a different paradigm to examine the ability of full-term neonates to remember a recently heard two-syllable word after a two-minute delay. This long delay between familiarization and testing distinguishes this experiment from the oddball paradigm, which involves novelty detection over much shorter intervals, typically only one or two seconds.

Instead, this new paradigm probes a qualitatively different aspect of the neonatal brain having to do with short-term working memory. A two-syllable word is repeatedly played to the infant for six minutes. Then, after the two-minute delay, either the same word or a different two-syllable word is played for another three minutes, though it is the response to the first word in this test series that is the most revealing. If the familiar word is played, a transient increase in oxygenated hemoglobin occurs in several areas of the frontal, temporal, and parietal lobes of both hemispheres. If a novel word is played, the response is a decrease in oxygenated hemoglobin in these areas. The response is unaffected if instrumental music is played to the infant during the two-minute delay, but it is significantly degraded if speech sounds are played during that time. This indicates that the response depends on a form of short-term memory specifically related to speech sounds because speech during the delay interferes with it but music does not.[50]

In another variant of this experiment, the test word either had the same vowels but different consonants as the familiar word or the same consonants but different vowels. In this test, the change in consonants alone went unnoticed by the infants: they responded as if hearing the familiar word. The change in vowel, however, elicited the same response as a completely novel word. This shows that it is mainly the vowel sounds of the word that newborns are remembering. This makes sense, considering that vowels carry most of the emotional prosody of speech, especially the exaggerated prosody of motherese. The "familiar vowel" response was also more anatomically specific than the "familiar word" response, being strongest in the right frontal cortex, but also present in the right parietal and left temporal cortex. Interestingly, the areas of parietal and frontal cortex involved are ones commonly activated in tests of working memory in adults.[51] The importance of vowels and prosody seen in this experiment is also consistent with the study I described near the end of chapter 4, in which NIRS imaging revealed a preference in the orbitofrontal cortex of newborns for motherese over adult-directed speech.[52]

The preference for motherese also shows up elsewhere in the neonatal auditory system. An fMRI study of sleeping newborns examined the responses of their brains to the sound of a woman reading a fairy tale in motherese and to artificially simplified versions of that sound. The simplifications changed the sound either into "hummed speech" (preserving prosody but stripping acoustic features needed for the recognition of distinct syllables) or "flattened speech" (preserving phonemes but eliminating prosody). Natural motherese activated the brain far more than either of the simplified stimuli, but the big surprise was that essentially all of the cortical areas activated by speech in adults are activated by motherese in the brains of sleeping newborns. There are, however, some interesting differences. As in the NIRS studies just described, the response in newborns is greater in the right hemisphere, in contrast to the left-hemispheric dominance for language in adults, and the connectivity between different speech areas within each hemisphere is much less in neonates. I discuss the significance of hemispheric dominance in chapter 12. The important points here are that the neonatal brain is highly attuned to the conjunction of prosody and phonemic content and that a cortical network of adultlike extent responds to speech in a newborn—albeit in an immature way.[53]

Beyond the auditory system, neuroimaging experiments reveal widespread activity throughout the neonatal cortex in response to stimuli of other sensory modalities. The orbitofrontal cortex is activated by the salient aromas of breast milk and colostrum, the pleasant smell of vanilla, and various unpleasant odors encountered in a hospital setting.[54] The somatosensory cortex of preterm neonates responds to wiping the skin with antiseptic, but it responds much more to a painful needle stick.[55] Gentle vibration of the hand excites broad regions of cortex beyond the primary somatosensory area, including temporal and parietal areas.[56]

A completely different approach to measuring cortical activity is simply to do neuroimaging while the subject is resting. In the absence of any specific sensory stimulus or motor task, the adult brain at rest undergoes slow changes in spontaneous activity that can be visualized using fMRI. Over periods of tens of seconds, different patterns of activity in multiple, widely separated regions of the brain come and go, but the patterns are not random. Instead they tend to fall into a small number of fairly consistent categories, each involving a characteristic set of communicating brain regions.[*57] The brains of sleeping newborns also show these slow transitions among different resting-state networks, though the networks appear to be simpler and fewer in number than in adults. Even so, together they encompass the sensory cortices, parietal and temporal areas, and the prefrontal cortex.[†58]

The visual system is arguably the least mature cortical system in neonates, but even in vision there is evidence of a surprising degree of innate cortical competence. In chapter 2 I summarized evidence for the innate visual perception of faces by newborn infants, including such elaborate aspects as a preference for eye contact, rudimentary gaze following, sensitivity to the Thatcher illusion, and the face-inversion effect—capacities suggestive of cortical involvement.[59] I also cited the case of a boy who suffered neurological damage to the face-perceptive cortex of the temporal lobe at birth and experienced a profound and permanent deficit in face recognition as a result.[60] This stunning lack of plasticity in the

---

* Despite their general consistency across many individuals, there is no universally agreed-upon way to subdivide and count these resting-state networks. A recent cluster analysis found them to be hierarchically organized, with two, seven, eleven, and seventeen clusters evident at successively greater levels of detail in the hierarchy (Lee et al. 2012).

† Fransson et al. (2009) found five distinct cortical resting-state networks in full-term newborns, plus a sixth that involved the basal ganglia.

face-recognition system reveals it to be a brittle, inflexible, special-purpose cortical network dedicated to one function, despite the fact that that function—representing the identities of thousands of faces—requires a vast amount of learning
throughout life. That the adjacent cortical areas could not assume this function
is reminiscent of the severe cases of acquired sociopathy that result from damage
to the orbital and medial prefrontal networks in infancy.[61] In both examples,
there appears to be some innately specified information about the nature of the
problem to be handled in that part of the cortex. Although solving the problem
requires learning, that learning evidently cannot happen without the innate
neural foundation.

The evidence I have reviewed in this section argues for humility in the
interpretation of the abilities and limitations of the neonatal cortex. Obviously
the myelination, synapse formation, and neural pruning yet to come imply that
the cortex of a newborn is both anatomically and functionally immature. But
the more we learn about the physiology of the neonatal cortex, the more active
and responsive it appears to be. At this point we should be open to the possibility that an innate neural model of the mother could be partly, and perhaps even
largely, cortical. And even if that model is not fully functional in the cortex at
birth, its elaboration over the first weeks and months of life could nonetheless
be partly innate.[62]

It was the neural basis of the feeling of certainty that prompted this digression into the interplay between innate and learned information in the cortex.
Interestingly, that interplay shows up in another, much stranger phenomenon
that also touches on the problem of short-circuit certainty. In contrast to the
tests of word recall and certainty of beliefs, this one may have more direct relevance both to the neonatal mind and to the illusion of a divine presence.

The Sensed Presence as a Phantom Mother

One of the most striking examples of high-confidence false percepts in all of neurology is the phantom limb—the illusion experienced by an amputee that the missing
limb is still present. Nearly all amputees who experience phantom limbs are aware
that the sensation is an illusion, but it is a sufficiently compelling one that they
may spontaneously try to use the phantom to answer a phone, to shake hands, or
to break a fall. Some patients have a vivid sense that they can move the phantom,
but for others it may be perceived as paralyzed, especially if there had been paralysis
prior to amputation. Many experience pain in the phantom that is real enough to
affect their quality of life. The perception of a phantom limb may appear within
a matter of hours after surgery and may endure for decades, though with time the
phantom often becomes physically distorted or may vanish entirely.[63]

I bring up the subject here, not only because there appears to be something very like short-circuit certainty involved in the perception of a phantom limb, but also because there may be deep commonalities in the neural bases of the phantom limb and the illusory sensed presence. The phantom limb is a manifestation of embodiment: the persistence of the neural representation of a part of the body that has been physically removed. Embodiment is so essential, so fundamental to our existence, that a feeling of certainty about bodily presence appears to be intrinsic to it. As I emphasized in the previous chapter, however, embodiment of the self is deeply connected with the perception of the presence and intentions of others, and the temporoparietal junction is essential for both kinds of neural computation. Just as a phantom limb results from the neural expectation that the limb exists, so may the illusory sensed presence arise from a neural expectation of the existence of another person.

A similar analogy was made decades ago, in psychological terms, by psychiatrist Colin Parkes. He suggested that the phantom limb results from denial on the part of the amputee—a refusal to accept an emotionally unbearable loss—which he compared to a widow's inability to accept her husband's death and her consequent sense of his presence.[64] I am suggesting something different, more neurological than psychological in origin,* and the kind of phantom I have in mind is rather different from that experienced by the typical amputee—much as the amorphous illusion of God's presence differs from a widow's highly specific feeling that her deceased husband is nearby. A better analogy is the kind of phantom experienced by a small fraction of those people born with missing limbs. These cases suggest that the neural representation of the body is partly innate and therefore appear to have more in common with the illusory sensed presence that I have argued arises from an innate neural model of the mother. There are alternative explanations for congenital phantom limbs, but, as I show in what follows, the weight of the evidence favors some degree of innate embodiment.

About one in five people with congenitally missing limbs report phantoms.[65] An especially interesting case was recently described by neurologists Paul McGeoch and V. S. Ramachandran. Their patient was a fifty-seven-year-old woman whose right arm was greatly foreshortened at birth, with only a rudimentary thumb, no index finger, and shortened and immobile middle and ring fingers. Only the little finger of the right hand was normal. At age eighteen her right hand was injured in a car crash and subsequently amputated. Prior to that time she had no phantom limb sensations from her right arm or hand, but after

---

* In an important sense, the distinction between psychological and neurological symptoms is one of degree, not of kind: all phenomena of mind originate in the brain. Parkes, however, tried to explain both the phantom limb and the deceased husband's ghost in purely mental terms (denial of a painful loss), rather than as conscious manifestations of the expectations of neural circuits that something (one's arm or one's spouse) exists.

the amputation she experienced a phantom right hand that included all five digits, with the thumb and index finger perceived as being about half normal length. McGeoch and Ramachandran argue that the phantom thumb and index finger resulted from an innate neural representation of these digits, the conscious experience of which had been inhibited during the patient's first eighteen years by sensory feedback, both tactile and visual, from her deformed hand. With the removal of this feedback following the amputation, the neural image of the complete hand emerged as a phantom. That the phantom thumb and index finger were shorter than normal and that the phantom hand as a whole lacked normal mobility shows that sensory experience with her deformed hand also affected the phantom. Interestingly, McGeoch and Ramachandran trained the subject to perceive the phantom hand as anatomically normal and fully mobile using a visual illusion. They placed her normal left arm into a box containing a mirror oriented such that she could see both her left arm and a mirror image of it where her right arm would be if it were of normal length and shape. She was told to open and close her left hand and right phantom in synchrony while watching this illusion. After a few weeks of these sessions, she perceived her phantom as having normal proportions and mobility, and her phantom pain was relieved—dramatic evidence of the interplay between nature and nurture in phantom limbs.[66]

Perhaps the most compelling and thoroughly studied case of this kind, however, was described by neurologist Peter Brugger and colleagues. Their subject, AZ, forty-four years old at the time of the study, was a college-educated woman who had been born without arms or legs and had experienced vivid phantom sensations of all four limbs for as long as she could remember. That her phantom limbs were the result of a neural representation of complete limbs rather than mere confabulation or wishful thinking was cleverly demonstrated by asking AZ to identify drawings of hands or feet as left or right and measuring her reaction time. Normal subjects take longer at this task when the drawing shows a hand or foot with the digits pointing down—evidence that they determine left or right in the drawing by mentally rotating the visual image and comparing it to their own hands or feet. The task can be done more quickly when the mental rotation is unnecessary. If AZ did the task by reference to a neural representation of her phantom hands and feet, then she should show the same delay in response as control subjects when mental rotation was required, and this was in fact the case.[67]

There were, however, subtle but revealing differences between the neural representations of AZ's phantoms and those of typical amputees. When an amputee attempts to move his phantom fingers, neuroimaging reveals activity in the corresponding hand area of the primary motor cortex, essentially the same as in cortex that represents an intact hand.[68] Similarly, when this area of cortex in an amputee is artificially stimulated via rTMS, the subject feels a localized muscle twitch in the phantom limb, much like the real muscle twitches that this evokes in control subjects.[69] In AZ, however, no change in activity was seen in her primary motor cortex when she was moving her phantom hand or fingers, though increased activity was seen in the supplementary motor area of her prefrontal cortex and in a region of her parietal cortex bordering the temporoparietal junction. In intact subjects these areas are involved in planning and executing complex bodily movements, in contrast to the more localized control of individual muscles seen in the primary motor cortex. Unlike amputees who have had years of experience with real limbs prior to their amputation, AZ apparently lacks this low-level representation of individual muscles in her phantom limbs. Similarly, stimulation of the hand area of her primary motor cortex via rTMS did not evoke phantom twitches, but instead evoked the perception of more complex phantom hand movements felt with a significant delay following the stimulus. These sensations probably resulted from the conduction of signals from the primary to higher order cortical areas, as similar sensations resulted from direct stimulation of her parietal and supplementary motor areas.[70]

The authors of the study concluded that congenital phantom limbs arise from neural representations of embodiment that are partly innate but also partly learned. In subjects with intact limbs, the detailed representations of the hands in primary somatosensory and motor cortices are heavily dependent on experience. In subject AZ, these areas represent the existing stumps but not her phantoms limbs. These appear instead to be represented in higher order prefrontal and parietal cortices, where innate representations of embodiment may be embellished and enriched in part through the experience of watching other people use their complete limbs.[71] The neurons capable of this are the so-called mirror neurons that respond not only during execution of a specific movement, but also when the subject sees someone else execute that movement. This seems likely, considering that amputees with phantom limbs report phantom sensations merely from watching the limbs of other people being touched.[72]

It even has been suggested that the mirror neuron system can entirely explain congenital phantom limbs, such as those of subject AZ, without recourse to any innate representation.[73] This idea, however, merely relocates the problem: how could AZ have mirror neurons that represent hands and fingers without some innate spatial representation of hands and fingers? In other words, how could the experience of seeing another person's finger movements lead to a phantom sensation of AZ moving her own fingers if her brain has no concept of the kinesthetic sense of finger movements? A subject with intact hands could conceivably acquire this sense through the association of proprioception with vision while watching his own hand movements, but AZ could not. That objection aside, the phenomenon of neonatal imitation strongly suggests that the mirror neuron system is itself at least partly innate, as I explained in chapter 2.[74]

Yet another line of evidence supporting innate embodiment concerns the experience of phantom sexual organs in transsexuals. In 2007, Ramachandran and McGeoch argued that people who seek gender reassignment surgery may be telling the truth when they describe themselves as women in male bodies or vice versa.[75] They suggested that a dissociation between actual bodily gender and its innate neural representation somehow occurs in these individuals during embryonic development. As a result, some people are born with the innate sense of embodiment appropriate for the opposite sex and feel that the sexual aspects of their bodies are somehow alien to them. This hypothesis is empirically testable because it predicts that male-to-female transsexuals should experience phantom penises at a significantly lower rate than men who undergo removal of the penis for other reasons. It also predicts that female-to-male transsexuals would be likely to report the experience of a phantom penis when none is physically present and that they would be less likely to experience phantom breasts after gender reassignment than women who undergo mastectomy for other reasons. A year later, Ramachandran and McGeoch published a study confirming these predictions.[76] Interestingly, something similar to the desire for gender reassignment is seen in subjects with a rare and subtle dysfunction of the right parietal lobe that causes them to experience a normal and healthy limb as being somehow alien to their bodies—so much so that they seek its amputation.[77] Like transsexuals, these individuals appear to be driven toward drastic surgery by a feeling of short-circuit certainty, a feeling so compelling that it persists in the face of bodily evidence to the contrary.

In the previous chapter and in chapter 2, I argued that an infant's innate model of its mother is intimately related to its innate sense of body ownership. The link between the two is most evident in neonatal imitation, a behavior through which a newborn establishes a connection with another person through the common language of embodiment. Here I have presented additional evidence that neural embodiment is partly innate and gives rise to a compelling feeling of certainty, a striking manifestation of which is the phantom limb. In this light, the illusion of God's presence can be seen as a phantom mother—not an imagined ghost of a person's real mother, but the vaguely specified loving and nurturing *other*, the neural representation of which is the innate model of the mother I described in chapter 1. The intimate relationship between this representation and that of the infant's own innate embodiment may in part explain the feeling of short-circuit certainty that typically attends the illusion of a mystical presence and contributes to the extraordinary tenacity of religious belief. The God thus perceived is therefore best understood as a supernormal phantom, not a supernatural spirit.

The Trick

The brain is not trying to deceive us. It has evolved to shape our behavior—and our behavioral flexibility—in ways that improve our reproductive success. We are learning machines, but our capacity to learn is neither boundless nor aimless. Our evolutionary history has shaped our brains such that we know some essential things at birth—what tastes good, smells good, feels good—and the most important of these things, I suggest, is that *another being exists*, one who brings love and comfort. This innate knowledge persists into adulthood as the foundation of our social nature—the direction in which evolution has aimed our learning. For some people it may never reappear as its anomalous neonatal essence—the illusion of a sensed presence—but the circuitry of that neonatal essence appears to persist in all of us. We see it in the response of orbitofrontal neurons to primary reinforcers, in the addictive nature of religious emotion, and in the overwhelming feeling of certainty without evidence that characterizes faith. In its essence, this trick of the brain may be merely an accidental consequence of evolution's use of nonmodifiable synapses to implement the core of human nature.

* * *

That concludes our guided tour of the mystical parts of the brain, guided not by mysticism, but by the idea that spiritual experience has its basis in the neural circuitry of neonatal expectation. Some neuroscientists, however, *have* tried to use various aspects of mysticism as their guide. In part III, I consider some of their findings, and I use insights gleaned from part II to suggest some new experiments.

Part III

NEUROTHEOLOGY MEETS NEUROETHOLOGY

CHAPTER 8

The Image of God

In the 1980s, a small group of neuroscientists arrogated for itself a new field of "neurotheology" which has become— not to put too fine a point on it—an embarrassment.— neurologist and Anglican minister Alasdair Coles[1]

What a difference the placement of one letter makes. Neuro*ethology* explores how the brains of various animals generate highly specialized natural behaviors that serve well-understood roles in their reproductive success. My approach in the preceding chapters was essentially neuroethological, not merely because I referred to prominent stars of the field like songbirds and prairie voles, but mainly because the proposed connection between human neonatal feeling and the illusion of a divine presence is an ethological one. If this hypothesis is right, then seeking the neural basis of an innate model of the mother may lead us more directly to the neural basis of the mystical. It might also clarify some of what neuro*theology* has already discovered. In this and subsequent chapters I examine a small sample of those findings in this different light, and I suggest a new direction for research of this kind.[2]

Neurotheology has a less than sterling reputation mainly because of the vagueness of what it studies.[3] If we wish to use modern neuroimaging methods to see what happens in the brain during religious or spiritual experience, whose religion and what kind of experience should it be? Meditation? Reading the Bible? Saying the Rosary? Speaking in tongues? How are we to interpret the results? For practical reasons, such experiments must focus on spiritual experiences that can be elicited on demand and with some reliability, but the most profound and life-changing mystical experiences—like the sensation of God's presence—are not so cooperative. Another problem concerns the appropriate control or baseline state to which the experience is to be compared. If a subject

109

is listening to a reading of the Bible while in the brain scanner and we see activity in the auditory cortex, that does not necessarily mean that the auditory cortex is the seat of mystical experience. Despite these and many other difficulties, several neuroscientists have charged ahead.

Sacred Recitation

In one of the first studies of this kind, scientists recorded images of neural activity while subjects recited the first verse of Psalm 23—a passage that depicts God as a source of guidance, protection, and nurturance. For comparison, images were also obtained while subjects recited a familiar nursery rhyme and read instructions for using a phone card. The study examined six fundamentalist Christians and six people who were not religious.[4] The pattern of neural activity in response to the religious recitation differed between the believers and nonbelievers, in that believers had greater activity in two areas of the dorsal prefrontal cortex and one in the medial parietal lobe. These are areas involved in reasoning and short-term memory,[*5] and the authors specifically noted the absence of activation in emotion-related areas like the amygdala and orbitofrontal cortex—two regions I identified in chapter 4 as probably involved in the innate model of the mother. They argued that, contrary to conventional wisdom, religious experience is not primarily emotional, but instead an "attributional cognitive phenomenon." This may be a fair description of reciting a familiar sacred verse, but is that act a good approximation of mystical experience?

Belief

Another study did not even try to elicit religious experience, but instead merely looked for a difference in the way the brain processes belief of religious versus nonreligious statements. It used fMRI to examine neural activity in fifteen committed Christians and fifteen nonbelievers as they responded "true" or "false" to statements like these:

- The Bible is the most important book we have.
- The Bible is a highly imperfect book.
- The U.S. Constitution is a very famous document.
- The U.S. Constitution was first written in Chinese.

---

* The area activated in the medial parietal lobe was the precuneus, an interesting but poorly understood region that appears to be involved in self-centered mental imagery and the retrieval of memories of personal experiences. See Cavanna and Trimble (2006).

Many such statements were used, and they were all deliberately balanced in this way—half religious, half secular, and with equal numbers of true and false responses expected—but they were presented in random order. In both groups of subjects, the neural responses to belief were essentially indistinguishable for the religious versus nonreligious statements. Belief alone turns out not to be a useful probe for revealing circuits that are in some way specifically involved in religious thought or feeling.[6]

The investigators did, however, extract other comparisons from the data, two of which bear some interesting if speculative resemblance to results I discussed in chapter 7. Responses to secular statements, relative to responses to religious ones, tended to activate memory-related structures in the medial temporal lobe—much like the responses to high-confidence true memories we saw in the word recall experiment in chapter 7.*[7] Similarly, blasphemous statements may have evoked a little short-circuit certainty in the subjects because they activated the inferior parietal lobe in or near the temporoparietal junction, along with the medial prefrontal network—a pattern reminiscent of the high-confidence false memory response in the word recall study.[8] Such similarities of response are merely suggestive, not conclusive, because other interpretations are possible.[9]

Meditation

A completely different way to see the mind in a mystical state is to scan the brains of subjects as they meditate. Not all practitioners consider meditation a religious experience, but some do, and it has the advantage of being inducible on demand. It is often described as an altered mental state in which the usual stream of conscious thought is deliberately suppressed, and this aspect of it has been explored using neuroimaging.†[10] The goal of the meditator may be relaxation, detachment from mundane troubles, or even an altered perception of embodiment, but one form of Buddhist meditation also can be an exercise in compassion. This immersion in empathetic compassion, to the exclusion of all other thought, may evoke the emotions of the mother-infant bond—albeit in a way more maternal than infantile.

That similarity is evident in an fMRI study of fifteen highly experienced Buddhist monks and fifteen novice meditators. Neural activity was recorded while the subjects meditated and while they rested in a non-meditative state.

---

* This contrast activated more areas in the brain than did the high-confidence true memory contrast in the word recall experiment—notably several prefrontal areas and retrosplenial cortex. Compare figure 2b and table 4 of Harris et al. (2009) with figure 5a and table 4 of Kim and Cabeza (2007).

† Experienced meditators return more quickly to the meditative state following a brief distraction than do novices. See Pagnoni et al. (2008).

During both conditions they would occasionally hear recorded sounds of other people: a woman in distress, a baby laughing, or the emotionally neutral sound of people talking at a party. The responses to these sounds were greater during compassionate meditation than during the resting state, and they occurred in many of the areas we explored in previous chapters: the anterior cingulate cortex, anterior insula, temporoparietal junction, and the medial prefrontal network, with greater activation in the right hemisphere for most of these.[11] The brains of expert meditators showed greater activation than those of novices in several regions, notably the amygdalae and temporoparietal junction on both sides, but the difference was greater on the right.[12] The authors note that these brain regions are part of the social brain—the network activated when experimental subjects infer the mental states of other people. As I suggested in part II, they may also be the substrate of an infant's innate neural model of its mother.

Prayer

During infancy, a critical function of that innate circuitry is crying for mother, the analog of which in religious experience may be supplicatory prayer. A Danish group used fMRI to examine neural activity during two common forms of prayer: habitual and personal. Both of these in some way express supplication, but they evoked strikingly different feelings and patterns of neural activity in the twenty adult subjects—all devout Protestants.[13]

For the habitual prayer, subjects were asked to recite silently the Lord's Prayer—a regular and frequent ritual in their denomination. For the personal prayer, they were asked to pray silently in a spontaneous and unrehearsed way. After the experiment they wrote from memory samples of their personal prayers, which typically included requests for help and protection for themselves and loved ones, contrition for sins, pleas for forgiveness, and expressions of gratitude. For nonreligious control conditions, the subjects silently recited a familiar nursery rhyme (habitual and structured, like the Lord's Prayer), and they silently and spontaneously expressed their list of Christmas wishes to Santa Claus (informal and supplicatory, like personal prayer). Neuroimaging was done in all of these conditions and in a baseline control condition in which the subjects counted backward from one hundred.

Although several of the subjects warned the experimenters that they would probably not see any response to the Lord's Prayer because it was habitual, a response was evident in the caudate nucleus—a part of the basal ganglia that is involved in sequential and repetitive behaviors.[14] The caudate lies just above the nucleus accumbens, which we explored in chapter 5 because of its role in mother-infant bonding, appetitive reward seeking, sexual pair-bonding, and drug addiction. Although addictive drugs initially exert their tenacious grip by raising dopamine levels in the nucleus accumbens, their influence gradually expands with repeated use to encompass the nearby caudate nucleus, where they reinforce the repetitive, habitual, and ritualistic aspects of drug-using behavior.[15] The caudate is also severely affected in Parkinson's disease, in which a deficiency in dopamine degrades repetitive and sequential behaviors that are so well learned as to be almost automatic. In chapter 11 we return to this intriguing connection between ritual and the effects of dopamine, but our main interest here is in personal prayer.

Personal prayer, relative to making wishes to Santa, activated four components of the social brain I discussed in part II in the context of the sensed presence—the temporoparietal junction, precuneus, temporal pole, and part of the medial prefrontal cortex. The activations were all in the left hemisphere, and a similar pattern was seen when personal prayer was contrasted with recitation of the Lord's Prayer.[16] The authors interpret the results as reflecting their subjects' concept of God as a real person, one who cares about and responds to their needs—in other words, an attachment figure. The study also demonstrates the importance of choosing the right kind of prayer—personal and spontaneous versus ritualized recitation—if the goal is to reveal the neural basis of personally meaningful religious thought and feeling.

Mystical Union

There may, however, be even more revealing forms of prayer or mystical states of mind that can be voluntarily induced. One of these is a form of Christian meditation called mystical union, during which the practitioner attempts to experience the presence and enfolding love of God. An fMRI study recruited fifteen Carmelite nuns who regularly practice mystical union, but they warned the experimenters that "God can't be summoned at will."[17] The nuns were therefore asked to lie in the scanner with eyes closed and to remember and relive the most intense mystical experience they ever felt in their lives as members of the Carmelite Order. As a control condition they were also asked to remember and relive an experience of intense closeness with another person. Also, their baseline neural activity was recorded while they simply rested with eyes closed.

After the experiment, the nuns described their experiences in the scanner as feelings of peace, joy, unconditional love, and sensations of the presence of God. Remembering a mystical union with God, compared with remembering closeness to a person, activated areas I discussed in chapters 1 and 2 in the context of innate knowledge and moral intuition—the orbital and medial prefrontal networks—along with the neural alarm system involved in infantile crying, the anterior cingulate cortex. Several regions in the temporal and parietal lobes were also activated, including but not limited to the temporoparietal junction.[18] The authors felt that the main limitation of the study was that the subjects were trying to recall a past mystical experience rather than fully achieving such an experience in the scanner.

Michael Ferguson and colleagues tried a similar fMRI experiment with practitioners of a different Christian sect, the Mormon Church.[19] They recruited nineteen young adult subjects, all of whom evidenced serious commitment to the faith though voluntary mission work, regular church attendance, and frequent prayer. The goal of the study was to discover what parts of the brain are active during a specific kind of religious experience called "feeling the Spirit"—a central aspect of Mormon theology and practice. From my brief sampling of online testimonies about this experience, they appear to be typical examples of the more persistent way of feeling God's presence that I described in chapter 1. Mormons are encouraged to seek these feelings daily, using prayer and reading Scripture to elicit them—something Ferguson, once a Mormon himself, surely appreciated.

During the fMRI session, each subject spent scheduled blocks of time praying, reading Scripture verses, watching short videos produced by the Mormon Church, and reading inspirational quotes attributed to famous religious leaders. By pressing buttons, the subject could indicate when and how strongly she was feeling the Spirit. She could also indicate how meaningful each experience was. Neuroimages acquired during the moments when subjects were maximally feeling the Spirit were compared to those from periods without such feelings, thus revealing neural activity specific to the spiritual experience.

One of the areas most consistently activated, in both the left and right hemispheres, was the nucleus accumbens, the source of reward, motivation, and addiction we explored in chapter 5 because of its importance in mother-infant attachment in rats. There I suggested that nucleus accumbens should be active during the illusion of God's presence in adult humans, a prediction I first published in 2016.[20] The neuroimaging data from Ferguson's group, published two years later, was the first evidence directly implicating nucleus accumbens in religious experience. Other areas activated when the Spirit was maximally felt included the anterior cingulate cortex (which I discussed in chapter 3 for its role in infantile crying) along with other prefrontal areas involved in selective attention, mainly in the right hemisphere. Of course you might expect a personal encounter with the Holy Ghost to catch your attention, but in chapter 10 I offer a more specific suggestion about the role of attention in mystical experience, one that is consistent with the evidence just described.

The neuroimaging studies of Carmelite nuns and devout Mormons are probably the closest we have yet come to seeing what happens in the brain during the illusion of God's presence, but even with these expert subjects, the experiences in the scanner fell short of the sudden and most intense variety spawned by utter despair and helplessness. Unfortunately, that kind of experience is fleeting and unpredictable and tends to occur under extremes of physical danger or emotional turmoil. If only we had a reliable way to elicit it on demand. . . .

CHAPTER 9

The Helmet of God

> Until science can teach us to reproduce such phenomena
> at will, science cannot claim to have explained them; and
> they can be judged only by their fruits.—T.S. Eliot, on the
> mystical experience of mathematician and scientist Blaise
> Pascal[1]

In chapter 6 I described two epileptic patients who experienced the illusory presence of another person in response to direct electrical stimulation of the temporoparietal junction. This suggests at least the possibility of eliciting such experiences on demand in healthy subjects. For this to be practical, however, the stimulus would need to be delivered in a noninvasive way.

Psychologist Michael Persinger claimed to have developed just such a technique by using an unconventional form of repetitive transcranial magnetic stimulation (rTMS). He gained some notoriety for inventing a device known in the popular media as the "God helmet": a modified motorcycle helmet adorned with numerous electromagnetic coils, each of which is driven by computer-generated electrical signals that vary in complex ways over time. His subjects wore this while sitting blindfolded in a soundproof chamber—a condition of partial sensory deprivation. In a double-blind study of this kind, forty-eight subjects were each given a twenty-minute session in the chamber, during which they received either no stimulation or one of three different patterns of magnetic stimulation to the temporoparietal area: slightly stronger on the left, slightly stronger on the right, or equal on both sides.[2]

After each such session, the subject reported what he or she felt during the experiment by selecting items on a form listing twenty subjective feelings—a collection of typical reports Persinger had received during many years of such research. The items included "sensed presence" and several other illusions related to embodiment and sense of agency.* The groups receiving right-biased or equally balanced stimulation reported sensed-presence experiences, but the left-biased group did not. Curiously, the group that received no magnetic stimulation also reported sensed presences, almost as often as the right-biased group.

Magnetic or Suggestive?

A few years after this study, a Swedish group led by Pehr Granqvist tried to replicate Persinger's results but could not. Although some of Granqvist's subjects reported feelings of a sensed presence, these reports were not in any significant way correlated with the magnetic stimulation. They were, however, significantly correlated with measures of the suggestibility of the subjects. Granqvist noted that the magnetic fields used in Persinger's method are much weaker than those used in conventional rTMS experiments—about a millionfold weaker than fields that are just barely strong enough to elicit muscle twitching when applied to the motor cortex or illusory spots of light when applied to the visual cortex. He argued that Persinger had not used proper double-blind protocol and that the sensed presences experienced by Persinger's subjects probably resulted from their suggestibility and from the peculiar conditions of sensory deprivation.[3] Persinger claimed that the Swedish group had miscalibrated their equipment.[4] He stood by his results until his death in 2018.

Granqvist's interpretation is consistent with some informal but intriguing experiments done in the early twentieth century by James Leuba, a psychologist who was interested in religion and the illusory sensed presence.[5] Each of Leuba's subjects sat blindfolded for ten minutes in a large, quiet, dimly lit room. Several experimental assistants sat behind the subject, twenty-five feet away. The subject was told that one of the assistants might come and stand near him, behind the chair. He was asked to "assume an attitude of passive expectancy" and to raise a hand whenever he became aware of a presence. At irregular intervals an assistant silently approached the subject, stood behind him for a few seconds, and then withdrew. The room had a thick rug, so the subjects detected only about half of these approaches. Occasionally an assistant deliberately made subtle sounds as if preparing to approach the subject but without doing so. After each session the subject gave a detailed account of the experience. From Leuba's description it is not clear how much the subjects knew about the purpose of the experiment or about Leuba's interests, but they were all psychology graduate students and so probably knew something.

---

* For details see appendix B.

About half of the subjects reported perceiving not only the assistant—by sound or movement of air—but also—at other times and without sensory cues—a qualitatively different sense of presence. At least some of these occurred with no one near the subject and often involved strong emotion—fear in some cases—and a highly specific feeling of spatial localization of the presence or of being surrounded by it. One subject apparently had a vestibular illusion as part of the experience, describing it as "the impression of a rhythmic motion. It was a rather restful feeling." In summarizing the results, Leuba emphasized not only the intense emotionality often associated with the sense of presence, but also the attendant feeling of certainty that he called "intensity of assurance":[6]

> With very rare exceptions our subjects found no difficulty in separating the inference of a presence, made on the basis of perceived sounds, from what they called a Sense of Presence. An inferred presence left our subjects more or less indifferent, while the Sense of Presence involved emotions varied in character and usually intense, and it carried with it also an intensity of assurance lacking in the mere inference. It must be emphasized that, however convincing the experience, the nature of the Presence remained extremely vague.

The study used only seven subjects and so carries little statistical weight. It shows, however, that a compelling illusion of a sensed presence can be elicited merely by suggestion, isolation, and partial sensory deprivation and therefore lends some support to Granqvist's case—as do more rigorous studies of sensory deprivation. Even without specific suggestion, sensory-deprived subjects often report elaborate hallucinations, including sensed presences and other illusions of embodiment, though it may take many hours for such effects to appear.[7]

I sincerely wish Persinger's magnetic stimulation had the effects he claimed for it because it held out the promise that the ephemeral feeling of a sensed presence could be nailed down and studied in a laboratory. That hope had been Granqvist's original motivation for trying Persinger's technique. Independent replication of results is, however, the essential currency of scientific credibility, so when Granqvist's paper was published, the hope appeared greatly diminished, if not completely dashed.

Even so, there are some intriguing aspects of Persinger's method and results that may warrant further investigation. Neurons are, among other things, oscillators: their membrane voltage tends to vary rhythmically, and extremely weak rhythmic stimulation applied at just the right frequency could conceivably have significant effects—rather like the shattering of a champagne glass by Ella Fitzgerald's long and resonant note.[8] Persinger's magnetic fields roughly mimic a prominent cortical rhythm, and so they might plausibly elicit more specific and interesting effects than the neurally disruptive blunderbuss that is conventional rTMS.[9] There is also some unusual anecdotal evidence in the reports from some of his subjects, like this account from psychologist Susan Blackmore:

> I was wide awake throughout. Nothing seemed to happen for the first ten minutes or so. Instructed to describe aloud anything that happened, I felt under pressure to say something, anything. Then suddenly my doubts vanished. "I'm swaying. It's like being on a hammock." Then it felt for all the world as though two hands had grabbed my shoulders and were bodily yanking me upright. I knew I was still lying in the reclining chair, but someone, or something, was pulling me up.
>
> Something seemed to get hold of my leg and pull it, distort it, and drag it up the wall. It felt as though I had been stretched half way up to the ceiling. Then came the emotions. Totally out of the blue, but intensely and vividly I suddenly felt angry—not just mildly cross but that clear-minded anger out of which you act—but there was nothing and no one to act on. After perhaps ten seconds, it was gone. Later, it was replaced by an equally sudden attack of fear. I was terrified—of nothing in particular. The long term medical effects of applying strong magnetic fields to the brain are largely unknown, but I felt weak and disoriented for a couple of hours after coming out of the chamber.[10]

Blackmore likened this not to a sensed mystical presence, but to the illusion of alien abduction—another effect sometimes reported by Persinger's subjects.[11] What I find most intriguing about Blackmore's account is that it contains many of the qualities associated with direct electrical stimulation of the temporoparietal junction: the vestibular illusion of swaying, the sense of external agency pulling and stretching her, gross distortion of embodiment, and even some incongruous emotions such as might attend moral judgment.

Against this fascinating anecdote must be weighed Granqvist's negative results, as well as Persinger's long but dubious history as a scientist. My overall impression from reading his book,*[12] some of his journal articles, and the abstracts and titles of dozens more, is that he was a prolific pseudoscientist with strange ideas about magnetism.[13]

---

* Some of the ideas in his book (Persinger 1987) resonate with mine, like his emphasis on the infantile aspects of religious emotion, but many others are bizarre and unsupported by literature citations.

Virtual Lesions

I began this chapter with Persinger's story mainly because of its notoriety, but it also makes a nice jumping-off point for a discussion of techniques other than neuroimaging. In the hands of careful and more conventional scientists, magnetic fields have robust and reproducible effects on the brain, effects that might illuminate the neural basis of religious experiences even if they can't precipitate one at the flick of a switch. I am referring, of course, to repetitive transcranial magnetic stimulation (rTMS), which we first encountered during our visit to the temporoparietal junction in chapter 6. There I emphasized that rTMS temporarily scrambles the normal functioning of the part of the brain that receives it. If this degrades a specific perception or behavior in a reproducible way, as happened with moral judgment in the temporoparietal junction, then we know that the targeted brain region is necessary for that perception or behavior. In effect, rTMS is a virtual lesion: temporary brain damage that can be safely used with human subjects.[14]

A few studies have tried to use this approach to probe the neural basis of religiosity. The experiments are difficult because it is not obvious where in the brain to probe or how best to measure short-term changes in religiousness. Cosimo Urgesi and colleagues[15] chose to probe the temporoparietal junction (specifically, the inferior parietal lobule) and the dorsolateral prefrontal cortex because prior studies had associated these brain regions with changes in spiritual transcendence.[16] To assay the religiousness of their subjects, they used a word association test that was cleverly designed to reveal, through objective measures of response times and percentage of correct responses, how much each subject associates religious or spiritual words with himself or herself.* A brief rTMS treatment was applied, coincident with the presentation of each word pair, at one of the two brain regions of interest or at a control site at the top of the head. The study found a significant increase in implicit religiosity and spirituality with a virtual lesion of the temporoparietal junction. They interpreted this as a weakening of the representation of the body in space (a well-established function of the temporoparietal junction), which might enhance feelings of self-transcendence. The virtual lesion of the dorsolateral prefrontal cortex had no significant effect on the implicit measures of religiosity and spirituality.[17]

In a different study, Colin Holbrook and collaborators chose to target the part of the brain we explored in chapter 3, the anterior cingulate cortex, with rTMS.†[18] Their interest was not its role in infantile crying or an innate model

---

* For details see appendix B.

† Because of the depth of the cingulate cortex in the brain, the cortex directly above it (dorsomedial prefrontal) was also affected by the rTMS. See Holbrook et al. (2016) for details.

of mother, but instead its more general function as a neural alarm system that responds to threat. Psychological studies have shown that thinking about one's own death can strengthen religious beliefs, presumably as a defensive mechanism against a perceived threat.[19] Similarly, a perceived insult or criticism against one's social group tends to increase hostility toward out-groups.[20] Holbrook's group sought to examine the role of the anterior cingulate cortex in these reactions by using rTMS as a virtual lesion.

They used a stimulus pattern designed to reduce cortical excitability in and above the anterior cingulate for at least an hour following the rTMS treatment.*[21] Half of the subjects received the rTMS, and the other half received a sham stimulus as a control. Following the treatment, the subjects did a writing task that evoked thoughts of their own deaths. Then they read essays purportedly written by two recent immigrants to the United States, one praising and the other criticizing their new homeland. Next, they filled out questionnaires about their religious beliefs, with items that probed both positive (God, angels, heaven) and negative (devil, demons, hell) aspects of belief in an afterlife. They also answered questions about how favorably they viewed each of the two immigrants who wrote the essays about America.

If the anterior cingulate were essential for enhancing religious belief in response to contemplating one's own death, then a virtual lesion in that part of the brain should result in relatively lower endorsement of religious beliefs in the rTMS group than in the sham group. Also, this effect should be specific to positive beliefs about the afterlife, because belief in hell should do little to assuage death anxiety. That was, in fact, the result: subjects in the rTMS group reported 32.8 percent less conviction in positive religious beliefs, on average, than subjects in the sham group, and there was no significant difference between the two groups with respect to negative religious beliefs. Similarly, all subjects gave less favorable ratings to the immigrant who criticized America than to the one who praised it, but the sham group was significantly more disapproving of the critic than was the rTMS group. The conclusion is that the targeted cortex is necessary for these responses, both religious and ethnocentric, to the perceived threats.[22]

The Holbrook study probed both dimensions of religiosity that I described in chapter 1. Hostility toward out-groups in response to a perceived threat against the in-group is clearly a feature of the social root of religion, though it was national rather than religious affiliation that was threatened in this experiment.†[23] In contrast, seeking a loving God of salvation in the face of inevitable

---

* This stimulus pattern is known as continuous theta-burst; see Huang et al. (2005).

† The ethnocentric reaction elicited in the Holbrook experiment was not, however, purely a consequence of the implicit threat posed by the immigrant's criticism of America. The reminder of mortality also likely contributed: in many psychological studies, cues of mortality salience magnify not only religious conviction, but also in-group loyalty and out-group derogation (Burke et al. 2010).

death is just as clearly a manifestation of the neonatal root. That these two aspects of religiousness derive, at least in part, from common neural circuitry suggests that they may not be completely independent of one another. This may partly explain why no religious or spiritual movement has been able to make itself all of one and none of the other, despite many attempts in recent history.[24]

As for the Urgesi study, the word association test for religiosity and spirituality is not so easily decomposed into neonatal and social dimensions. Instead, the significance of this experiment is that a virtual lesion of the temporoparietal junction affects this subtle measure of religiousness. I explored this part of the brain in chapter 6 for completely different reasons: its early maturation during infancy, its connections to the medial and orbital networks of the frontal lobe, its role in embodiment and sensing the presence of others, and its involvement in imitation and inferring the emotions of others from nonverbal cues—all necessary capacities for an innate model of mother. I do not, however, have any specific explanation for the direction of the effect (i.e., increase in implicit religiousness with suppression of the temporoparietal junction), apart from the observation that permanent lesions in this part of the brain have caused illusions of a sensed presence in some patients.[25]

There is, however, one important methodological caveat I should point out about the Holbrook study. In every other rTMS study I have cited in this book, the magnetic pulses were delivered either simultaneously with the perceptual or behavioral task under study (as in Urgesi's experiments) or prior to it over a prolonged period (typically ten or twenty minutes) at a low frequency (typically one pulse per second). The latter protocol is called offline rTMS, because the treatment happens before, not during, the experimental test of its effects. Offline rTMS depresses cortical activity for tens of minutes following the treatment, during which time the experimental task can be done.[26]

By contrast, the Holbrook study used a newer version of the offline protocol, called continuous theta burst stimulation (cTBS), in which the pulses are delivered at a higher frequency but for a much shorter duration (forty seconds in the Holbrook experiments).*[27] The cTBS protocol has been widely used since its introduction in 2005, when a study showed that its depression of cortical activity lasts for more than an hour following the treatment.[28] In recent years, however, the reliability and reproducibility of theta burst protocols have come into question.[29] The issue likely will be resolved by validation of TMS protocols using neuroimaging methods,[30] but for now, studies using theta burst protocols should be interpreted with caution.

---

* A total of six hundred pulses were delivered over forty seconds as a series of bursts of three pulses at fifty hertz. There were five such bursts per second.

Real Lesions

Of course there is no such concern about the potency of a lesion if it is real and permanent rather than virtual and temporary. Real brain damage—whether caused by strokes, tumors, bullets, or degenerative disease—has long been a source of insight into the workings of the human brain. In chapter 4, I told the sad story of Keith Jordan, whose frontotemporal dementia transformed him from a kind, considerate, and loving father into a person so childish, impulsive, vulgar, and irresponsible that he nearly drowned his own daughter on a whim. The shriveling of his frontal and temporal lobes obliterated his sense of morality, social norms, and parental attachment. The effect on his religiousness, if any, was not described.[31]

We know from other reports, however, that frontotemporal dementia can indeed warp religiousness. The anecdotes are few, fascinating, and tragic—painful reminders that everything about our humanity flows from the activity of neurons. The condition is rare, about three cases per one hundred thousand in the general population,[32] and only a small fraction of those have symptoms of altered religiosity. The change is usually described as hyper-religiosity and most often occurs in cases in which the degeneration is mainly in the right hemisphere. The religious anomalies are almost always accompanied by other, more common symptoms of the disease, such as memory problems, social disinhibition, obsessive and ritualistic behavior, aggression, depression, altered food preference, inability to recognize familiar faces, loss of empathy, overeating, or behavioral rigidity.[33]

The hyper-religiosity of these patients is seldom described in much detail, but when it is, it seems not merely amplified, but also strangely distorted. One woman, a Lutheran since childhood, converted to Catholicism at age sixty-nine and made donations to the church. She later falsely claimed that she and the priest were lovers. Within six months of her religious conversion, she was profoundly demented.[34] A fifty-six-year-old Greek woman who had previously been religiously indifferent suddenly became obsessed with religion. She prayed at home, read Christian magazines, and often attended church. She collected personal health folders and decorated them with religious pictures.[35] A Protestant pastor in North Carolina became strangely fixated upon his idiosyncratic interpretation of ambiguous biblical passages. His extreme views and rigidity of thought caused his congregation to dwindle, and he was shuffled among successively smaller churches over a period of ten years. His odd theological ideas were compounded by his growing social disinhibition. He would approach total strangers and reveal intimate personal and financial details about his family. His daily routine became increasingly obsessive and compulsive. His occupation had long masked the pathological nature of his hyper-religiosity, but eventually he was diagnosed with degeneration of the right temporal lobe.[36]

The brain damage in these patients is too diffuse and extensive to be of much use in localizing the neural source of their hyper-religiosity, especially by the time the symptoms are severe enough for a clear diagnosis. The cases are so rare that they yield little statistical insight, yet they demonstrate that brain damage can spawn dramatic changes in religious behavior, and they hint at the importance of the frontal and temporal lobes in religious belief.

More specific and localized damage, like the surgical removal of a brain tumor, can be more informative. Erik Asp and colleagues used this approach to look for changes in religiousness associated with damage in a specific part of the prefrontal cortex. They had theorized that this area is important for generating doubt, and so patients who had lesions there would have greater credulity for extreme beliefs. More specifically, they expected the patients to score higher than control subjects on psychometric tests for authoritarianism and religious fundamentalism.[37] The study included ten patients with lesions in the ventromedial prefrontal cortex, an area roughly equivalent to what I called the medial and orbital networks in chapter 4.*[38] There were also twenty-six control subjects: ten with damage elsewhere in the brain plus sixteen others who had recently experienced serious medical problems that did not involve brain damage. All of the subjects reported affiliation with moderate or liberal Christian denominations except for one control subject who belonged to a fundamentalist Mennonite group. No subjects reported a change of religious denomination after their medical event.

Consistent with the authors' predictions, the subjects with damage in the ventromedial prefrontal cortex all scored significantly higher than controls on authoritarianism and religious fundamentalism. Their scores were also significantly higher than those of the general American population. The experimental group also had some but not all of the symptoms typical in frontotemporal dementia, including impaired autonomic responses to emotionally charged pictures and deficits in empathy, guilt, and insight. Their general intelligence and working memory, however, were mostly intact. This study gave much cleaner results than the anecdotal evidence from frontotemporal dementia, mainly because of the smaller and well-defined lesions. The authors' "doubt deficit" hypothesis, however, is not the only possible explanation for the observed changes in religiousness—a subject to which I return near the end of chapter

---

* The term "ventromedial prefrontal cortex" has no precise and consistent definition in the scientific literature. As used in this study, it includes most of what I called the medial and orbital networks in chapter 4 except for the more lateral portion of the orbitofrontal cortex, which was not damaged in the experimental subjects. Most of the lesions also included the rostral poles of the frontal lobes, which are neither ventral nor medial; a few extended slightly into the dorsal and lateral prefrontal cortex. For a composite map of the lesions, see figure 1 of Asp et al. (2012). Four of the lesions were from strokes and six from the surgical removal of tumors.

13. Also, the number of subjects was small, mainly because patients with lesions highly localized to this part of the brain are rare.

There is, however, a different approach to lesion studies that can localize the behavioral effects of lesions with high spatial precision, even when the lesions themselves are large and not restricted to the region of interest. If the lesions need not be similar in size and location, then many more patients are eligible as subjects. The technique is called voxel-based lesion behavior mapping (VLBM),* a voxel being a tiny unit of brain volume analogous to a pixel in a two-dimensional image. The basic idea is to map onto a standardized 3D computer model of a human brain the lesions of patients who vary not only in the geometry of their lesions, but also in a specific symptom of interest—like greater religious fundamentalism. If there are enough subjects of sufficient variety, then it becomes possible to estimate the likelihood, for each voxel, that damaging that voxel in a person's brain would cause the symptom. There are, of course, many subtleties and caveats involved in the statistical analysis, but the technique has offered fresh insight into many problems in neurology.[39]

Jordon Grafman and collaborators tried to use VLBM to get at the neural basis of religious fundamentalism in a group of Vietnam veterans with lesions caused by penetrating traumatic brain injuries. Following the lead of Erik Asp's group, they expected the ventromedial prefrontal cortex to be involved, but their working hypothesis was somewhat different. Their idea was that this part of the brain represents diverse religious beliefs; damage to it therefore causes a narrowing of religiousness, which would manifest as fundamentalism. They also predicted that decreases in cognitive flexibility and openness would be associated with this change.

Their first finding was a methodological disappointment: their VLBM analysis yielded no statistically significant results. They then tried a more traditional statistical approach, dividing their patients into three groups: twenty-four patients with lesions mainly in the ventromedial prefrontal area; thirty-one with lesions mainly in the dorsolateral prefrontal area; and thirty-seven patients—the "posterior" group—whose lesions did not affect the prefrontal cortex at all. They also included thirty healthy control subjects in their analysis. With respect to religious fundamentalism, the only statistically significant difference among these groups was greater fundamentalism in the ventromedial patients relative to the ones with posterior (non-prefrontal) lesions. In a separate analysis of the same subjects, they found that deficits in cognitive flexibility and openness were significant predictors of religious fundamentalism.[40]

A more recent study from the Grafman lab had better success with the VLBM technique. As in the earlier experiments, their subjects were Vietnam

---

* Voxel-based lesion behavior mapping (VLBM) is sometimes also called voxel-based lesion symptom mapping (VLSM).

vets, but this time the behavioral measure of interest was a psychometric test for having a personal relationship with God. The VLBM analysis found a statistically significant effect in the ventromedial prefrontal cortex of the right hemisphere: damage there was associated with having a stronger personal relationship with God. Also, subjects who had such damage had a greater sense of control over their personal lives than did control subjects.[41]

The authors pointed out an apparent contradiction in their findings. They noted that their study, like the earlier ones I have already described,[42] found that damage to the ventromedial prefrontal cortex was associated with an increase in religiousness, yet neuroimaging studies had found greater activation in this part of the brain during highly emotional religious experience in subjects with no brain damage.[43] If the ventromedial prefrontal cortex is specifically activated during religious experience (as seen in the imaging studies), then you would expect damage there to cause a *decrease* in religiousness—the opposite of what was found in the lesion studies. The authors tried to explain this by suggesting that the ventromedial prefrontal cortex normally inhibits other parts of the brain that produce religiousness, so damage in the ventromedial prefrontal cortex would release these downstream areas from inhibition and so increase behavioral or psychometric measures of religiousness. Unfortunately, however, that argument doesn't eliminate the contradiction: if the ventromedial prefrontal cortex inhibits religiousness, then its activity should *decrease*, not increase, during neuroimaging studies of intense religious experience.

By the same reasoning, the lesion results appear to contradict the argument I made about the medial and orbital networks in chapter 4. There I emphasized that this part of the brain becomes active in mothers when they see their infants[44] and in infants when they interact with their mothers.[45] On the basis of this and other evidence, I suggested that the orbitofrontal cortex is part of a widely distributed network that constitutes an innate neural model of mother in the brain of an infant and that, in some adults, this model is largely responsible for the illusion of God's presence and the infantile aspects of religiousness. If I am right about that, then we should expect neuroimaging studies of subjects experiencing this kind of religiousness to show activation of the orbitofrontal cortex—as has indeed been reported.[46] We should not, however, expect an increase in this kind of religiousness when the orbitofrontal cortex is damaged.

This raises an obvious question: what kind of religiousness is being measured in these lesion studies? Also, do the findings truly represent an increase in the religiousness of the brain-damaged patients, or is the change more one of the *quality* of their religiousness? I suggest the latter. In two of the studies, it was an increase in religious fundamentalism that was associated with damage to the ventromedial prefrontal cortex.[47] If the medial and orbital networks are part of the neural basis of belief in an unconditionally loving god, then their destruction in a specific patient would be expected to shift his or her religiousness more toward belief in the cruel and judgmental god of the social root—a shift toward fundamentalism and authoritarianism. The study that found a closer personal relationship with God in the patients with ventromedial prefrontal lesions might seem to contradict this interpretation, but again, what kind of relationship was being assayed by the psychometric test? Luckily, the study included the test as an appendix. None of the items probe God's unconditional love, nurturance, or protection—essential attributes of the god of the neonatal root. In my view, the two that come closest to the infantile dimension are "God lifts me up" and "I feel warm inside when I pray." Most of the others concern the subject's perceived quality of communication with God during prayer.*

Some of these ambiguities were clarified in a lesion study led by Cosimo Urgesi,[48] whose experiments with virtual lesions I described earlier in this chapter. He and his collaborators used an assay of spirituality, the self-transcendence test, which is a bit more closely aligned with the neonatal root of religion than those just discussed. It comprises three parts: self-forgetfulness (versus self-conscious experience); transpersonal identification (versus self-isolation); and spiritual acceptance (versus rational materialism). The first two mainly emphasize connectedness with nature and other people. Only the spiritual acceptance part touches on some aspects of religiousness, and none of those is fundamentalist in nature.†

Another distinction was that they measured spirituality in their subjects both before and after their surgical lesions, which is probably a more specific and sensitive experimental design than comparison with unlesioned controls. There were eighty-eight patients in the study comprising four groups:

1. twenty-four patients with high-grade (rapidly growing) glioma,
2. twenty-four with low-grade (slowly growing) glioma,
3. twenty with recurrent gliomas who had previously had other brain tumors removed, and
4. a control group of twenty patients with brain meningiomas (tumors on the surface of the brain that do not require removal of neural tissue).

Each of these was equally divided into four subgroups by location of the lesion: left anterior, right anterior, left posterior, and right posterior, where "anterior" lesions were mainly in the frontal or temporal lobes, and "posterior" ones were mainly in the parietal, temporal, or occipital lobes.

In contrast to the other lesion studies I have discussed, a voxel-based lesion behavior mapping (VLBM) analysis of these patients found no significant

---

* See appendix B for details.
† See appendix B for a five-item sample.

associations between spirituality and lesions anywhere in the frontal lobes. In fact, there were only two sites in the entire brain where significant associations of this kind were found: the right and left temporoparietal junctions.*[49]

In addition to testing the patients for self-transcendence before and after surgery, the authors also evaluated them prior to surgery for self-reported reli- giosity (including frequency of prayer and church attendance) and for mystical experiences. Among all eighty-eight patients, those with posterior lesions rated themselves significantly more religious than did those with anterior lesions. This was also true for two of the subgroups: patients with rapidly growing tumors and patients with recurrent ones. The only other significant differences to emerge from these preoperative interviews concerned mystical experiences. Patients with recurrent tumors in posterior regions of the brain had significantly more mysti- cal experiences relative to those with recurrent tumors in the anterior regions. The authors described the experiences in some detail:

> These consisted mostly in experiencing the presence of God or visions during prayer, while one patient reported a single-event feel- ing of a presence. No patient reported frank out-of-body experiences, although some patients reported undefined illusory bodily sensations before the occurrence of epileptic seizures.[50]

That this difference was significant only in the recurrent group makes sense: these were the only patients who had had lesions prior to the interview, typically months before. That the illusions were more numerous in the posterior cases also makes sense, considering the role of the posterior parietal cortex and the temporoparietal junction in embodiment—one of the main lessons of chapter 6. Similar reasoning could explain the association between lesions in the tem- poroparietal junction and the postsurgical increase in self-transcendence scores found in the VLBM analysis.

A Network of Lesions

In the years since Urgesi's study was published, new approaches to voxel-based lesion mapping have been developed. One of the most promising of these is lesion network mapping, in which lesions are mapped onto a 3D computational

---

* More precisely, the VLBM analysis found associations with self-transcendence scores in the left inferior parietal lobule and the right angular gyrus, both of which are generally considered part of the tem- poroparietal junction. The other subscales of the Temperament and Character Inventory were also used in VLBM analyses, but none of these yielded statistically significant associations with lesions in the brain. There were, however, a few significant associations between specific patient groups and some of these personality subscales, but they were unrelated to the surgeries. See Urgesi et al. (2010) p 313 for details.

model of the brain as before, but then the neural connections of the lesioned areas are added to reveal the complete network of brain regions affected by the lesions. This connectivity information comes from the human connectome,[51] a 3D database of neural pathways in the human brain, and possibly also from patterns of functional connectivity derived from fMRI experiments.[52] Normally what emerges from this approach is the discovery of some critical point of intersection in the brain to which all or nearly all of the lesioned areas connect, and which therefore is likely to be important to the symptom or behavioral change associated with the lesions. In some cases, this essential junction may not itself be damaged in any of the mapped lesions, yet it suffers from loss of the information it would normally have received from the lesioned sites.[53]

Recently Michael Ferguson led a collaboration involving Urgesi, Grafman, and others to reanalyze and extend their earlier findings using lesion network mapping.[54] They began with the data from Urgesi's study of the surgical patients whose spirituality had been measured using the self-transcendence test.[55] Rather than use the complete self-transcendence scores in the new study, however, they decided to analyze only scores from the spiritual acceptance component of the test—presumably because it more directly measures belief in the supernatural. They found that lesions affecting this measure of spirituality map onto a distributed neural network that converges on the periaqueductal gray, a relatively small structure in the core of the midbrain that had not previously been considered relevant to spirituality or religiousness. They also did a validation study using data from Vietnam veterans as Grafman's group had done in their prior experiments. With these subjects they used as their behavioral assay the simple question, "Do you consider yourself to be a religious person?" The religiousness of these subjects was significantly associated with the spirituality expected from their lesion locations and from the lesion network map of spirituality generated using the independent dataset of Urgesi's surgical patients.*[56]

This discovery does not mean that the periaqueductal gray is itself the source of religious experience. It only means that the parts of the neural network that in some way contribute to spiritual acceptance all connect with the periaqueductal gray. Even so, as the essential anatomical hub of the spirituality network, the periaqueductal gray should tell us something about the essence of spirituality— and it does.

The periaqueductal gray is a structure of ancient origin, common to all vertebrates.[57] Its most primitive function, also likely common to all vertebrates, was probably to coordinate the response to life-threatening danger: fight, flee, or freeze.[58] In large-brained animals with complex behaviors, it has taken on a wide variety of functions, but they all share the common threads of urgency, drive,

---

* Ferguson et al. (2021) did additional validation tests and examined other correlations I have not described here; see their paper for full details.

motivation, survival, and reproductive success.[59] Just to name a few examples, the periaqueductal gray is critical for the regulation of breathing,[60] pain,[61] urination,[62] and courtship vocalization.[63] But what does any of that have to do with religion or spirituality?

As Ferguson and colleagues point out, the periaqueductal gray does other things that align well with some aspects of religiousness—what I call its neonatal dimension. In neuroimaging studies of human subjects, the periaqueductal gray is active during feelings of unconditional love[64] and empathy for others in pain.[65] It is more active in mothers when they watch silent video clips of their own infants than when they see similar clips of unfamiliar babies,[66] and direct recording of its electrical activity in adult subjects reveals a short-latency response specifically attuned to the sounds of human infants babbling, laughing, or crying.[67]

The role of the periaqueductal gray in attachment and mother-infant bonding is even more firmly established in nonhuman mammals. Lesions of the periaqueductal gray severely impair nursing behavior in mother rats,[68] as does the infusion of opioid drugs into this part of the brain.[69] Similarly, ultrasonic isolation cries are reduced in rat pups that have received lesions to the periaqueductal gray.[70] Consistent with these behavioral findings, oxytocin-containing fibers from the hypothalamus innervate the periaqueductal gray in mice,[71] and oxytocin is found in the periaqueductal gray of postmortem human brains.[72]

The periaqueductal gray is the final common pathway to attachment behavior for the cortical and subcortical areas we explored in part II of this book. It mediates between low-level behavior-generating circuits of the brain stem and spinal cord, on one hand,[73] and high-level centers of emotion, aversion, and reward-seeking on the other, including the medial and orbital networks, amygdala, insula, and hypothalamus.[74] To the extent that religiousness derives from the illusion of a divine presence and an innate neural model of mother, it makes sense that the periaqueductal gray would also be a final common pathway for the neural circuitry of religion.

As we have seen, the search for the neural substrate of religiousness via lesions, whether virtual or real, has sometimes produced inconsistent or ambiguous results. Yet real progress is being made as experimental and analytical techniques improve, and there is plenty of room for further improvement. We need better tests for religiousness in these studies. In particular, sensitive and specific psychometric tests that distinguish between the neonatal and social dimensions of religiousness might help.*

---

* See appendix B.

The Helmet Revisited

We have come a long way from Michael Persinger and the God helmet, but his goal of evoking mystical experiences in a laboratory was not completely crazy. Although difficult to produce on demand, the illusion of a sensed presence is compelling and common enough that there is significant literature on the subject.[75] A psychometric test for it has recently been published.[76] It might never be possible to study the illusion via neuroimaging, but, as we have seen in this chapter, there are other ways to map neural functions onto the brain. One possibility is to build on Leuba's informal experiment of nearly a century ago[77]— using subtle sensory cues and suggestion to evoke the illusion—and some pilot experiments of that kind have recently been done.[78]

But the most promising approach, ironically, might be to follow Persinger's protocol—not because the helmet excites neurons with its weak magnetic fields (it probably doesn't), but because it excites the imagination and floods the mind with suggestion.[79] A fascinating pilot study of this kind was published in 2014 by a group of scientists led by Uffe Schjoedt in Denmark. Their working hypothesis was essentially the same idea of top-down perception that I laid out near the end of chapter 1: perception comes not only from sensory input, but also from what the brain expects. More specifically, the experimenters predicted that subjects who were more experienced in spiritualistic or New Age thinking would have greater top-down expectation of a mystical experience and so would report more sensed-presence events in a mock-up of Persinger's experiment with the God helmet. They recruited twenty-three subjects: eight members of a spiritist group who claimed prior experience of contact with spirits, seven with meditation experience recruited from a New Age store, and eight controls from the general population. The investigators deliberately used the trappings of a research hospital to amplify the effects of suggestion:

> To add credibility to the induction context, participants were received in the basement of the neurological section at Aarhus Hospital. After a short welcome, the participants were given a tour and shown a range of equipment associated with neurology and neuroscience including two MR-scanners. The study itself was carried out in the section's EEG laboratory. . . .
>
> All participants were then given a highly suggestive instruction (identical across participants). The instructions explained what was likely to happen in the room; that similar studies had been effective in eliciting unusual sensory experiences, more precisely, experiences of another being: a sensed presence. . . . Participants were then equipped with EKG sensors and a respiration-belt to boost the effect of the laboratory setting. Participants were instructed to

push a response button whenever the feeling of something unusual ended ("not during but after the experience"). . . . The participants were then seated in a comfortable chair in a sensory-deprived room, blindfolded and equipped with earplugs while wearing the helmet. To indicate that the helmet was activated, we inserted a power plug into the helmet and informed the participants that the helmet was active.[80]

Of the twenty-three subjects, eleven reported sensed-presence experiences, and eighteen reported some other kind of "unusual experience." The effect of prior mystical experience on the number of reported sensed-presence events was statistically significant, with the greatest number of events in the spiritist group and the least in the inexperienced controls, as predicted.[81]

The authors noted that many other experimental questions could be addressed with this protocol, and I agree. For me the most enticing possibility would be to combine this procedure with real rTMS: give some interesting part of the brain a virtual lesion with a genuine rTMS coil just before the session with the sham helmet. If that part of the brain is relevant to such illusions, there should be a significant difference in the number of sensed-presence events in treated subjects relative to untreated controls. In chapter 13 I offer a few specific predictions about this kind of experiment. For now, however, there is a completely different way of inducing mystical experience that merits our attention.

CHAPTER 10

The Flesh of God

> The mind is attached as by an elastic cord to the vagrant senses.—R. Gordon Wasson, describing the mystical state induced by hallucinogenic mushrooms of the genus *Psilocybe*[1]

The Aztecs called them *teonanacatl*—God's flesh—and in 1957 Gordon Wasson captured the imagination of readers of *Life* magazine with the story of his encounter with these hallucinogenic mushrooms in a remote Mexican village. Even after centuries of repression by the Catholic Church, the shamanic practice of eating *Psilocybe* mushrooms had persisted among a small group of Mixeteco Indians, who introduced Wasson and his companions to the ritual. The hallucinations he described were mainly visual, usually beginning with abstract patterns that gradually evolved into complex but meaningful scenes, as in this account from his first experience:[2]

> [T]he visions came whether our eyes were opened or closed. They emerged from the center of the field of vision, opening as they came, now rushing, now slowly, at the pace that our will chose. They were in vivid color, always harmonious. They began with art motifs, angular such as might decorate carpets or textiles or wallpaper or the drawing board of an architect. Then they evolved into palaces with courts, arcades, gardens—resplendent palaces all laid over with semiprecious stones. Then I saw a mythological beast drawing a regal chariot. Later it was as though the walls of our house had dissolved, and my spirit had flown forth, and I was suspended in mid-air viewing landscapes of mountains, with camel caravans advancing slowly across the slopes, the mountains rising tier above tier to the very heavens.

Abnormal but Not Supernatural

It is completely understandable that those who first stumbled upon these effects attributed them to supernatural forces and beings. We now know, however, that these seemingly magical visions are entirely the work of *psilocybin*, a molecule that mimics in an unusual way the neuromodulator serotonin—the same molecular mimicry that makes hallucinogens of LSD, mescaline, and DMT.[3] Other hallucinogens, like the dissociative anesthetics ketamine and phencyclidine, block an important kind of glutamate receptor and have different subjective effects—notably distortions of embodiment and illusory out-of-body experiences—but these, too, are often given supernatural interpretation.[4] Yet another plant revered by Mexican Indians for its mind-altering properties, the mint *Salvia divinorum*, yields *salvinorin-A*—a potent hallucinogen that acts on one type of opiate receptor.[5] The amphetamine derivative MDMA, commonly known as ecstasy, affects several different neurochemical systems and induces a euphoria that seems mystical to some users.[6] Collectively such drugs have been labeled entheogens, meaning "that which creates God within,"[7] and the study of their subjective effects constitutes a major branch of neurotheology.[8] All of this raises an obvious question: is God's flesh a more practical alternative to the God helmet? Are these drugs a reliable and reproducible path to the sensation of a divine presence?

The short answer, regrettably, is no. What a person experiences with an entheogen depends not only on its neurochemical properties, but also on the setting in which it is taken and on the expectations of the subject. As Wasson's description suggests, what one sees in these visions is at least partly influenced by conscious thought, and one's interpretation of it even more so. The abstract geometric patterns in Wasson's hallucinations reminded him of carpets and wallpaper, but to a terminally ill cancer patient taking LSD, a glowing diamond-shaped figure was a sign of the comforting presence of God.[9] More to the point, most religious people who avoid these drugs do not report vivid and overwhelming sensory hallucinations as part of the experience of their faith—not even when they feel the presence of God. Perhaps the word entheogen has too narrow a meaning to be a good label for these drugs, as their effects seem to overlap only partly with the domain of religious experience.

Hallucinations the Hard Way

In a deeper sense, however, hallucinations—whether drug-induced or not—have much to teach us about the workings of the human mind in general and about the illusion of God's presence in particular. The lesson encompasses not only hallucinations and mystical states, but also such mundane aspects of mental experience as perception, attention, and visual search. By way of introduction, I present an extremely mundane example from my own experience.

In springtime I engage our local weeds in hand-to-root combat, not so much for appearances, but mainly to shift the balance of power in favor of our beleaguered California native plants. The invasive non-native species are mainly European mustard, thistles, and foxtails. During each session I concentrate on the species that is closest to dispersing its seeds, so I may spend hours searching for, and trying to eradicate, only one kind of plant. The work is demanding, not only physically, but also cognitively. It is a challenging visual search task because the weeds are intermingled with, and closely resemble, some of the native plants. Subjectively I feel that I am exerting attentional effort, deliberately not trusting my first impression that an area is free of weeds, and scanning it repeatedly from different angles. I imagine the shape, color, and arrangement of the leaves of my quarry. I feel that my visual perception is sensitized by the expectation that I will soon see what I imagine. There is probably also some reinforcement learning involved in this, as each successful discovery and removal of a weed provides some satisfaction, and my performance seems to improve over time.

After many hours of this I go indoors, covered with sweat and allergens, and eagerly dive into the shower. When I close my eyes to wash my face, I often see a canonical image of the target of my visual search—an idealized and symmetrical thistle, for example. This is not a retinal afterimage, because it is not in any way related to what I was seeing immediately prior to closing my eyes.* On another day when I pursued foxtails, it is a clump of foxtail grass that I see when I close my eyes in the shower. These are authentic closed-eye hallucinations, less interesting, perhaps, but otherwise essentially like those elicited by appropriate doses of the hallucinogens MDA or atropine.[10]

That I have these uninspiring visions does not necessarily mean that my weeds exude hallucinogens or that I am psychotic. At least I am in good company. Psychologist Nelson Hanawalt and his wife had closed-eye hallucinations of blackberries after picking them all day.[11] Similarly, when neurophysiologist David Ingle closed his eyes at bedtime, he saw the waveforms of the neuronal action potentials he had been scrutinizing all day on the screen of an oscilloscope. Shortly before his death in 2006, he published several papers describing the phenomenon and experiments he had done to learn more about it. He found that a few percent of subjects experience these closed-eye visual illusions far more readily than others—after only a few seconds of looking at a visual target—and

---

* A retinal afterimage is a negative image seen with the eyes closed after staring at a high-contrast scene long enough to cause significant adaptation of retinal photoreceptors. See https://en.wikipedia.org/wiki/Afterimage (accessed 22 December 2021).

that the apparent position and size of the illusory object could be affected by grasping and moving it by hand.[12]

Expectation and the Streams of Perception

What these and other observations show is that hallucinations are intimately related to the neural mechanisms of attention, learning, and perception. Although hallucinations are not fully understood, a scientific consensus is forming around the idea that they somehow arise from an unusual disturbance in the normal balance between the two streams of information that contribute to perception.[13] One of these streams constitutes sensory information flowing into the brain from the bottom up—that is, from the sense organs into the thalamus and cortex. The other is the flow of perceptual expectations from the top down—that is, from higher cognitive and memory areas of the brain into the lower-order sensory areas.[14]

In any perceptual task, what the brain is really trying to do is to determine what is out there in the environment by choosing from those familiar percepts supplied by the top-down stream the one that most closely matches the pattern of sensory input arriving from the bottom up. Our perception is therefore never purely objective but is always to some degree biased by our expectations. When our sensory input does not match anything in our repertoire of expectations, the mismatch spawns a new percept that gets added to the repertoire. Conversely, when a specific expectation is unusually important, sensory areas become more sensitive to it—and less receptive to distractions. Learning is therefore an essential aspect of perception.[15]

This strategy greatly improves our performance in difficult pattern recognition tasks, especially when the pattern we seek is degraded by noise or obscured by distracting stimuli. When all but a small part of one leaf of a thistle is hidden by another plant, the top-down information in my brain fills in the missing parts of the image. I imagine the whole plant to which that partial leaf must belong, I see in my mind's eye where it must be rooted, I part the leaves of the obscuring plant, and another thistle bites the dust.

In more formal terms, perceptual tasks like weeding are problems in signal detection, in which two different kinds of error are possible. I can fail to notice a weed where one is present—a false negative response—or I can think I see one where there is none—a false positive response. In weeding, a false negative error is costly because any weed that I miss will disperse its seeds, and next year's weeding ordeal will be worse. A false positive error, however, is no big deal. If I mistake a harmless plant for a weed at first glance, I will look at it carefully before uprooting it, and I usually catch the error in time. It therefore makes sense for me deliberately to adjust the bias in my perception such that I am more sensitive to anything that looks remotely thistlelike, even if it means that I make some false positive errors.[16] This is one way of exerting attentional effort, and, after many hours of it, this top-down information has subtly reshaped my visual system. Through changes in its synaptic connections, my visual cortex has come to expect the image of a thistle in whatever I am seeing. This top-down bias is not so great as to override sensory input that is not plantlike. I do not mistake my shower faucet for a thistle. But when I close my eyes, the bottom-up information vanishes, and the top-down bias becomes evident as the ultimate false positive response: a compelling hallucination of an idealized thistle.

Attention and Acetylcholine

The neural mechanisms that underlie all of this have been explored in animals trained to do attention-demanding tasks. In a typical experiment of this kind, a thirsty rat is placed into a Skinner box equipped with two levers, a small light just above the levers, and a tube that can supply water.[17] The small light is normally off, and the whole box is illuminated from above by a larger lamp that is always on. At random times, a brief audible tone cues the start of a four-second time interval, during which the rat must press one of the levers to get a drink of water. In some trials, the small light briefly flashes one second before the tone; in others there is no flash. If there was a flash just before the tone, the rat must press the left lever to get a reward. If there was no flash, it must press the right lever. Pressing the wrong lever during the four-second interval or pressing either lever at any other time has no effect. After they learn this task, rats will perform it for many consecutive trials.*[18]

In contrast to my weeding task, there is no advantage for the rat to bias its perceptual expectation toward the presence of the flash signal—assuming the flash and non-flash trials occur with equal probability. There is, however, an advantage to paying attention to both the light and the tone. Also, the task can be modified so as to demand greater attentional effort from the rat. For example, the flash could be shorter in duration and therefore less noticeable, or it could be obscured by distracting stimuli like a flickering of the overhead light, or the tone could be obscured by irrelevant sounds. Even without these distractions, the task requires gradually increasing attentional effort over time because the rat becomes fatigued. Most importantly, the brain of the rat can be manipulated

---

* Minor variations on this task design have been used; for example, using food instead of water as the reward or insertion and withdrawal of the response levers to indicate the trial period rather than the audible tone (McGaughy and Sarter 1995). Other task designs are also effective for this kind of research; see for example Robbins (2002).

during these experiments. The activity of its neurons can be recorded, drugs can be directly injected into tiny neural structures, and changes in the concentration of neurotransmitters and neuromodulators can be measured in critical circuits while the rat is behaving.

The most important discovery from this research is that the interplay between the top-down and bottom-up streams of information and the effects of attentional effort on that interplay are mediated mainly by acetylcholine.[19] Acetylcholine is a small molecule that acts as a neurotransmitter in the peripheral nervous system—where it causes contraction of skeletal muscles and adjusts heart rate, among many other functions—but in the cortex it acts both as a fast-acting neurotransmitter and as a slow-acting neuromodulator that has various and complex biochemical effects on its target neurons.[20] The neurons that release it lie below the cortex in an area called the basal forebrain, but their axons innervate the entire cerebral cortex.[21]

These cholinergic neurons of the basal forebrain are activated when the animal needs to pay more attention.[22] For the thirsty but distracted rat in the Skinner box, this happens when it starts making more errors and misses its rewards. A direct signal from its prefrontal cortex excites the cholinergic neurons, which then release acetylcholine into various parts of the cortex, including the prefrontal areas that began the process.[23] The pattern of acetylcholine release is not uniform or random but is directed to specific cortical regions that are relevant to the problem at hand. For example, if the rat is trying to pay more attention to the signal light and the cue tone, the concentration of acetylcholine will be disproportionately increased in the visual and auditory cortex.[24] This focused release of acetylcholine changes the responsiveness of neurons in the target areas—for example, by increasing the excitability of auditory neurons that are tuned to the pitch of the cue tone, while a relative deficit of acetylcholine elsewhere depresses neuronal responses to distracting stimuli.*[25]

In part these changes in responsiveness are effected through changes in the strength of synaptic connections—the essence of most forms of learning. The mechanisms of synaptic plasticity are complex and multifaceted, and acetylcholine is not the whole story—in previous chapters we saw similar if more specialized roles for dopamine, oxytocin, and vasopressin. Acetylcholine affects synaptic plasticity through two different kinds of receptors. A fast-acting one, called nicotinic, is important for learning the precise timing of neuronal signals and is therefore likely to be involved in mastering sequential patterns of sensation or behavior.[26] A slow-acting acetylcholine receptor, called muscarinic,

---

* The effects of acetylcholine on the receptive field properties of sensory neurons have been studied in a variety of sensory systems. For an interesting sample of this research, see Ego-Stengel et al. (2001), Herrero et al. (2008), Kilgard and Merzenich (1998), Murphy and Sillito (1991), Roberts et al. (2005), Sato et al. (1987), and Weinberger (2007).

triggers a cascade of biochemical reactions that can affect synaptic strength in several ways, including changes in the sensitivity of glutamate receptors[27] and in the physical shape and size of dendritic spines[28]—the sites of modifiable synaptic connections.[29]

The acetylcholine system is therefore doing something more elaborate than merely shifting the balance between the top-down and bottom-up contributions to perception. It is using top-down expectations to tune and shape the responsiveness of the lower-order sensory cortex to incoming bottom-up information, thus imposing a perceptual bias that is helpful in a specific task. Through its effects on synaptic plasticity, it can make this bias an enduring one. In humans, the consciously directed focusing of attention—such as occurs in my brain when I am weeding—is driven by areas in the lateral prefrontal cortex and by parietal areas, including the temporoparietal junction, mainly on the right side.[30] That the right temporoparietal junction has a role in this should come as no surprise, considering the profound deficit in attention to the left side of the body and surrounding space that results from damage to this part of the brain—the hemispatial neglect syndrome I described in chapter 6.

Expecting the Primordial Savior

But attentional effort can also be driven by emotion, and emotion-related parts of the brain—especially those that I identified in part II as likely components of an infant's innate model of its mother—directly innervate the cholinergic basal forebrain. These include the amygdala[31] and orbitofrontal cortex,[32] which store innate knowledge of fundamentally pleasant and aversive stimuli, including the primordial social sensations of the mother-infant bond. But another cortical area that sends a dense input to the cholinergic basal forebrain system and that often shows up in neuroimaging studies of attention in humans is the anterior cingulate cortex[33]—the neural alarm system that drives infantile crying.

This, too, should come as no surprise. As I explained in chapter 3, the anterior cingulate cortex is activated by a bewildering variety of conditions, but what they all seem to have in common is something gone awry, something wrong, something urgent, something painful, something in need of attention. When the anterior cingulate cortex of an infant detects separation from mother, it triggers not only crying and the autonomic stress response, but probably also activation of the cholinergic basal forebrain system. In doing so, it presumably is biasing the sensory cortex with an expectation of mother.*[34] For a newborn infant, this

---

* The anterior cingulate cortex is specifically activated during states of emotional expectation. See Bermpohl et al. (2006).

expected sensory image can only be its innate image of her—that she has a face, that she is warm, that she provides nourishment, that she has a melodic voice of feminine pitch, that her presence is accompanied by rhythmic vestibular sensations, that she is another being, a loving agent who can and will help.

The vision of a human newborn is functional but significantly worse than that of an adult with respect to acuity,[35] contrast sensitivity,[36] and color perception.[37] It completely lacks stereoscopic depth perception.[38] Its olfactory and auditory thresholds are relatively high.[39] All of its senses have previously had only the extremely limited experience of life in the womb. With its naive and suboptimal sensory systems, a newborn is clearly in a position to benefit from the contribution of top-down information to perception. Just as it helps me find thistles in a landscape of visual noise and distractions, it would help a newborn recognize contact with mother, despite its sensory limitations, and thereby promote the emotional contact that is reinforcing both to mother and infant. As I explained in chapter 4, the innate image of the mother appears to be the primordial model—the collection of social primary reinforcers—that forms the essential foundation for learning to function as a social animal through reinforcement. This may be the single most important selective pressure that made that innate image necessary. If it exists, as I have suggested, then it is only natural that other neural systems, like those of sensory perception, would also exploit its existence.

If that innate model also persists into adulthood, then the illusion of God's presence can be understood as a compelling hallucination of the sense of agency. This is especially so when the illusion is elicited by extreme stress, helplessness, and emotional pain—conditions so extreme, so primordial, that the anterior cingulate and orbitofrontal cortices bias perception to expect the primordial savior. This would be a hallucination not primarily in visual, auditory, or other low-level sensory cortices, but in the temporoparietal junction and related structures, like the medial prefrontal network, which distinguish self from other and infer the presence and intentions of other people.

Since the cholinergic basal forebrain system is at the heart of this perceptual biasing, the foregoing speculations predict that: (1) cholinergic drugs should influence the susceptibility to religious experience of this kind; (2) the cholinergic basal forebrain system is present and functioning in the brain of a newborn; and (3) at least some cholinergic drugs should be hallucinogenic. I elaborate on the first prediction in chapter 13. The second and third have already been abundantly confirmed.

The cholinergic basal forebrain system appears early in human fetal development[40] and is essentially complete—including its axons that innervate cortex—by the end of the second trimester.[41] By contrast, the cortex is much less mature by this time, which suggests that the basal forebrain system plays a role not only in postnatal cortical function, but also in prenatal cortical development.[42] The entire cholinergic system is present at birth, including the acetylcholine receptors in the cortex, though the density of cholinergic innervation changes throughout the lifespan.[43]

As for hallucinations and the acetylcholine system, plants that contain the muscarinic cholinergic blocker scopolamine and related compounds have been used in shamanic rituals since antiquity because of their hallucinogenic properties.[44] Unlike hallucinations from LSD, which tend to involve surrealistic distortions of real percepts, those induced by scopolamine are usually complex, familiar, seemingly real but wholly nonexistent objects, animals, persons, or scenes. In a study of 158 healthy male volunteers given scopolamine injections, subjects grasped at imaginary objects, smoked nonexistent cigarettes, conversed with hallucinatory friends and family members, and witnessed complex and panoramic scenes, like a reenactment of a battle from the Civil War.[45] It is partly because of this compelling realism that scopolamine hallucinations are still sought by some recreational drug users, despite the potentially lethal side effects of the drug.[46]

Exactly how scopolamine creates these visions is not fully understood, but by blocking the normally attention-directing effects of acetylcholine on sensory perception, it appears to be diminishing the influence of bottom-up information on conscious perception—much as closing my eyes in the shower brought a nonexistent thistle into my vision. It also, however, evidently admits into consciousness prospective top-down percepts that normally would only enter conscious perception if their presence were confirmed by sensory input. To use Wasson's metaphor, scopolamine indiscriminately stretches that elastic cord connecting the mind and senses and makes the senses vagrant. The psilocybin he ingested, like LSD and several other classical hallucinogens, presumably also does this, though through a different neuromodulatory mechanism.[47] In the absence of hallucinogens, the acetylcholine system does the opposite: focusing attention and biasing perception in some meaningful or urgent way, thus selectively and tightly coupling the mind to sensory input.

If this interpretation is correct, then the illusion of God's presence seems less like a scopolamine or psilocybin hallucination and more like my hallucinatory thistle in the shower. Like the illusory thistle, the feeling of God's presence arises from prior hyperactivity of the cholinergic basal forebrain system, not from its blockade. It comes from an unconscious perceptual bias gradually brought on by overwhelming guilt, fear, loneliness, pain, or defeat—but above all, by helplessness. And just as eye closure reveals the thistle, a diminution of sensation seems to help trigger the illusory sensed presence—closed or weeping eyes, a face buried in the hands, Leuba's quiet, darkened room, or Persinger's sensory deprivation chamber. Sleep deprivation may play a similar role in some of these experiences, as circuits in the brain stem struggle to disengage the senses from conscious perception as a prelude to sleep.[*48] The most effective of these triggering sensory diminutions, however, is probably social isolation, whether real or imagined. The illusion arises because the brain has been primed to expect a percept in the social modality—the presence of a loving agent—and the absence of any agent, like my absence of vision in the shower, may manifest that neural bias as an illusory sensed presence.

This interpretation is similar to the one I developed in chapter 5 for the maternal potentiation behavior of rat pups. There I suggested that when an isolated rat pup briefly experiences the presence of its mother and is then isolated again, its neural state becomes one of appetitive reward seeking. Its greatly exaggerated crying may reflect not increased stress, but an increased expectation that its mother is present nearby and will soon retrieve it. The essential role of the nucleus accumbens in maternal potentiation supports this view, as this part of the brain is a neural epicenter of appetitive reward seeking.[49] If this view is right, however, then maternal potentiation should also increase cortical acetylcholine levels so as to impose that expectation of mother on the sensory cortex.

To my knowledge, this prediction has not yet been directly tested.[†50] The ideal test would be to monitor the concentration of acetylcholine in the sensory cortex of a rat pup throughout a maternal potentiation experiment; the prediction is that a significant increase in cortical acetylcholine should accompany the pup's potentiated ultrasonic crying during its second period of isolation. The experiment would be technically difficult, though feasible.[‡51]

Other lines of evidence, however, are at least consistent with the hypothesis. Several experiments show that the nucleus accumbens is critically involved in the control of the cholinergic basal forebrain system.[52] In one of these, rats were trained to drink a mildly unpleasant-tasting liquid to obtain a highly appealing tasty food. This unusual task requires tolerating a slightly aversive primary reinforcer—the bad-tasting drink—as a way to obtain a highly rewarding one—the delicious food—and so presumably involves both attentional effort and appetitive reward seeking. Normally this task causes an increase in acetylcholine levels in the medial prefrontal cortex, but that increase never occurs if the nucleus accumbens is temporarily inactivated by a microinjection of a neurotoxin.[53] In another series of studies, cortical acetylcholine levels of rats were greatly

---

* About a quarter of the general population reports occasional hallucinatory experiences just before the onset of sleep. See Ohayon (2000).

† Although it does not directly test the idea that maternal potentiation increases cortical acetylcholine, the study by Ricceri et al. (2007) has relevance here. I discuss it in some detail in chapter 13.

‡ The acetylcholine concentration in somatosensory barrel cortex might be the most relevant for two-week-old pups (Khazipov et al. 2013). The traditional microdialysis technique used for measuring cortical acetylcholine concentrations in studies of attentional effort might not have adequate temporal resolution for the maternal potentiation experiment, but new techniques using biosensors should work (Cifuentes Castro et al. 2014, Jing et al. 2020, Parikh and Sarter 2008).

increased by injecting into their nucleus accumbens the drug NMDA—a molecule that activates a specific kind of glutamate receptor that is important for synaptic plasticity.[54] The effect was enhanced by simultaneous activation of the D1 variety of dopamine receptors in the nucleus accumbens,[55] but it also could be *completely blocked by activating D2 dopamine receptors* there.[56]

I emphasize that last arcane point because the drug used to activate the D2 dopamine receptors in that experiment—quinpirole—is the same one that Harry Shair and colleagues used in the nucleus accumbens of rat pups specifically to block maternal potentiation.*[57] This common sensitivity to quinpirole in the nucleus accumbens is at least consistent with the idea that an increased release of acetylcholine into the cortex accompanies a rat pup's high expectations of its mother's presence.

The experiment also suggests that the NMDA type of glutamate receptor is somehow involved in the interplay between the top-down and bottom-up contributions to perception, and this has been confirmed.[58] NMDA receptors are abundant in the brain and are required for many forms of learning. Because learning and perception are so deeply intertwined, drugs that block NMDA receptors—like PCP and ketamine—cause such profound distortions of consciousness that they, too, are considered entheogens. They are also considered psychotomimetic[59] because their effects can mimic the paranoid delusions that bring some people into highly disturbing relationships with aliens, demons, or God—a subject for the next chapter.

---

* I discussed this in chapter 5.

CHAPTER 11

The Madness of God

> The mystical experience is very much that of the schizophrenic,
> except that the mystic knows how to swim in the waters in
> which the schizophrenic drowns.—Joseph Campbell[1]

Schizophrenia is probably the most familiar of several neural maladies that can dramatically affect religious thought, feeling, or behavior. Others include Parkinson's disease, temporal lobe epilepsy, obsessive-compulsive disorder, and various forms of dementia. To the extent that these conditions can be linked to specific neurochemical systems or circuits in the brain, they may provide clues to the neural basis of religiosity in healthy subjects and so have attracted the attention of neurotheologists.[2]

Schizophrenia and the Misperceived Self

Twin studies show that schizophrenia is largely a heritable condition, with more than 80 percent of its variance attributable to genetic factors.[3] A survey of schizophrenia research found that religious delusions and religiously themed hallucinations are common in schizophrenics of all cultural groups examined.[4] For example, a study of 295 schizophrenic patients in predominantly Catholic Lithuania found that 64 percent had religious delusions, with women often believing themselves to be saints and men often believing themselves to be God.[5] Of the 295 patients, 88.5 percent described themselves as religious, which appears to be more than in the general population.*[6] Although the tendency to

---

* A different study found that only 49 percent of Lithuanians claim to believe in God, but an additional 36 percent believe in "some sort of spirit or life force" (European Commission 2005, p 9). Some

have delusions is heritable in schizophrenia, the tendency to give them a religious interpretation is largely cultural in origin.[7]

Whether religious or not, the delusions of schizophrenics almost always involve some kind of misinterpretation, not only of what they perceive, but also of what they themselves do and think. They may feel that they are the victims of some grand conspiracy, or that the mental voices they experience are not their own thoughts, but are instead messages from gods, demons, aliens, counselors, or enemy conspirators. They may be convinced that their thoughts and actions are being remotely controlled by some external force or person.[8]

These symptoms strongly suggest that schizophrenia is at least in part a disturbance in the sense of agency—especially in the discrimination between self and other—a faculty that I have argued is centrally involved in the illusory sensed presence. In chapter 6 I emphasized the role of the temporoparietal junction in this discrimination especially because of the way it uses signals from the motor system to distinguish one's own movement from that of another person. In essence, these signals are copies of the commands sent to muscles to execute the behavior.[*9] The copies sent to the temporoparietal junction anticipate the sensory consequences of the movement—for example, that the sight and sensation of curling fingers should be apparent because a command to close the hand has just been issued. The precise timing of these motor signals is essential for distinguishing self from other, as demonstrated by experiments in which healthy subjects watched slightly delayed video displays of their own hand movements.[10]

In schizophrenics, something appears to be wrong with the generation or processing of these anticipatory signals.[11] In healthy brains, copied signals of this kind are evidently issued not only in conjunction with behavior, but also along with virtual behavior like silent mental speech—the internal linguistic thought that tends to dominate our stream of consciousness. In those schizophrenics who hear intrusive alien voices, the signals that should identify their internal verbal thoughts as their own are absent or so distorted that they experience these thoughts as coming from someone else.

The most direct evidence for this comes from experiments in which waves of electrical activity from the auditory cortex are evoked by brief sounds and are detected by electrodes on the scalp. In healthy subjects, a specific component of this neural response is diminished if the subject is speaking during the presentation of the sounds—evidence that a copy of the motor signals from speech-generating cortex is altering the receptivity of the auditory cortex in anticipation

---

caution is warranted in comparing these results to those of Rudalevičienė et al. (2008) because the two studies used different questions to assess religiosity.

* More precisely, they are predictions of the sensory consequences of the self-generated action. The computation that transforms copied motor signals to anticipated sensory signals occurs unconsciously, probably in the cerebellum. See Synofzik et al. (2008); Wolpert et al. (1998).

of self-generated sounds. Interestingly, the same diminishment occurs when the subjects speak silently to themselves during the presentation of the sounds—evidence that verbal thought alone also generates these anticipatory signals. The most interesting result from these experiments, however, is that during both experimental conditions—actual speech and silent verbal thought—the electrical signature of the anticipatory signals is missing in schizophrenic patients.[12] Although this experiment shows that there is something abnormal about the processing of self-generated thought and action in schizophrenics, it does not directly assess their conscious experience of body ownership or their subjective sense of agency—two critical aspects of embodiment that could be affected by this deficit in neural signaling.

The Rubber Hand and Weak Embodiment

One way to assay the feeling of body ownership is an intriguing illusion of partial disembodiment known as the rubber hand illusion. The subject sits with one arm resting on a table in front of him but hidden from view behind a screen. On the visible side of the screen, a lifelike rubber forearm is placed on the table near the real arm and in the same orientation, with a cloth covering the elbow of the dummy arm and the subject's upper body, so as to suggest that the dummy is connected. While the subject watches the dummy hand, an experimenter uses two small paintbrushes to stroke both the dummy and the subject's real hand in synchrony. After a few minutes of this, and despite their objective knowledge to the contrary, healthy subjects report the feeling that the rubber hand is their own, and their subjective sense of the location of their real hand is shifted toward the position of the dummy.*[13]

Several studies report that schizophrenic subjects feel this illusion sooner and more vividly than healthy ones, and they perceive a significantly greater shift in the position of their hand toward the dummy.[14] This is consistent with a weaker or more flexible sense of embodiment in schizophrenia—so much so that, in one subject, the rubber hand illusion triggered a complete, if illusory, out-of-body experience. This patient had a history of out-of-body experiences, usually occurring either near the onset of sleep or associated with a psychotic episode.[15]

Even healthy subjects report weak illusions of complete disembodiment in a clever elaboration of the rubber hand experiment called the full body illusion. While wearing virtual reality goggles, the subject watches a 3D image of

---

* You can watch a video of the rubber hand illusion at www.youtube.com/watch?v=sxwn1w7MJvk (accessed 18 December 2021).

himself, filmed from behind by video cameras, while an experimenter strokes the subject's back with a rod. The video image is either synchronous with the actual stroking of the subject, or, in other trials, it is delayed. In the synchronous condition, subjects identify the person seen as themselves and misperceive their physical location in the room as being closer to the self seen in virtual reality—much like the distortion in perceived hand position in the rubber hand illusion.[*][16]

Considering the greater sensitivity of schizophrenics to the rubber hand illusion, one might expect them to have a similarly enhanced response to the full body illusion. Surprisingly, however, in the only such study I have found, the response of schizophrenic patients to the full body illusion was not significantly different from that of healthy controls. The authors noted that the studies of schizophrenic subjects using the rubber hand illusion used small sample sizes and so had low statistical power. They also reanalyzed the data from three such studies that were similar enough in design as to qualify for meta-analysis and found no significant effect of schizophrenia on sensitivity to the rubber hand illusion.[17] More experiments will be needed to resolve this apparent discrepancy, but for now the link between schizophrenia and a deficiency in body ownership appears dubious at best. A much stronger case can be made, however, for a deficiency in the sense of agency.

Disturbances in Agency Detection

Many experiments reveal striking psychosis-related anomalies in the sense of agency, and some of them confirm the essential role of the temporoparietal junction in the perception—whether real or delusional—of an external agent. In these experiments, subjects must do some task with their hands, like point at a target, manipulate a joystick, or remove pegs from a board while viewing their actions on a video display that in some way distorts their visual feedback. The distortion can be something adjustable, like a time delay or a change in the angle of movement, or it can be a display of a similar task being done by another person. In healthy subjects, the greater the distortion—and the greater the subjective sense that another person is controlling what the subject sees—the greater the activation of the right temporoparietal junction.[18] Schizophrenic patients are less likely than healthy subjects to be aware of the sensory distortion,[19] they are more likely to infer external agency in these experiments,[20] and even in the absence of sensory distortion—when they are clearly watching the consequences of their own actions—their brains show abnormally high activity in the temporoparietal junction.[21]

---

* You can watch a video of the full-body illusion at: www.youtube.com/watch?v=ee4-grU_6vs (accessed 18 December 2021).

This confused sense of agency is also evident in studies of the verbal thought and speech of schizophrenics. These experiments involve cleverly designed word recall tests that require the subject to remember not only a list of words, but also the source from which they experienced each word. During the training phase, the subject sees each word on a computer screen for three seconds, immediately preceded by one of two written instructions: "hear this word" or "imagine hearing this word." In the "hear" case, the subject hears the experimenter read the word aloud when it appears on the screen. In the "imagine" case, there is no sound, and the subject must imagine the word being read aloud. After sixteen of these trials (eight of each kind randomly intermingled), the subjects are given a list of the sixteen test words, along with eight distractor words that were not seen during the training phase. The subject must identify each of these as heard, imagined, or new.

The reasoning behind this experimental design is that, if schizophrenics hear hallucinatory voices because they incorrectly attribute their own verbal thoughts to an external agent, then they should do worse than healthy subjects at remembering whether a word was heard or imagined in this test. But this result could also occur merely from a general memory deficit in the schizophrenic patients, so the test includes the distractor words as a control. If the schizophrenics do as well as healthy subjects at distinguishing the distractors from the test words, yet still have a deficit in recalling the source of each word, then the deficit specifically involves the attribution of agency.

That is, in fact, the result obtained. A series of such experiments reveals that unaffected siblings of schizophrenics do significantly worse than healthy subjects at this test, non-hallucinating schizophrenics do worse than the sibling group, and hallucinating schizophrenics do worst of all. There is, however, no significant deficit in any group at distinguishing test words from distractors. That unaffected siblings are deficient in this test suggests that it is sensitive enough to detect vulnerability to schizophrenia, even before symptoms of the condition appear.[22]

In chapter 6 I described experiments in which the temporarily disruptive effects of transcranial magnetic stimulation were applied to the right temporoparietal junction to show that this part of the brain is essential for the perception of biological motion, for the discrimination between self and other, and for the moral judgment of the behavior of other people. When hallucinating schizophrenics are given rTMS treatments to the left temporoparietal junction over a period of five days, their performance at the word recall test significantly improves.[23] Since chronic rTMS tends to depress neural function in the stimulated area,[24] the implication is that aberrant or hyperactive circuitry in the temporoparietal junction of schizophrenics contributes to their misattribution of agency, and temporary suppression of this activity relieves the symptoms.

This interpretation is consistent with the neuroimaging experiments that reveal hyperactivity in this part of the brain in schizophrenics[25] and with the abundantly replicated finding that rTMS to the temporoparietal junction temporarily relieves auditory hallucinations in some of the most intractable cases of schizophrenia.[26] Interestingly, rTMS to the temporoparietal junction also diminishes some effects of the rubber hand illusion.[27]

In the previous chapter I suggested that the illusory sensed presence is a vivid hallucination of the sense of agency, primed by the intensely emotional prior expectation of a loving agent, precipitated by overwhelming helplessness, triggered by sensory or social deprivation, and likely to involve the temporoparietal junction. The delusions of schizophrenia arise from a far more profound disturbance in neural function, one that involves multiple genes, multiple neurochemical systems, and a partial breakdown in the normal cooperation of different brain regions that is essential for the distinction between self and other.[28] Despite these differences, the two phenomena share some revealing commonalities. Both involve anomalous prior expectations of external agency, in one instance arising from emotional need and the latent circuitry of infantile expectation and in the other from deficiencies in the expected sensory consequences of self-generated thought or action. Both result in the incongruous perception of another being—albeit with greater sensory and verbal content in psychotic hallucinations—and in both cases the incongruity is often rationalized in religious or spiritual terms.

There are other interesting commonalities and differences at the level of neural hardware. There is good evidence that the temporoparietal junction is centrally involved in both, though probably more on the right side for the sensed presence and more on the left for the verbal hallucinations of schizophrenics. This difference reflects the typical hemispheric specialization in this part of the human brain. Although the left temporoparietal junction is typically more specialized for language comprehension and the right more for embodiment and the inference of the intentions of others from nonverbal cues, each side has some of these abilities, and both subserve theory of mind—the interpretation of the minds of others that is the hallmark and foundation of human social behavior.[29]

The Neurochemistry of Delusion

If the neurochemistry of mother-infant bonding is as relevant to the illusory sensed presence as I have suggested, then that illusion and the delusions of schizophrenia both involve some of the same neuromodulatory systems, though in different ways. In chapter 5 I summarized evidence that dopamine mediates an infant's appetitive longing for its mother, particularly through its effects in the nucleus accumbens. Similar transient dopaminergic effects in the nucleus accumbens probably also attend the desperate longing that precipitates the sensed presence. Schizophrenics, by contrast, tend to have a deficiency of dopamine in the prefrontal cortex but a chronic excess of it in the nucleus accumbens. Long-term abuse of amphetamines—which flood the nucleus accumbens with dopamine—often induces psychotic delusions much like those of schizophrenia, and a major class of antipsychotic drugs blocks dopamine at D2 receptors.[30]

Probably more important, however, are the neuromodulatory systems central to attention, perception, and synaptic plasticity. Embodiment and the sense of agency depend on the precise timing of anticipatory signals from the motor system that must cohere with the actual sensory consequences of a movement— sensations such as sight, touch, acceleration, and the feeling of limb position. This is merely a special case of the reconciliation of top-down expectation with bottom-up sensation that I discussed earlier in the context of hallucinations—in this case the motor system generates precisely timed top-down expectations. Synaptic plasticity is essential for this reconciliation because the most robust way to achieve precision in any complex system is through adjustability and fine tuning.

In the previous chapter I emphasized the role of acetylcholine in the synaptic plasticity that underlies the effects of attention on perception, and I argued that hyperactivity of this system might set the stage for the illusion of a sensed presence. A deficiency in cholinergic function may contribute in a different way to the hallucinations of schizophrenia. A common genetic deficit in schizophrenia affects nicotinic acetylcholine receptors that are involved in a special form of plasticity—one that is sensitive to the precise timing of signals at the synapse.[31] Although this deficit is only part of the problem in schizophrenia, it is enough of a problem that cholinergic drugs are effective antipsychotics for some patients.[32]

But a different neurochemical system—one that is even more crucial for timing-dependent plasticity—has recently emerged as the primary focus of research in schizophrenia. This is the ubiquitous NMDA receptor, a kind of glutamate receptor that is especially well suited to detecting the coincidence of activity in the two neurons on either side of a synapse.[33] Although many different genes and their various neurological consequences contribute to schizophrenia, a good case can be made that nearly all of them impinge in some way upon the NMDA receptor system.[34] NMDA receptors are abundant in essentially all of the putative circuits of the innate model of the mother I discussed in part II. They are essential in various forms of imprinting,[35] reinforcement learning,[36] autobiographical memory,[37] sensory perception,[38] and the fine-tuning that subserves embodiment and the sense of agency.

The last of these roles is demonstrated by drugs that specifically block NMDA receptors, like ketamine and phencyclidine. As I mentioned earlier, these drugs induce profound distortions of embodiment, including illusory out-of-body experiences, along with distortion of the sense of time, memory lapses, delusions, and other psychotic symptoms that mimic the early stages of schizophrenia—though typically without verbal hallucinations.[39] Ketamine even exaggerates the effects of the rubber hand illusion.[40]

Susannah Cahalan, a reporter for the *New York Post*, provides a chilling account of an extreme episode of NMDA receptor dysfunction, caused not by ketamine abuse, but by a rare autoimmune disorder in which the immune system attacks NMDA receptors in the brain.[41] This mysterious, profound, and chronic collapse of her NMDA system plunged her into a nightmare of severe mood swings, paranoid delusions, memory loss, distortions of embodiment, and hemispatial neglect. Luckily one of her neurologists made the right diagnosis and rescued her from what could easily have been a fatal disease.[42] Prior to her recovery, Cahalan also suffered severe epileptic seizures—a consequence of uncontrolled waves of electrical activity in the brain that can have many different causes but that often arise, as in Cahalan's case, from some kind of imbalance in NMDA receptor activity.[43]

Epileptic Psychosis and the Temporal Lobe

Epilepsy figures prominently in the literature of neurotheology mainly because of a possible link between extreme religiosity and epilepsy of the temporal lobe.[44] Neurologists have described a set of peculiar behavioral symptoms that seem to occur together in some of these patients when they are not having seizures: intense religious devotion or delusions, diminished sexuality, and a compulsion to write hundreds or even thousands of pages, often about their unique religious or philosophical insights.[45] Others have described intense mystical experiences or religious conversions that come just before, during, or in the aftermath of a seizure.[46] These papers are often embellished with colorful speculation about epileptic seizures that may have spawned the revelations of prominent saints and prophets, including a few so historically distant that their existence—let alone their neurological symptoms—may be apocryphal.*[47] Neurologists with theistic leanings tend to embellish in a different way, criticizing a caricature of these ideas—a straw man who claims that all mystics had temporal lobe epilepsy, that all religious sentiment can be dismissed as neural misfiring, and that the temporal lobe is the sole neural substrate of religious experience.[48]

One source of this controversy is the fact that only a small fraction of temporal lobe epileptics, a few percent or less, are unusually religious.[49] They tend to be those whose epilepsy has led to psychosis,[50] often with severe atrophy in

---

* For examples see Devinsky and Lai (2008), Saver and Rabin (1997), and references therein.

the medial temporal lobe of the right hemisphere.[51] Like schizophrenia, epileptic psychosis sometimes spawns delusions and hallucinations—including illusions of a sensed presence[52]—that easily fit the mold of religious thought. The one major difference between them is that delusions of being controlled by an external agent are rare in epileptic psychosis.[53]

If my tentative mapping of the circuitry of infantile and mystical experience proves correct, then it is not surprising that epilepsy of the temporal lobe can spawn religious delusions, because it straddles two critical components of that circuit. At its front end lie the amygdala and temporal pole, which are involved in innate knowledge, reinforcement learning, identification of specific other persons, and the setting of interpersonal boundaries. At the other end lies the temporoparietal junction, which is essential for embodiment, perception of biological motion, discrimination between self and other, moral judgment, and inference of the presence and intentions of other people. In between are networks for recognizing faces and voices, along with memory-related circuits that are especially vulnerable to epilepsy. But an equally relevant correlate of temporal lobe epilepsy and one far more prevalent among epileptics than extreme religiosity, is obsessive compulsive disorder (OCD).[54]

Cleansing and Moral Absolution

What is relevant here about obsessive compulsive disorder—along with similar psychopathologies that involve impulsive or compulsive behaviors—is that many of the peculiar behaviors they engender strikingly resemble religious rituals.[55] Although religiosity does not cause OCD, religious scrupulosity is a common manifestation of the disease in religious cultures.[56] People with this condition are plagued by unwanted and intrusive thoughts about contamination, cleanliness, and purity—both physical and moral—and feel compelled to relieve these anxieties through rituals of washing, dietary restriction, and avoidance of the unclean. The equation of physical with moral cleansing is not unique to OCD but appears to be a fundamental aspect of human nature. Cleansing rituals are ubiquitous in religion—with such prominent examples as baptism in Christianity, washing before worship in Islam, and purification in the sacred Ganges in Hinduism—but the connection can also be seen in nonreligious contexts.

One example is known to psychologists as the Macbeth effect. The reference, of course, is to Shakespeare's Lady Macbeth, who was obsessed with washing her bloodstained hands as a means of purging her guilt. The Macbeth effect is the equivalence between physical cleansing and moral absolution that appears in clever psychological tests of healthy subjects. In one such experiment, half the subjects were asked to recall an unethical act from their past, and the other half an ethical one, and all subjects were asked to describe their feelings about the recalled event. Then each subject did a word completion task in which three of the word fragments were W _ _ H, SH _ _ ER, and S _ _ P. These can be completed to make words that are cleansing-related (wash, shower, soap) or that are not (wish, shaker, step). Subjects who had recalled an unethical act were significantly more likely to use the cleansing-related words, which suggests that their feelings about the immoral act somehow made those words more readily accessible. In another variation of this experiment, the word completion task was replaced by a simple choice between two possible free gifts: an antiseptic wipe or a pencil. Those who had remembered an unethical act were more likely to choose the antiseptic wipe, which suggests an increased desire for cleansing. Yet another variation more directly demonstrated the equivalence of cleansing and absolution. All subjects recalled an unethical event from their past, after which only half of them washed their hands with an antiseptic wipe. They were then asked if they would be willing to volunteer, without pay, to participate in another experiment to help out a desperate graduate student. Those who had washed their hands were less likely to volunteer, suggesting that physical washing reduced their need for the compensatory absolution that comes from doing a good deed.[57]

This deep connection between physical and moral cleansing also shows up in neuroimaging studies. Just as evolution has co-opted the circuitry and neurochemistry of physical pain in representing the social pain of isolation or ostracism, so too has it evidently co-opted the neural substrate of physical disgust in representing the moral disgust evoked by social transgression. Just as physical disgust motivates aversion to the ingestion of substances that cause disease or death, moral disgust motivates aversion to acts of betrayal, dishonesty, or unfairness, and spawns the related emotions of guilt and shame when we commit these wrongs or outrage when we see them in others. Both physical and moral disgust, as well as their positive emotional opposites, specifically activate the insula,[58] a cortical area I described in chapter 6 as being involved in the conscious perception of sensations from the body and gut, the sense of taste, physical pain, vestibular sensations, and the conscious cravings of addiction. The insula is also consistently activated when healthy subjects perceive expressions of disgust in photographs of faces,[59] and this response is exaggerated in OCD patients.[60]

As I mentioned in chapter 6, the insula is heavily interconnected with the anterior cingulate cortex, the neural alarm system that perceives threats and generates appropriate responses—one of which is infantile crying. Together the insular and anterior cingulate cortices detect and respond to potential threats—usually generating a state of anxiety that persists until the threat is resolved—and a defect in this system appears to be part of the problem in OCD.[61] Neural activity in these areas tends to be abnormally high in OCD patients, especially under conditions that evoke compulsive behaviors,[62] but also in challenging cognitive tasks like the rapid recognition of a specific letter in a string of distracting ones.[63] For severely affected patients who do not respond to drug therapy, small surgical lesions in the anterior cingulate cortex reduce the symptoms of OCD in about a third of cases.[64] These lines of evidence show that OCD can be partly explained, in mechanistic terms, as an exuberant alarm system with a reluctant "off" switch. A deeper explanation, however, may lie in another aspect of the condition.

Vigilance, Doubt, and Certainty

The threat of contamination is only one of several themes that appear in OCD. Another prominent one is the fear that some critical act has been forgotten—an anxiety that leads to compulsive checking that the stove is turned off, the doors are locked, or the children are safe. Anxiety and doubt lie together at one end of a spectrum, the other end of which harbors security, comfort, safety, satiety, assurance, and certainty. We have seen this emotional dimension before, especially in the isolated and helpless infant whose anxiety and doubt transform to comfort and certainty in the arms of its mother. I have suggested that there lies between these extremes an emotional transition state, fostered by the innate expectation of the mother's presence and the short-circuit certainty of her existence and characterized by a mood of appetitive reward seeking.

The same emotional dimension is evident on the other side of the mother-infant bond, where mothers seek their young—literally, as addicts seek drugs—and also seek assurance and certainty through behaviors of nest building, checking, and vigilance that are much like those carried to pathological excess in OCD.[65] The high lifetime prevalence of OCD, about two percent,[66] may be an evolutionary by-product of the intense selective pressure for unusually strong emotional bonds in humans—not only between infant and parent, but also between sexual partners.[67]

Of course the management of anxiety and the transition from doubt to certainty are also central to religion, and in this light religious ritual appears to be a behavioral manifestation of the need for assurance. A series of experiments from the laboratory of psychologist Ian McGregor show that threats to personal certainty increase religious zeal. In one of these experiments, psychology undergraduates were asked to read and summarize a highly technical and virtually incomprehensible paragraph that described a complex and obscure statistical method. A control group did the same for a clearly written paragraph about a simple statistical procedure. Because psychology students tend to be insecure about mathematical statistics, the first group experienced a significant threat to their certainty about their educational path and career choice. After reading and summarizing the paragraph, each student responded to an eight-item questionnaire that measures religious zeal. Those whose personal certainty had been threatened scored significantly higher on religious zeal than the control group. In a similar experiment, non-Muslim students who reflected on an uncertain relationship in their personal lives expressed higher levels of agreement with statements that derogate Islam, relative to controls who reflected on a friend's uncertain relationship.[68]

McGregor and his colleagues see religion as a zone of safety and unimpeachable certainty to which believers return when their certainty about other aspects of life is challenged.[69] They also liken it to the ritualistic, self-oriented behaviors that some animals do when anxious or severely stressed—seemingly pointless scratching, grooming, tail chasing, licking, or yawning.[70] One of the most heartbreaking of these is the self-rocking seen in monkeys raised without maternal contact and in severely neglected human children. In both cases, self-rocking appears to be a means of self-comforting. Interestingly, it also bears a striking resemblance to many religious rituals, like the repetitive rocking of orthodox Jews in prayer, Catholic priests swinging censers, Buddhist monks in meditation, Muslims reciting the Koran, and Shakers in ritual dancing.[71] Psychological studies show that rituals like praying the Rosary and reciting psalms tend to reduce anxiety.[72] In part because of its simple, repetitive, and predictable nature, ritualistic behavior appears to evoke feelings of comfort and familiarity in times of anxiety and uncertainty.

Helplessness, Magical Thinking, and the Need for Control

An important, special form of anxiety-provoking uncertainty is frustration or lack of control in the pursuit of a goal. Several physiological and psychological experiments have shown that experiencing a lack of control increases the anxiety of experimental subjects and triggers a compensatory increase in their belief in a controlling God.[73] Other studies show that an experimentally induced feeling of lack of control increases the tendency of healthy subjects to accept conspiracy theories or superstitious belief, to perceive illusory objects in images of random visual noise, and to perceive with greater sensitivity real objects in visually degraded images.[74] These results support the suggestion, which I made in chapter 10, that a feeling of helplessness increases attentional effort, increases the influence of top-down biases on perception, and thus lowers the threshold for illusory perception. There is, of course, no more extreme example of lack of control than the profound helplessness of a human infant, but a close second is the feeling of spiritual brokenness that often precipitates the illusion of God's presence. Another example is the learned helplessness of physically abused children, who as adults may be unusually prone to superstition and magical thinking.[75] Not surprisingly, patients with OCD also fit this pattern, with a propensity for magical thinking and a profound need for control.[76]

These and other more cognitive aspects of OCD may explain the involvement of other parts of the brain in this disorder. Unusual activity in the temporoparietal junction shows up in some neuroimaging studies of these patients, and—because of the role of this part of the brain in the redirection of attention—this has been interpreted as a correlate of the rigid fixation of attention commonly seen in OCD.[77] But the parts of the brain that are most consistently implicated in OCD, using both structural and functional neuroimaging, are the orbitofrontal cortex and associated parts of the basal ganglia, notably the caudate nucleus.[78]

Cognitive Rigidity

We explored the orbitofrontal cortex in chapter 4, where I emphasized its roles in representing stimuli that are innately rewarding or punishing and in learning to associate these primary reinforcers with other stimuli, thus making them secondary reinforcers. Neurons of the orbitofrontal cortex quickly adapt to changes in the reward value of a secondary reinforcer, especially when that value completely reverses—becoming a punishment after having been a reward, or vice versa.[79] A deficit in reversal learning of this kind is one aspect of the cognitive rigidity that characterizes patients with OCD.[80] In chapter 4 I also emphasized the importance of the social primary reinforcers in the orbitofrontal cortex that are the foundation for acquiring moral behavior through reinforcement learning. In that light, it is not surprising that the obsessions of many patients with OCD involve fears of harming or neglecting other people or fears of personal contamination—a proxy for moral purity.

The cognitive rigidity of OCD can be measured by means of a card game in which the subject draws one card at a time from any of four decks. Each card pays a reward, but some decks pay higher rewards than others. Also, a few cards in each deck carry penalties that cost the player some of his winnings. After drawing a few dozen cards, most healthy subjects learn that the two decks that pay the highest rewards per card sometimes exact such severe penalties that they should be avoided—their perceived reward value has reversed—whereas the other two decks remain more rewarding than punishing over the long run. Patients with OCD,[81] like those with bilateral damage to the orbitofrontal cortex,[82] are deficient at this task, tending to persist at drawing cards from the decks that initially rewarded them the most, despite the severe subsequent punishments that deplete their winnings. A neuroimaging study revealed abnormally low activity in the orbitofrontal cortex of OCD patients as they performed a similar reversal learning task, and it even found this abnormality in their unaffected close relatives.[83] A deficiency in serotonin in the orbitofrontal cortex is associated with poor performance at reversal learning,[84] and drugs that increase the availability of serotonin reduce the symptoms of OCD for many patients.[85]

Dopamine, Compulsion, and Ritual

Other OCD patients, however, respond better when given an additional drug that blocks dopamine receptors,[86] and this effect is thought to occur in the caudate nucleus—a part of the basal ganglia where an excess of dopamine may exacerbate the symptoms of OCD.[87] We encountered the caudate in chapter 8 as the part of the brain specifically activated when Christians recited the Lord's Prayer.[88] There I noted that it is also involved in addiction, where it reinforces the ritualistic aspects of drug-taking behavior. The action of dopamine in the caudate gradually transforms addictive behavior from what is initially a voluntary act of pleasure seeking into a habit and ultimately into a compulsion.[89] Its role in OCD appears to be similar: more concerned with compulsive behavior than with anxious obsession.

Its role in religious ritual appears to be similar as well. In chapter 5 we explored a related part of the brain just below the caudate—the nucleus accumbens—which is more involved in the early phase of addiction and in appetitive reward seeking more generally. I suggested that religion to a recovering addict may be a spiritual methadone, an emotional bonding behavior that taps into the circuitry of reinforcement and reward—more specifically, the circuitry of mother-infant bonding—and thereby gradually displaces and weakens the tenacious grip of the addictive drug on that same part of the brain. As the former addict's spiritual commitment grows, the rewards of seeking religious social support and the comfort of God's presence transform into habitual, ritualistic, and ultimately compulsive behavior. Presumably something like this happens in anyone, whether previously addicted or not, who becomes deeply immersed in religious or spiritual belief. Even participants in the informal and nondenominational meetings of twelve-step groups have rituals, like the reading aloud of the Serenity Prayer, the announcement of first name and confession of addiction prior to speaking, the greeting in unison by the group that follows this announcement, and the use of a conventional name for the subsequent speech: *sharing*.

Abnormalities of dopamine regulation in the caudate, nucleus accumbens, or other parts of the basal ganglia may also contribute to other compulsive or impulsive behavioral disorders, including compulsive gambling, hair pulling, skin picking, nail biting, hoarding, binge eating, compulsive shopping, kleptomania, and pathological sexual excess.*[90] Often these can be treated with drugs that block dopamine receptors, and some of them appear as symptoms of chronic addiction to drugs that increase dopamine levels in the caudate and nucleus accumbens. Methamphetamine addicts, for example, are called tweakers because of the compulsive skin picking that commonly afflicts them.[91] At the opposite extreme are unmedicated Parkinson's patients, in whom a severe depletion of dopamine produces profound lethargy, apathy, tremors, and other movement deficits.[92]

This picture changes dramatically after Parkinson's patients begin taking drugs that increase the concentration of dopamine in their brains. At first their mood and motivation rise, and their movement disorders subside.[93] Over subsequent months and years, however, unpleasant side effects of dopamine replacement therapy appear. One of these is punding, which refers to compulsive, repetitive, and pointless rituals that typically involve dismantling household appliances or other machines and spending hours arranging and inspecting the parts.[94] Others include addictive abuse of dopamine replacement drugs, compulsive gambling, compulsive shopping, hypersexuality, hallucinations, sensed-presence illusions, and psychosis.[95] Although dopamine is not the whole story, these observations support the idea that dopamine and its effects in the basal ganglia are central to motivation, appetitive reward seeking, and habitual or ritualistic behavior—all hallmarks of religious behavior.[96]

In this and the three previous chapters I have considered some challenging and complex questions at the intersection of neuroscience and religion, none of which is fully resolved. I conclude the discussion with one that appears to be much simpler, though appearances can be deceiving. Can we reliably attribute the neural substrate of mystical experience more to one side of the brain than the other, and, if so, which?

---

* Evidence for involvement of dopamine and/or the basal ganglia is stronger for some of these disorders than for others. See Fineberg et al. (2010), Grant et al. (2006), and Saxena (2008).

CHAPTER 12

The Handedness of God

> On my bed when I think of you,
> I muse on you in the watches of the night,
> for you have always been my help;
> in the shadow of your wings I rejoice;
> my heart clings to you,
> your right hand supports me.—Psalm 63:6–8 (NJB)[1]

If you hand a woman a pillow and ask her to hold it to her chest, she will most likely clutch it symmetrically along her midline. If, however, you also tell her to imagine that the pillow is a baby, she will probably cradle it against her left side.[2] Most people write with their right hand, and, if forced to use the other, produce a crude scrawl like that of a small child. Most strokes that affect speech occur on the left side of the brain.[3] The movements involved in emotional facial expressions tend to be greater on the left side of the face,[4] and most people are better at perceiving the emotion in faces seen on the left side of their visual field.[5] Why are we lopsided in these ways, and are we equally lopsided in our perception of God?

The literature of neurotheology seems confused and conflicted regarding this question, even in the small sample of it I have considered in this book. The hyper-religiosity associated with epilepsy of the temporal lobe tends to occur more often with pathology of the right hemisphere[6] or with problems on both sides of the brain.[7] By contrast, the hallucinations of schizophrenics are typically associated with abnormal activity of the left hemisphere.[8] The neural correlates of Buddhist compassionate meditation are greater in the right hemisphere,[9] but those of spontaneous Christian prayer are greater in the left.[10] At least one reviewer of this literature has concluded that no consistent hemispheric bias for religious behavior can yet be discerned.[11] Perhaps there is none, but if there is, it might be more apparent from the perspective of neuro*ethology*.

Why Lateralization?

The vertebrate body plan is one of bilateral symmetry, which means that we have left and right halves that are essentially mirror images of one another. This symmetry extends to some of our internal organs, including the brain. For the most part, each side of the brain receives sensations from the opposite side of the body—or of surrounding space in the case of vision—and controls muscles of the opposite side as well. This left-right division of labor ceases to make sense, however, at higher levels of the nervous system that deal with increasingly abstract concepts and behavior. There is, for example, no intrinsically meaningful way to divide into left and right halves such things as grammar, syntax, and vocabulary. Linguistic computation could be equally apportioned in both cerebral hemispheres, but without intrinsic constraints that favor such symmetry, evolution and development are free to try asymmetrical alternatives and evidently have done so.

This asymmetry of neural function is called lateralization, and anatomist Paul Broca's discovery of the left-side dominance for language in 1861 was the first clear example of it.[12] For the next century, lateralization of the brain was thought to be unique to humans, in part because of reasoning like that expressed in the previous paragraph. Language is abstract and symbolic, and it is easy to imagine the computational advantages of putting most of it into one hemisphere. Delays that result from shuttling signals from one hemisphere to the other could be eliminated. The corresponding part of the other hemisphere could be specialized for a different function—perhaps one that must be done in parallel with speech, like expressing emotion in tone of voice without regard to the meaning of words. Having discovered this trick, the human brain could apply it elsewhere, as in the specialized control of the right hand for highly skilled tasks like making and using tools. Or perhaps tool use and right-handedness evolved first, with the consequent lateralization in the brain setting the stage for language. Either way, lateralization came to be seen as a critical evolutionary breakthrough in the origin of humans.[13]

There are, however, some serious problems with this tidy story. One is that the supposed computational advantages of lateralization cannot explain why the language-dominant hemisphere is on the left side for most individuals in the population. If computational efficiency were the only selective pressure, right-side dominance for language would work just as well, and the expected result would be a roughly equal mix of linguistically right- and left-lateralized individuals.

Another problem is that abstract cognition is not as uniquely human as we once thought. Chimpanzees, dolphins, elephants, and European magpies all have enough self-awareness that they spontaneously recognize their image in a mirror as a reflection of themselves.[14] Apes can be taught to communicate using the rudiments of symbolic language,[15] and starlings can be trained to recognize recursive patterns in a series of syllables—an aspect of syntactic comprehension long thought to be uniquely human.[16]

But the most serious problem with the idea of lateralization as the distinctly human spark is that it turns out not to be uniquely human. During the last few decades, studies of our distant nonhuman cousins have found that lateralization of the brain occurs in all classes of vertebrates and is more the rule than the exception.[17] Typically it involves aspects of perception and behavior for which left and right do have meaning, like foraging or predator avoidance, which shows that lateralization need not be restricted to abstract thought. A toad is more responsive to the threat of a snake approaching from the toad's left side than from its right.[18] Domestic chicks are better at detecting aerial predators with the left eye, but better at distinguishing edible grain from pebbles with the right.[19] Lizards of the genus *Anolis* use mainly the left eye during aggressive territorial disputes with others of their species.[20] Some species of fish more often turn left than right to escape a predator, but other species have the opposite bias.[21] Rats are better at learning to discriminate complex sequences of tones heard through the right ear than through the left,[22] and there is good evidence that chimps, like humans, tend to be right-handed.[23]

This is an active area of research, and many questions remain unanswered.[24] From the examples that have so far come to light, lateralization appears to be a complex aspect of neural development that occurs in varying degrees with different behaviors. In some cases the bias is evident only at the level of the individual animal, whereas in others, most of the population is biased to rely on the same side of the brain. Lateralization is probably controlled by multiple interacting mechanisms, including environmental, genetic, and epigenetic factors that affect the developing brain.[25] The relative contributions of these, however, are unknown for most examples of lateralized brain function.

Emerging Vertebrate Patterns

Despite these gaps in knowledge, a few somewhat consistent patterns have begun to emerge in those behaviors most commonly lateralized at the population level.[26] Generally the left hemisphere is specialized for dealing with familiar stimuli and self-motivated behaviors, like feeding, whereas the right is more vigilant for new, unexpected, or threatening stimuli that demand immediate response. Also, the right hemisphere is better at seeing large-scale spatial relationships (the forest), whereas the left is tuned for spatial detail (the trees).[27] Detection of predators and aggressive interactions with conspecifics typically involve the right hemisphere. Less urgent behaviors that require detailed analysis of sensory input, like the foraging of a chick trying to distinguish between grain and pebbles, are more often a function of the left hemisphere. Visual recognition of specific individuals occurs mainly in the right hemisphere, but interpretation of the vocalizations of other individuals is done mainly in the left. Lateralization in humans is consistent with several of these patterns, which strongly suggests that it is inherited from a basic pattern that appeared early in vertebrate evolution—though how early is still a matter of debate.[28]

A more vexing question is *why* it evolved. A toad that is more responsive to predators on its left side is more vulnerable to predators on its right. If most toads in the population have the same bias, then in principle predators could exploit this weakness. Any snakes that have a heritable tendency to approach their prey from the victim's right will be more successful at catching toads, and the genes that made them so will propagate throughout *their* population. Although it is not yet known whether toad-eating snakes are biased in this way, the thought hints at some of the subtle interactions that can occur in interacting populations of lateralized individuals—not only between predator and prey, but also between individuals of the same species. In nearly all known examples of lateralization at the population level, there is a small but significant minority of the population that has the opposite or no lateralization. *Why?*

Frequency-Dependent Selection and Evolutionary Stability

The animal for which we have the best answer is probably *Perissodus microlepis*, a fish that eats the scales of other fish in Lake Tanganyika. It attacks its prey by sneaking from behind and snatching scales from the victim's flank. Its mouth is distorted, permanently turned to one side at the direction of its genes. Both possible forms—right- and left-facing mouths—occur in the population. Fish with right-facing mouths always attack their prey from the victim's left side and vice versa. The prey are sufficiently adept at evading these attacks that the predator's success rate is only about 20 percent, so even a slight advantage can result in large consequences for the predator's survival.

That fact is evident in a record of the frequencies of right- and left-facing mouths in the population sampled annually over eleven years: this ratio oscillates with a period of about five years, during which right-facing mouths are more abundant, then the two become about equal, then left-facing mouths dominate, and so on. When right-facing mouths are in the majority, the prey are more attentive to attacks coming from their left. This means that they are less attentive to attacks from the minority of predators with left-facing mouths, so the minority always has an advantage. Intuitively it seems that the population should reach equilibrium with equal numbers of right- and left-facing mouths, and this 50:50 ratio is in fact the point around which the system oscillates. The oscillations probably arise because there is a delay between the production of the next generation of predators and their influence on the behavior of the prey—the predators eat plankton, not scales, when they are small.[29]

The important point is that the numbers of right- and left-lateralized scale-eating fish are determined by a selective pressure that varies with their relative abundance. This is called frequency-dependent selection, the same aspect of evolutionary dynamics that figured prominently in my interpretation of religious cults in chapter 1. With respect to lateralization, mathematical models show that it can give rise not only to oscillatory behavior, but also to stable states in which one lateralized form remains in the minority. This can result from interactions between predator and prey when social grouping of the prey is to their advantage.[30] Another model shows that a stable state can arise from the effects of lateralization on cooperation and competition among members of the same social species.[31]

The latter model assumes that an individual with the same lateralization as the majority of the population has a fitness advantage over the minority in *cooperative* social behavior. That individual may, for example, do better in cooperative pack hunting or in using the same tools as others of the group. The model also assumes that an individual with the minority lateralization has a fitness advantage in *competition* against other members of the species because its behavior is less predictable. The evolutionary behavior of the model depends on the relative magnitudes of these advantages and on their sensitivity to changes in the frequency of each kind of lateralization. If the advantage of social cooperation is relatively small, then both forms of lateralization occur with equal frequency in the stable state. If the advantage of cooperation is relatively high, then the population stabilizes in one of two ways: either fully right- or fully left-lateralized. Between these extremes, however, the population reaches an evolutionarily stable state in which one form of lateralization remains the majority and the other the minority.*[32]

There is some evidence that handedness in humans can be at least partly explained in this way. About a tenth of humans are left-handed, a state that has probably been stable for millennia.[33] The model predicts that left-handers should, on average, have an advantage over right-handers in competitive interactions—like fighting or interactive sports—and several studies confirm this. In sports that require the anticipation of an opponent's behavior, like boxing, tennis, fencing, basketball, ice hockey, or baseball, the fraction of left-handers

---

* Lateral bias at the population level is just one of many ways an evolutionarily stable state can arise from frequency-dependent selection. For a more general discussion see Maynard Smith (1982).

among participants is significantly greater than in the general population—both in athletic college students and especially in elite professional athletes. The extreme case is fencing, in which the fraction of left-handers reaches 50 percent for some world champion teams. By contrast, the proportion of left-handers in noninteractive sports, like gymnastics, bowling, darts, or javelin throwing, does not differ from that of the general population.[34]

The model also predicts that the fraction of left-handers should be greater in more violent cultures, but the evidence on this point is mixed. One study of left-handedness and homicide rates in eight traditional societies found that the two are positively correlated, with the fraction of left-handers being about 3 percent in the most pacifistic societies and about 20 percent in the most violent.[35] A recent study of the most violent of these cultures, however, used more direct tests of handedness and did not find a high fraction of left-handedness,[36] so there may be other selective pressures that maintain this trait in humans. The important point is that, whatever those pressures turn out to be, they are probably frequency dependent.[37]

Laterality of the Sensed Presence

The foregoing digression into evolutionarily stable states and the lateralization of behavior in nonhuman animals is in keeping with a major theme of this book: seemingly unique aspects of human thought, feeling, and behavior often turn out to have antecedents in other species. In chapter 1 I made this point for the feeling of God's presence,[38] and throughout this book I have argued that the neural circuitry that underlies such feelings might best be found by seeking the neural substrates of infantile emotion and expectation. If that approach is valid, then there is a clear pattern of lateralization in the circuitry that produces the illusion of God's presence. A striking right-hemispheric bias is evident in the activity of most of the brain areas that I identified as likely components of an infant's innate model of its mother. Most of this evidence comes from studies of these neural networks in human adults experiencing social emotion or cognition, but the same bias appears in experiments done with infants.

In adults, the right-hemispheric bias is especially evident in the social and embodiment-related functions of the temporoparietal junction,[39] in the vestibular responses of the temporoparietal junction and insula,[40] and in the face-recognition area of the inferior temporal lobe.[41] The bias is more subtle in the amygdala, which also responds to faces but is more generally dedicated to emotional perception and behavior. In most neuroimaging studies of human responses to emotionally charged stimuli, activation of the left amygdala is reported more often than of the right, but this may be an artificial bias arising from limitations of experimental technique.[42] In experiments designed to measure rapid responses, the right amygdala is as responsive as the left, but only to the onset of the stimulus. The left amygdala, by contrast, responds as long as the stimulus is present, and so predominates in experiments that measure responses averaged over longer blocks of time.[43] This is in keeping with a general pattern of lateralization seen in other vertebrates: the right side is more concerned with urgent responses to unexpected events, while the left does slower, more detailed analysis. There may also be a right-side bias in the amygdala for unpleasant emotions. In patients with damage to the amygdala on only one side, those with lesions on the right were less able to recognize facial expressions of fear.[44] This, too, is consistent with the common vertebrate pattern of a right hemisphere specialized for predator avoidance.

In the prefrontal parts of the brain we explored in chapters 3 and 4, the right-side bias is evident as a physical lopsidedness of neural tissue. The anterior cingulate cortex, for example, is typically larger on the right than on the left, and this difference tends to be greater in women than in men. Interestingly, whether in men or in women, this right-side bias in the size of the anterior cingulate cortex correlates with greater harm avoidance as a personality trait.[45] This is consistent with the interpretation of this part of the brain as a neural alarm system,[46] with its roles in obsessive compulsive disorder[47] and maternal behavior[48] and, once again, with the typical vertebrate bias for predator avoidance in the right hemisphere.

There is a similar anatomical bias in the orbitofrontal cortex, with the right side larger than the left,[49] and several neuroimaging studies implicate the right[50] or bilateral[51] orbitofrontal cortex in a mother's response to her infant. As I explained in chapter 4, the orbitofrontal cortex represents innately rewarding and punishing stimuli that constitute a foundation for reinforcement learning, and damage to it causes severe deficits in social behavior and decision making. The right-side bias may be important in this regard: adolescents who are genetically at high risk for alcohol addiction have a significantly lower right-side bias in their orbitofrontal cortex than other teens,[52] and patients with damage to the right orbitofrontal cortex are especially impaired at empathetic behavior relative to those with left-side damage.[53]

Lateralization in Human Infants

But if the illusion of God's presence can be traced back to an infant's innate expectation of its mother's presence, then it is the right-side bias in the brains of infants that is of greatest relevance here. The right hemisphere is dominant in newborns and young infants because it matures earlier than the left hemisphere.

We saw a hint of this in chapter 2, where I noted that human newborns tend to imitate finger movements using the left hand, and therefore the right hemisphere, regardless of which hand of the model they are imitating.[54] More direct evidence comes from a study of blood flow in the brains of quietly resting infants and small children. Cerebral blood flow, which is a good proxy for general neural activity, is significantly greater in the right than in the left hemisphere from about age one to three years—especially in the region that includes the temporoparietal junction. Beyond age three the asymmetry reverses, and the left hemisphere becomes dominant.[55] Although in this study the right hemispheric dominance in blood flow did not reach statistical significance for infants younger than a year old, other more sensitive measures reveal greater right- than left-side maturity during the first year and even before birth.

One of the most important studies of this kind concerns the acetylcholine-containing nerve fibers that figured prominently in my discussion of attention and hallucinations in chapter 10. There I emphasized that these cholinergic neurons focus attention and bias the sensitivity of the sensory cortex in accord with top-down perceptual expectations. I suggested that the illusion of God's presence begins with prior hyperactivity of this system—typically during a prolonged period of desperation and helplessness—thus priming the brain with the innate expectation of a maternal savior. The illusion then manifests as a sensed presence during diminished sensory input, social isolation, or sleep deprivation. I also argued that the cholinergic system plays a similar role in the brain of a newborn, where the biasing of the sensory cortex with the innate expectation of mother compensates for deficiencies in neonatal perception and promotes mother-infant bonding.

The study in question measured the density of cholinergic nerve terminals in two areas of cortex on both sides of the postmortem brains of fetuses and adults.[56] One area was in the primary motor cortex of the frontal lobe, the other in the superior part of the temporal lobe. In the left hemisphere of an adult, the superior temporal cortex is mainly concerned with the perception of spoken words. The corresponding area of the right temporal lobe is specialized for the perception of prosody in speech and the visual interpretation of movements of the lips, facial expressions, and other forms of biological motion. The visual functions are especially evident in its more posterior part, where it merges into the temporoparietal junction.

In the fetal brains, the study found that the density of cholinergic nerve terminals was significantly greater in the right temporal lobe than in the left. The opposite asymmetry was found in the adult brains, though the absolute densities of cholinergic terminals were the same in the right temporal lobes of both the fetal and adult brains. This means that the cholinergic fibers arrive early in the right temporal lobe and reach mature density there *before birth*, whereas they arrive later on the left, eventually surpassing the density on the right as the brain matures during childhood. In the motor cortex, by contrast, there was no hemispheric asymmetry in cholinergic innervation in either the fetal or adult brains. This underscores the special importance of lateralization in and around the temporoparietal junction, and the study as a whole suggests that the right temporoparietal junction plays a special role in human newborns.

That special role is evident in neuroimaging studies of infants younger than a year old as they watch changing facial expressions and similar social stimuli. As I described in chapter 6, the responses occur in or near the temporoparietal junction on both sides but more strongly on the right.[57] Similar right-hemispheric dominance is seen in studies that use auditory stimuli—tones, speech, or artificial speechlike sounds—and some of these experiments demonstrate the early functional maturation of the right hemisphere when the subject is still in the womb.

Such experiments are possible only because of magnetoencephalography (MEG), a technique that measures the weak magnetic fields generated by electrically active neurons.[58] Unlike the electric fields measured by EEG, magnetic fields are relatively undistorted by the tissue and fluid in the mother's abdomen. The experiment begins with ultrasonic imaging of the fetus to see which side of its head is against the uterine wall and where on the mother's abdomen the detector must be placed. Then short acoustic tones are applied to the mother's abdomen while magnetic fields from the fetal brain are recorded and averaged. The results show that the response of the fetal brain to sound is sluggish, meaning that it comes at a much later time relative to the sound than does the auditory response in an adult brain. This long latency of the fetal neural response becomes shorter as its brain matures, and that process continues long after it is born. The real discovery, however, is that a prominent component of the response occurs with shorter latency on the right side than on the left, meaning that the right hemisphere is more mature than the left, even before the baby is born.[59]

Other studies of young infants find right-side dominance in response to more elaborate sounds. An EEG study of premature infants found a right-hemispheric advantage in discriminating tones of different pitch.[60] A study of three-month-old infants using the NIRS method compared their neural responses to a sentence spoken with normal prosody and the same sentence artificially stripped of its emotional tone. This difference between the two responses was specifically localized in the right temporoparietal junction.[61] A similar study of sleeping newborns found a greater response in the right temporal lobe than in the left to a change in prosody of a two-syllable speech fragment. Intriguingly, the same study also found a greater response in the left temporoparietal junction of newborns in response to a change in vowel in the second syllable. This is the part of the brain that processes phonemes in the adult brain. The finding suggests that, long before language acquisition has occurred and even while the right hemisphere is dominant, the left temporoparietal junction is in some way specialized for phoneme discrimination.[62]

Cradling Bias and the Laterality of Face Perception

Given the overwhelming evidence for right hemispheric dominance in infancy, it may surprise you to learn that a typical newborn infant, when lying on its back, has a spontaneous tendency to turn its head to the *right*, as if its attention were being directed by the less mature, not-yet-dominant *left* hemisphere.[63] There is evidence that this bias may result from the typical orientation of a fetus during the third trimester and the asymmetrical effects of that on its developing vestibular system.[64] But that would only explain *how* the rightward head-turning bias comes about, not *why*. An answer to the *why* question may lie in a complementary behavioral asymmetry on the other side of the mother-infant bond.

In nearly all cultures so far examined, human mothers have a strong tendency to hold or cradle their babies on the left side.[65] The bias does not occur in the handling of inanimate objects—only with infants and a few other things that elicit similar emotions, like baby dolls and beloved pets.[66] As with many other aspects of human lateralization, this bias appears to be inherited: most great apes and some species of monkey also have it.[67] In humans, the bias is completely unconscious and unrelated to the handedness of the mother. When it is pointed out to them, right-handed mothers tend to say they do it that way so that their right hand is free to feed the baby, whereas left-handed mothers often explain that they are holding the baby with their stronger arm.[68] Several other explanations have been offered, and the issue is not yet settled, but the weight of evidence strongly suggests that left-side cradling is a consequence of right-hemispheric dominance for maternal affection, vigilance, and emotional connection.[69]

Earlier in this chapter I described the enlargement of the right anterior cingulate cortex in women, with its likely role in maternal vigilance, and the right-side dominance in the response of a mother's orbitofrontal cortex to the sight and sound of her infant. In chapter 6 I emphasized the role of the right temporoparietal junction in imitating others and in inferring the mental states of others from nonverbal cues—two essential aspects of early mother-infant interaction. Of special relevance here is the right-hemispheric dominance for communication of emotion through facial expressions.

When a mother cradles her infant on her left side, the baby's face is in her left visual field and thus well situated for perception by her right hemisphere.

The right hemisphere is better not only at perceiving the emotion in a face, but also at expressing it—which means that the left half of the face is more expressive.[70] When a mother holds her baby on her left side, it is mainly the more emotional left side of her face that the baby sees.[71] If the infant is lying on its back on its mother's left arm, its tendency to turn its head to the right not only exposes the most expressive side of *its* face to the mother, but also places the mother's face in the *infant's* left visual field and right hemisphere. For both mother and infant, this arrangement also favors the left ear, which is likewise better at perceiving vocalized emotion by virtue of its connection to the right hemisphere.[72] Of course, the right turn of the infant's head also has the practical advantage of directing its mouth to the mother's breast.

Several experiments support this interpretation of the role of facial perception in the lateral biases of infants and their mothers. One study measured competence at facial perception in adults who had a congenital cataract in one eye during their first few months of life. The experimenters showed them pairs of facial images, some of which were digitally edited to have various subtle differences, and asked the subjects to rate each pair as "same" or "different." The only problem detected was in those subjects who had been deprived of clear vision in their left eye during infancy. As adults they had significant trouble recognizing small differences in the spacing between parts of the face—a skill that is critical for expert facial recognition. This shows not only that the right hemisphere is uniquely and inherently specialized for this skill, but also that its development requires clear visual experience during the first few months of life.[73]

The relevance of this to the biases in maternal cradling and infantile head-turning is suggested by another study that found a significant right-hemispheric bias for the perception of female faces. Normal adult subjects were shown digitally edited images of faces, in each of which, the left and right halves of the face came from two different people—in most cases from those of the opposite sex. These so-called chimeric faces briefly appeared on a computer screen while the subject fixated on the center, so that each hemisphere of the subject saw only the opposite half face. The subjects were asked to identify each face as male or female. Like many similar studies before it, this one found a strong advantage in the right hemisphere for this discrimination. The surprise was that this proficiency turned out to be entirely attributable to better performance by the right hemisphere on female-left/male-right chimeric faces. In other words, the right hemisphere is especially attuned to the perception of female faces, which the authors suggest may be a consequence of the laterality biases that cause young infants to use their right hemispheres when watching their mothers.[74]

Chimeric faces can also be used to test more directly the idea that left-side cradling is related to better emotional communication between mother and infant. In these experiments, the two half-faces in each chimeric face express different emotions, and the subjects must report the emotional expression they see after each brief exposure to one of these dual-emotion faces. Not surprisingly, the emotion reported tends to be the one on the left—the part seen by the right hemisphere.[75] Interestingly, this bias for perceiving emotion in the left side of the face is reduced in adults who, as infants, had been cradled on the right.[76] An analogous experiment, called dichotic listening, can be done using the sense of hearing. A sentence is read aloud using two different emotional tones of voice, and these sounds are played simultaneously to the subject, one in each ear using headphones. Again, the emotion reported is usually the one heard by the left ear and right hemisphere.[77]

If a mother's cradling bias comes from her bias in emotional attention, then we should expect cradling bias to be correlated with emotional hemispheric bias as revealed in the chimeric faces and dichotic listening tests. In other words, a mother who cradles on the left should score higher on these tests using her left visual field and left ear; one who cradles on the right should do better with her right visual field and right ear. Several studies have tested for this correlation, but so far the results are mixed, and the issue remains unsettled.[78] Another caution-ary note is raised by a study of laterality biases in baboons, which found that infant baboons have a rightward head-turning bias, but their mothers have no significant cradling bias.[79]

Despite these caveats, the evidence for right hemispheric dominance during infancy is overwhelming. If an innate model of the mother gives rise to the illu-sion of God's presence in an adult, then we should expect greater neural activity in the right hemisphere during such an experience. But just as language is not exclusively a function of the left hemisphere, neither is emotional perception purely a function of the right. Unlike an infant sensing its mother, the typical adult sensing God's presence has a mature and dominant left hemisphere. In what way might it contribute to that illusion?

Magical Thinking, Theology, and Verbal Virtuosity

One intriguing possibility emerges from studies of the human response to stimuli that arrive in random sequence. The task is called *probability learning* and seems simple enough. You sit in front of a panel that has two lights, left and right. You are told that on each trial one light will flash, under the control of a computer, and at the start of each trial you must guess which side will flash. You are not told that the experiment is biased, such that one side will flash more often, maybe on 80 percent of trials and the other only 20 percent. After a few dozen trials, however, the bias becomes obvious.

When human subjects do this task, they typically end up adjusting their guesses such that they place about 80 percent of their bets on the side that flashes 80 percent of the time. Monkeys, pigeons, and rats also can be trained to do this task, but they come to a radically different solution. They eventually learn to place nearly all of their bets on the 80-percent side, a strategy that pays off significantly better than the human one of matching the frequencies of the flashes.[80] Why are these animals better than humans at this task?

Hoping to answer that question, psychologist John Yellott came up with an especially revealing variant of this experiment for his human subjects. He ran the experiment until the subject had clearly discovered the bias and adjusted the frequencies of his guesses accordingly. Then, without informing the subject, Yellott ran the next fifty trials under manual control, always making the flash match the subject's guess. After fifty trials of 100 percent success, he stopped the experiment and asked the subject to describe his impressions. During those last fifty trials, most of the subjects continued to guess in a way that matched the original bias, but in their comments they described elaborate and complex rules or patterns that determined which light would flash. They said it took them a while to figure it out, but their discovery explained their perfect performance at the end. Yellott described his subjects' behavior as superstitious. They had been seeking order and pattern where there was none.*[81]

Michael Gazzaniga and colleagues formulated a more elaborate version of this experiment for use with split-brain subjects—patients whose left and right cerebral hemispheres had been surgically disconnected from one another for relief of otherwise intractable epilepsy.[82] In effect, split-brain patients have two distinct and separable conscious minds, each with a different view of and knowledge about the person's surroundings—an observation that has profound implications beyond the scope of this discussion.†[83] Here my interest lies in the kind of superstitious behavior Yellott described. Gazzaniga had his subjects visually fixate a mark on the center of a computer screen and make guesses about which of two symbols, red or green, would appear on one side of the fixation point on the next trial. Some trials used only the left half of the screen, others only

---

* Yellott (1969) used a panel that displayed either "X" or "Y," not flashing lights, but I have glossed over that detail for simplicity. When the experiment is done with animals, the choice is usually between two levers the animal can press to obtain a food reward—if the chosen lever is the correct guess. In fairness to the human subjects in these experiments, the performance of the nonhuman animals may not be exactly comparable, because they were hungry and may have been more highly motivated. The human subjects were paid a fixed amount for participating, without regard for their success at the task.

† In chapter 14 of *The Illusion of God's Presence* I briefly explore some of these deeper ramifications of the split-brain procedure (Wathey 2016, pp 278–79). The physical splitting of the brain in these patients is never truly complete, because some communication between the hemispheres persists through unsevered connections in subcortical structures. In some tests, therefore, the two conscious minds of a split-brain patient are not completely distinct but show some residual unity. For a recent discussion of these and other subtleties, see de Haan et al. (2020).

the right, but in either case the choice was between red or green appearing on the specified side, and the frequencies of their appearance were biased as in the typical probability learning experiment. Guesses about the left half-screen were made by pressing buttons using the left hand; guesses about the right were made using the right hand. In effect, Gazzaniga was running two probability learning experiments, in parallel and independently, one for each cerebral hemisphere of the subject. Because the nerve fibers connecting the two hemispheres had been severed, the perception and behavior of one had no influence on the other.

The remarkable result was that the right hemisphere behaved more like the pigeons and rats, placing significantly more than 80 percent of its bets on the color that appeared 80 percent of the time, while the left hemisphere behaved in the more typically superstitious human pattern, distributing its guesses in a way that roughly matched the frequencies of appearance of the two colors. Similar results were obtained in separate experiments on patients who had suffered strokes in either the left or right frontal lobe. Those with intact right hemispheres performed about as well as pigeons and rats; those with intact left hemispheres performed like superstitious humans. Gazzaniga interpreted the results in terms of the human tendency to look for causal relationships in the events around us and argued that this may have been advantageous during early human evolution.

I suspect, however, that there may be something more fundamental going on here. To some extent this superstitious behavior can be seen as an extreme example of a pattern commonly seen in vertebrate lateralization: the left hemisphere misses the optimal strategy (the forest) because it is too concerned with the temporal and sequential details of the arriving stimuli (the trees). A glimmer of this specialization is evident in the rat, whose left hemisphere is better than the right at discriminating complex temporal sequences of sounds.[84] The human left hemisphere, however, actively seeks order and meaning even where there is none—not only in temporal sequences, but also in static images of random visual noise.[85] Like the right hemisphere's propensity to feel the presence of another being, this inclination of the left hemisphere toward magical thinking might have its basis in another peculiar computational demand of human infancy.

A human infant is bombarded by highly complex temporal sequences of sound, and its left hemisphere is innately driven to seek order, pattern, rules, and meaning in this apparent chaos. With good reason: there really is meaning there, and the propensity and drive to find it form the basis for human language acquisition. This tendency may be so fundamental that it endures throughout life, leading us to find not only useful and real causal relationships, but also misleading and illusory ones. When the right hemisphere of an emotionally distraught adult conjures that ineffable feeling of the presence of a loving being, it may be largely the left hemisphere that gives that being a name, puts the feeling into words, and interprets the experience as incontrovertible evidence of an unseen mystical or theological reality. Just as Yellott's subjects fabricated arcane rules to account for nonexistent order in a random sequence, so theologians like Thomas Aquinas and Anselm of Canterbury may have taxed their left hemispheres to formulate arcane verbal arguments for the existence of an invisible God—the real origin of which probably lies in the innate and infantile expectations of the right hemisphere.

Might there be a correlation between verbal virtuosity, on the one hand, and religiosity or magical thinking on the other? I have found no scientific study that explicitly addresses this question, though some indirect support for the idea emerged from a neuroimaging experiment that compared believers in the supernatural to more skeptical subjects.[86] The twenty-three subjects, twelve believers and eleven skeptics, were chosen for their extreme scores (highest and lowest 10 percent, respectively) on a psychometric test for paranormal beliefs. The subjects were told before the experiment that they would be given several short stories that described challenging life events. With each story they would also be given a photograph. Their task was to imagine themselves in the situation described in the story, as if they were walking along a street thinking about it, when suddenly they look up and see the image in the photograph as a poster on a building. After doing each such task in the fMRI scanner, the subject was asked to what degree he or she would think that the image contained a sign or a message about how the situation would turn out. They also rated their emotional reaction to the image as positive, negative, or mixed. The stories were about relationships, money, health, justice, or work life; all involved elements of uncertainty, loss, or foreboding. The photos were sharp, color images of lifeless objects or scenery containing no letters, numbers, animals, or people.

Not surprisingly, the believers reported seeing meaningful signs in the images about twice as often as the skeptics. Seeing signs was correlated with the strength of the emotional reaction to the image, without regard to valence of the emotion. As for neural activity, viewing the pictures activated the left inferior frontal gyrus in both groups; the striking difference was in the corresponding area of the *right* hemisphere, where neural activity was much greater in the skeptics than in the believers. What's interesting about the inferior frontal gyrus is that, in the left hemisphere, it contains Broca's area, the high-level cortex that generates speech and, to a lesser degree, interprets the speech of others. It and the cortex around it are primarily concerned with symbols, meaning, and messages. The authors interpret the role of the corresponding area in the right hemisphere as one of cognitive inhibition—restraining an overly enthusiastic Broca's area in the left hemisphere from finding more meaning than is really there. A similar role for the right inferior frontal gyrus has shown up in several other contexts.[87]

Although this study did not look for a link between verbal virtuosity and mysticism, the link it revealed between the speech cortex and magical thinking suggests that such a correlation may yet be found. Meanwhile I can offer only some intriguing anecdotal evidence. From Oprah's spirituality to M. Night Shyamalan's popular movie about crop circles and malevolent extraterrestrials,[88] the motto of modern mysticism is "There are no coincidences." The phrase may have its origin in this quote from novelist and prominent magical thinker William S. Burroughs II:

> In the magical universe there are no coincidences and there are no accidents. Nothing happens unless someone wills it to happen. The dogma of science is that the will cannot possibly affect external forces, and I think that's just ridiculous. It's as bad as the church. My viewpoint is the exact contrary of the scientific viewpoint. I believe that if you run into somebody in the street it's for a reason. Among primitive people they say if someone was bitten by a snake he was murdered. I believe that.[89]

Perhaps it is no coincidence that Burroughs is considered one of the most innovative novelists of mid-twentieth-century America, that Oprah is arguably the most successful conversationalist on the planet, and that Joseph Mezzofanti—legendary nineteenth-century hyperpolyglot who claimed to speak forty-five languages—was a cardinal in the Catholic Church.[90]

There is also evidence that the adult right hemisphere—despite its likely role in the illusion of God's presence—is in some sense more skeptical, more firmly grounded in reality, and better at using probabilistic knowledge than is the left. Gazzaniga's experiment speaks to the last point. Neurologist Oliver Sacks suggested the first after seeing his left-hemispheric stroke patients laugh at a televised speech by then-president Ronald Reagan, whose words they could no longer comprehend.[91] A more systematic study of stroke patients has since confirmed that those with left-hemispheric damage and severe deficits in speech comprehension are better at detecting lies of an emotional nature from prosody and facial expression than are healthy controls or patients with damage in the right hemisphere.[92] But my personal favorite among revealing neurological anecdotes is V. S. Ramachandran's description of his split-brain patient LB, whose left hemisphere believes in God but whose right hemisphere does not.[93] In those believers who manage to break free of religion, it may, ironically, be the nondominant right hemisphere—with its better grasp of simple probability and more holistic view of the forest—that releases the trickle of skepticism that ultimately extinguishes the fire of arcane religious rationalization in the left.

CHAPTER 13

Predictions

> Nothing ever satisfies her but demonstration. Untested theories are not in her line, and she won't have them.—
> Adam describing Eve in his diary, as conveyed by Mark Twain[1]

In the preceding chapters I have hinted at testable predictions that fall out of the neuroethological perspective on religious experience. Here I examine these in detail. Most are mechanistic questions about how some neurochemical system or part of the brain is involved in an infant's innate expectation of its mother, in an adult's illusion of God's presence, or both. The essential point—and the central hypothesis of this book—is that these two mental states, though not identical, originate from a common core of neural circuitry. The overarching prediction is that deep commonalities between the two should be evident at the level of the nervous system. If such commonalities are not found, then the hypothesis is wrong.

Before we dive into the details, however, a note of caution is in order. We do not fully understand the neural basis of the infant's part of the mother-infant bond, much less that of a sensed mystical presence, so in some sense this effort is premature. Yet we must start somewhere. I have tried to suggest experiments that rely on the most well-established current knowledge of the infantile brain. If some of that current knowledge turns out to be wrong or not applicable to humans, then predictions derived from it will be of little value. But as our neuroscientific knowledge grows in both domains—the neonatal and the mystical—a pattern of commonality should become obvious if the hypothesis is right.

The Infant Brain and Mother-Infant Bonding

It would help to know more about the brains of infants. Noninvasive imaging methods like NIRS and MEG could reveal much of the essential neural circuitry involved in human infantile isolation crying and in the comforting union with mother that accompanies cradling, rocking, and breastfeeding. Some simple drug experiments could be especially revealing, though in human infants great care must be taken to ensure the safety of the subjects, and most experiments of this kind could be done only in conjunction with administration of the drug for medical reasons unrelated to any research questions. Drugs affecting endogenous opiate systems are sometimes used with infants, so these may be good candidates. Do opiate blockers, like naltrexone, increase the intensity or duration of human infantile crying, as they do in some other mammalian infants? Does a low dose of a μ-opiate, like morphine, reduce or abolish crying in human infants?[2]

Is infantile maternal potentiation behavior really a good model for the illusory sensed presence in an adult human? To answer this question, we must first know more about the behavior in nonhuman infants and whether anything like this happens in humans. One promising line of research would be to explore the possible commonalities between the maternal potentiation behavior of rodents and the increased rate of crying—with concomitant reduction in stress—that occurs in isolated rhesus and squirrel monkey infants when they can see, but not get to, their mothers.[3] Does the level of the stress hormone corticosterone decrease in rat pups during maternal potentiation, as does that of cortisol in the infant monkeys? Is the increased rate of crying in the infant monkeys abolished by the dopaminergic drug quinpirole, which specifically blocks maternal potentiation in rat pups?[4] Affirmative answers to these questions would support the idea that maternal potentiation is a more general phenomenon that includes at least one primate species.

Another important question, which I briefly mentioned in chapter 10, concerns the possible role of the cholinergic basal forebrain system in maternal potentiation. In my discussion of hallucinations and the illusion of a sensed presence, I argued that the illusion might arise as a consequence of prior hyperactivity of the cholinergic system brought on by stress, perceived helplessness, and social isolation. If the adult illusion and infantile maternal potentiation are both manifestations of this underlying process, then cortical acetylcholine concentrations should be unusually high in the relevant sensory cortical areas of rat pups during maternal potentiation behavior and, to a lesser degree, during the initial isolation period preceding it. It should be lowest during the brief period of reunion with the mother. The most relevant cortical area to monitor for changes in acetylcholine would probably be the somatosensory barrel cortex, as it is relatively mature in the two-week-old pups typically used in such experiments.[5]

There are now highly sensitive techniques for detecting rapid changes in acetylcholine levels that could test this prediction.[6]

Similarly, blockade of the cholinergic basal forebrain system in rat pups, either by drugs like scopolamine or by specific immunotoxic lesion of the cholinergic neurons,[7] should affect maternal potentiation in some way, though these effects may be only subtle alterations in the attention or distractibility of the pups that could be difficult to detect.[*8] For this reason, I cannot make any specific predictions about the outcome of such experiments, but they are nonetheless worth pursuing. Only a few studies of this kind have been done with rat pups. In one of them, a drug that activates the muscarinic type of acetylcholine receptor reduced the rate of ultrasonic cries in isolated pups, though maternal potentiation was not examined.[9] Laura Ricceri and her colleagues have done a few studies of the effects of immunotoxic lesions of the cholinergic basal forebrain system in rat pups, but the results thus far are difficult to interpret. In two of the studies, the cholinergic lesions reduced the rate of ultrasonic cries in the rat pups,[10] but a subsequent study found no significant effect of cholinergic lesion on isolation cries or on maternal potentiation.[†11] The discrepancy may have to do with a lower efficacy of the lesion in the latter study or other methodological differences.[‡12] But another apparent discrepancy concerns the experiment with the cholinergic drug, which reduced the rate of isolation crying in unlesioned pups. In the most simpleminded interpretation of these experiments, a drug that activates acetylcholine receptors should have an effect *opposite* to that of a lesion that eliminates acetylcholine from the cortex, yet all of these experiments had the *same* effect: reduction of the rate of isolation cries. Perhaps simplemindedness does not suffice here. Just to mention one possible confounding factor: the cholinergic neurons of the basal forebrain are not the only cholinergic cells involved in rat vocalizations. The midbrain periaqueductal gray receives cholinergic input from the brain stem, which in adult rats is necessary for their 22 kilohertz ultrasonic alarm cries.[13] Whether something like this is involved in infantile isolation cries is unknown, but it underscores the need for more experiments.

For my purposes, however, the most important question about maternal potentiation is a simple behavioral one: does it occur in human infants? Some

---

* Neigh et al. (2004) found little effect on behavior in an attention-demanding task when task-related increases in cortical acetylcholine were inhibited by microinjection of a neurotoxin into the nucleus accumbens.

† Ricceri et al. (2007). The authors noted that a reduction in the magnitude of maternal potentiation in the lesioned pups, relative to controls, fell just barely short of statistical significance (the p value was 0.06). The efficacy of the lesion in this study was only about 50 percent (see next footnote), so a higher dose of toxin might have yielded a significant reduction in maternal potentiation.

‡ The density of cholinergic terminals in the cortex was reduced by only about 50 percent in the Ricceri et al. (2007) experiments, whereas earlier experiments from the same lab achieved about 70 to 90 percent decrease in cholinergic terminals with the same immunotoxic lesion.

anecdotal evidence suggests that it might,[14] but the question deserves systematic study. We know that in rodents the behavior emerges only after some imprinting has occurred,[15] so we should seek it in human infants at several different stages of development. The normal crying behavior of human infants varies across a spectrum, with minimal fussiness at one end and inconsolable colic at the other.[16] Infants from the middle of this range would probably be the most revealing for these studies, especially at ages younger than three months.

Of course the experiments with human infants could not exactly follow the protocols used with rats and mice, but the essence of the question could still be asked. When a crying infant is briefly picked up and comforted by its mother and then put down again, does it tend to cry more vigorously? At later ages, the experiments might more closely resemble those done with monkeys. When a toddler is separated from its mother and starts crying, does a brief reunion, followed by another separation, make the crying worse? If the crying toddler can see but not reach its mother—if, for example, her face appears through a small window in the wall—does the infant's crying intensify? Such experiments would have much in common with Mary Ainsworth's Strange Situation test for evaluating mother-infant attachment.[17] Perhaps good evidence for maternal potentiation in humans already exists in the raw data collected for attachment studies.

The Adult Brain and the Sensed Presence

Ultimately the results of these and similar experiments on infants are prerequisite for testing the hypothesis in adults, but there are some interesting questions we could ask in the meantime. I have suggested that prayer is an adult manifestation of infantile supplication, but not all prayers are created equal. Intense, heartfelt prayers motivated by desperate need should bear the greatest neurological resemblance to the crying of a helpless infant. We need not place experimental subjects under unethically severe conditions of pain and duress to do these experiments; life does that for us—and all too often. We could, for example, recruit as subjects for our prayer studies the loved ones of terminally ill patients, whose prayers for the dying would probably have the appropriate qualities. Images of neural activity during these prayers should resemble those from the brains of crying infants, and drugs that affect infantile crying should have analogous effects on prayer. Naltrexone, for example, should increase the frequency and intensity of spontaneous prayer, and doses of morphine low enough not to cause sedation should diminish it.

As I emphasized in my discussion of Michael Persinger's research, however, it is the illusion of the sensed presence that I would most like to see brought under scientific scrutiny. In chapter 9 I described the controversy over Persinger's claim that his unconventional form of transcranial magnetic stimulation elicits the illusion. In my judgment, the evidence against this claim is overwhelming, but if any of Persinger's colleagues still believe it and want to be taken seriously, they will need to do some rigorous research on the biophysics of the method. It would help if extremely weak but rhythmic magnetic stimulation could be shown to affect the firing of cortical neurons in experimental animals. It might also help to apply those complex magnetic fields to regions of the human cortex in which the expected effects are simpler and more easily interpretable. For example, can visual perception be reproducibly altered in any way by stimulating the primary visual cortex in this way? What about movement, muscle tone, and the primary motor cortex? Finally, it may be that attaching the magnetic coils to a helmet causes variability of results for subjects with heads of different shapes and sizes. It might be better to place the coils stereotactically using fMRI data to establish equivalent sites of stimulation across multiple subjects.[18]

Probably none of that would pan out, and for some purposes it would not matter. As the Danish group beautifully demonstrated, we can use Persinger's protocol to elicit sensed presence illusions merely by power of suggestion and deprivation of the senses,[19] and many experimental variations are possible. Just before starting each sensory deprivation session with the helmet, we could use real rTMS, from a conventional TMS coil, to give half the subjects a virtual lesion in some interesting part of the brain. The other half would be controls who each get a sham rTMS treatment that looks and sounds the same but has no effect. Ideally the rTMS treatment should be an offline protocol that depresses cortical activity in the treated area for an hour or more—long enough for the entire session with the helmet. As I mentioned in chapter 9, however, the offline protocol that seems best suited to this purpose, continuous theta-burst stimulation, has credibility problems of its own.[20] These can and obviously must be resolved before the experiments I suggest here can be done.

But once we have settled on the best protocol for doing the virtual lesion, what parts of the brain would be the most informative targets? From the evidence I summarized in chapter 6, the obvious first choice is the temporoparietal junction of the right hemisphere. I predict that subjects given a virtual lesion there should have significantly more sensed-presence illusions during their suggestive sessions with the God helmet than will control subjects. Other areas related to embodiment and the distinction between self and other would also be promising targets, including the medial prefrontal cortex, the precuneus, and the extrastriate body area. Various combinations of these might be worth trying. For example, paired virtual lesions of the temporoparietal junction and extrastriate body area might be especially effective, as they are jointly active in some tasks related to embodiment.[21] Another promising pair would be the temporoparietal junction and medial prefrontal cortex, as their activity tends to be correlated during theory-of-mind tasks[22] and even at rest.[23]

Another strategy would be to choose targets that might alter the quality of the sensed presence in a predictable way. In chapter 4 I summarized evidence for the role of the orbitofrontal cortex in affiliative behavior, including mother-infant attachment. I also described the role of the amygdala in aversive behavior, fear conditioning, and maintenance of interpersonal space in adults, and I noted that it and the orbitofrontal cortex often act antagonistically. This suggests that a virtual lesion of the orbitofrontal cortex should increase the relative influence of the amygdala and make any sensed presence seem more threatening than loving. Likewise, a virtual lesion of the amygdala should tip the balance in the other direction, yielding sensed presences that are perceived as loving, benevolent, or at least benign.

And rTMS is not the only way to manipulate a sensed presence evoked by suggestion and sensory deprivation. The quality or rate of occurrence of the illusion in such experiments should be predictably alterable by making the condition of the subject more or less like that of a helpless infant. For example, if the hypothesis is right, then a subject whose life is more emotionally challenging and stressful should be more likely than control subjects to experience the sensed presence in a suggestive and isolated environment and more likely to experience that presence as benevolent. If the illusion arises from prior hyperactivity of the cholinergic system, as I have suggested, then a long session of intense prayer prior to entering the sensory deprivation chamber should increase the probability of the illusion, and that effect should be augmented by drugs, like physostigmine, that enhance the effects of acetylcholine.*[24] Conversely, several hours of prior partial blockade of the cholinergic system—by wearing a scopolamine patch, for example—should decrease the probability of the illusion.

Other neurochemical dimensions of the hypothesis could be tested in this way. For example, the potentiating effect of prior intense prayers of supplication should be enhanced by a concurrent dose of the opiate blocker naloxone. The onset of the illusion of a benevolent presence should be more likely after an intranasal dose of oxytocin.[25] If maternal potentiation can be demonstrated in human infants, if it has the same sensitivity to quinpirole that it has in rodents, and if it and the illusion of God's presence share a common neural substrate, then systemic quinpirole should block the illusory sensed presence in these experiments. That long string of "ifs" implies a more general warning. We need to establish the effects of these drugs on behavior and neural activity in human infants, especially for those neurochemical systems—like dopamine, oxytocin,

---

* Physostigmine enhances the effects of acetylcholine by blocking the enzyme that eliminates it from the synapse. The drug may have mixed effects in this experiment, enhancing attention in a general way but diminishing the effects of top-down expectation. See Bentley et al. (2004).

and vasopressin—that can have radically different effects in two closely related species or even at different ages or for different behaviors within the same individual animal.

Given the ephemeral and serendipitous nature of the illusion of a sensed presence, it might never be feasible to conduct such experiments in an fMRI scanner—apart from studies of self-induced states like the ones achieved by Carmelite nuns and devout Mormons, as I described in chapter 8. But with the right subjects and the right suggestive and pharmacological encouragements, the phenomenon might be sufficiently reliable that useful imaging data could be obtained with EEG, NIRS, or both. Probably the most obvious prediction of the hypothesis that could be seen with these methods would be a surge in activity in the temporoparietal junction, mainly in the right hemisphere, during the illusion itself.

Laterality of the Illusion

That last prediction suggests another—one that could be tested without suggestion, drugs, or sensory deprivation chambers. In the previous chapter I explored the special role of the right hemisphere in the neonatal human brain and, presumably, in the illusion of God's presence. Does this right-hemispheric dominance make any testable prediction about the laterality of the illusion itself? In many descriptions of this illusion, the presence is highly localized in space, and often specifically described as being on the subject's left or right side. Should we expect a bias in these reports that reflects the hypothesized right-hemispheric dominance in its neural origin?

For almost any cortical area other than the right temporoparietal junction—the likely substrate of the spatial aspects of the illusion—the answer would be, "Yes, of course—the presence should be biased toward the left side, opposite the right hemisphere." The right temporoparietal junction, however, represents all extrapersonal space, left and right, which explains why a stroke there, but not on the left, causes hemispatial neglect: the surviving left temporoparietal junction handles spatial attention only for the right side of space.[26] In principle, then, the right temporoparietal junction could place the illusory sensed presence anywhere in space, and reports of these illusions do appear to encompass a wide variety of spatial localization. Furthermore, most of these reports come from neurologically normal subjects with intact connections between their hemispheres, so the left hemisphere may contribute to the spatial aspects of some of these experiences, even if they are triggered on the right.

Despite those arguments, there remains a good reason to expect at least some left-side bias in the localization of the sensed presence. The left-side cradling bias of mothers and the right-side head-turning bias of supine neonates are good evidence that leftward-directed visual attention and orientation are the dominant bias of the mother-infant bond in humans. As I explained earlier, when mother and infant relate in this way, each presents the left half of the face to the other, and each is in the left side of the other's visual field. Interestingly, a similar left-oriented bias occurs when sexual partners kiss.*[27] If these biases arise from an asymmetry in emotional attention driven by the right hemisphere and if the illusion of God's presence has a similar neural basis, then that presence should, more often than not, be felt on the subject's left side when it is experienced as spatially localized.

A survey of lateralized sensed-presence experiences in the neurological literature has attempted to address this question.[28] All of these were, of course, illusions precipitated by some kind of neuropathology, and the number of cases was small, so the results may not apply to the general population. Of thirty-one such cases, twelve were felt on the left and nineteen on the right. Unfortunately, in nineteen of the cases (seven left- and twelve right-side presences), the pathology could not be localized to the left or right side of the brain. Of the five left-side presences for which the pathology was localized, three involved the right hemisphere and two the left. Of the seven right-side presences for which the pathology was localized, one involved the right hemisphere and six the left. We could add to these two more subsequently published cases of left-hemisphere epilepsy, in which electrical stimulation of the left temporoparietal junction elicited an illusory sensed presence on the right.[29] Perhaps the most meaningful conclusion that can be drawn from these results is that most of the presences were felt on the side opposite the neural disturbance. A real test of the hypothesis of right-hemispheric dominance in this illusion requires many more cases, preferably from a healthy population.

Luckily there is a large collection of such reports that could provide a meaningful test. The Alister Hardy Trust has solicited from the general public thousands of reports of religious and spiritual experiences over many years.[30] Not all of these describe sensed presences localized to the right or left, but probably many do.†[31] To my knowledge the collection has not yet been examined for a possible left or right bias, but it should be. My bet is on the left.

---

* The bias in kissing is often described as a rightward head-turning bias, but this refers to a lateral flexion of the head to the right in preparation for the kiss. This inevitably is accompanied by a rotation of the head about the spinal axis, toward the left, and each partner in such a kiss is mainly in the left half of the other's personal space. See van der Kamp and Canal-Bruland (2011) and references cited therein.

† In chapter 5 of *The Illusion of God's Presence*, I quoted one that described a right-localized presence (Wathey 2016, pp 75–76).

Religious Conversion

Finally, there is yet another strategy by which the infantile basis of religious experience can be tested—one that is less sensitive to the vicissitudes of the sensed-presence illusion. Rather than struggle to nail down that illusion in the laboratory, we could instead examine one of its chief natural consequences: religious conversion. The idea would be to recruit a large number of subjects who are not committed to any religious belief but who are actively exploring spirituality by attending the services of various religions. Adults age thirty-five or younger would probably be good subjects, as most Americans who leave their childhood faith and convert to another do so by that age.[32] The subjects would be given a drug or a placebo treatment prior to each service, and the effect of the drug on the likelihood of religious conversion would be evaluated by statistical analysis of the outcomes.

The drugs used would be those known to have specific effects in mother-infant bonding, and therefore predicted to have similar effects on conversion experiences precipitated by the feeling of God's presence. Based on our current knowledge, promising candidates include the drugs I have already discussed in the context of experiments on the sensed-presence illusion: scopolamine, physostigmine, naloxone, morphine, and quinpirole. Scopolamine would presumably be most effective if given continuously for several hours prior to each religious service, assuming that it is prolonged hyperactivity of the cholinergic system that primes the brain for religious experience. Oxytocin might facilitate conversion during a moment of religious fervor, but its effects wear off too quickly to be practical for experiments of this kind. Perhaps longer-acting synthetic drugs that block or activate oxytocin receptors will become available.[33] Obviously, the more we know about the neurochemistry of human infantile behavior and emotion, the more sensitive and specific these experiments can be.

I could list more tests, but what matters most is the direction of experimentation suggested here. The essential idea is to look for deep commonality in an infant's neural representation of its mother and an adult's neural representation of a mystical presence. Some of the predictions I have made here may need to be revised as our understanding of the neuroscience of human infancy improves, but the hypothesis must be discarded if that deep commonality cannot be found.

How Does God Become a Cruel Tyrant?

I began the first chapter of this book with a brief discussion of the cruel and judgmental side of the two-faced god, the god of the social root. Most of the book, however, has focused on God's other face—his nurturing and unconditionally loving side—and on the neural circuitry that brings such beliefs to life in the human mind. In this final chapter I have developed some mechanistic *how* questions from answers to various *why* questions that spring from the neonatal root of religion. But what about the social root? Can we find its mechanistic basis in the brain?

A central theme of part II of this book—that the neural substrate of the innate model of the mother is the skeleton of the adult social brain—suggests that some parts of the brain may be deeply involved in both biological roots of religion. For example, the conjunction in the amygdala of stimulus valuation with the identification of specific familiar persons may be critical for the social evaluation of other people[34]—an important aspect of reciprocal altruism. The pathologies of social behavior that result from damage to the medial and orbital prefrontal networks identify these structures as likely neural substrates for our innate moral intuition[35]—the emotional glue that holds a society together.[36] Neuroimaging studies implicate both the amygdala and orbitofrontal cortex in the expression of in-group bias,[37] and damage to these structures causes severe deficits in reasoning about the kind of social rules that are the foundation of cooperation.[38]

But how does the concept of God fit into this? Is there anything about the answer to "Why is God a cruel tyrant?" that suggests a promising mechanistic *how* question? Can we develop any suggestions for brain research from the idea that human reciprocal altruism is supported by belief in a cruel and punishing God who demands costly sacrifice?

I am hopeful that we can, and I already have one general suggestion for research along these lines. In chapter 1 I emphasized a possible link between the social root of religion, especially as manifest in religious cults, and narcissistic personality disorder (NPD), the symptoms of which include extreme self-centeredness, cruelty, lack of empathy, and an insatiable lust for costly sacrifice as proof of the devotion of others. I argued that the cruel gods of human religions were probably modeled after human personalities of this kind. This psychopathology is highly heritable[39] and persists in a small but significant fraction of the human population,[40] probably through frequency-dependent selection.[41] Its genetic and neurological bases are poorly understood, but the answers to those genetic and neurological *how* questions may reveal the mechanism, not only of narcissistic personality disorder, but also of some of the darkest aspects of religious belief and practice.

Some progress has already been made in this direction. One approach is to choose a trait that is deficient or dysfunctional in NPD subjects—empathy, for example—and use neuroimaging, lesion mapping, or any other relevant technique to work out what structures and networks in the brain are important for that trait. We can then look for anomalies in these neural circuits in NPD subjects. We must also, of course, be careful about how we measure empathy. Psychologists distinguish two components of empathy: emotional empathy is about experiencing the same feeling as another person who is displaying pain, anguish, joy, or some other strong and unambiguous emotion; cognitive empathy involves taking the perspective of another person and trying to read their thoughts from nonverbal cues or behavioral context.[42] A meta-analysis of lesion studies found that emotional empathy requires the inferior frontal gyrus, inferior parietal lobule (part of the temporoparietal junction we first considered in chapter 6), anterior insula, and anterior cingulate cortex (the neural alarm system from chapter 3). By contrast, cognitive empathy was most consistently affected by damage to the ventromedial prefrontal cortex (part of the orbital and medial networks from chapter 4), the medial temporal lobe, and the temporoparietal junction.[43]

Some of these empathy-related areas do in fact appear to be deficient in narcissistic subjects, but the results vary depending on the details of the experiments. The anterior insula, which is strongly associated with emotional empathy, was one of the most consistently implicated structures. When twenty-two healthy subjects did empathy-related tasks in an fMRI scanner, those with relatively high scores on a narcissism scale showed diminished responses in the anterior insula relative to those with low narcissism scores.[44] A structural MRI study using 103 healthy subjects found correlations between their narcissism scores and local variations in gray matter volumes in the insula, anterior cingulate cortex, and several prefrontal areas including the ventromedial and orbitofrontal cortex.[45] Another structural MRI study compared seventeen clinical NPD patients to seventeen healthy controls matched for age, gender, handedness, and intelligence. Relative to controls, the NPD subjects had significantly less gray matter volume in the left anterior insula, and that volume was positively correlated with the self-reported emotional empathy of all the subjects, including controls. The NPD subjects also had gray matter deficiencies in the right anterior cingulate, the left middle cingulate, and the middle and superior frontal gyri of both hemispheres.[46]

Another hallmark of NPD is an insatiable craving for attention and praise from others. A possible mechanistic explanation for this is a deficient connection in the brain between self and reward—more precisely, between areas that handle self-referential processing, like the medial prefrontal cortex we examined in chapters 4 and 6 and the nucleus accumbens, the epicenter of reward, motivation, and addiction we explored in chapter 5. This hypothesis was tested in a structural MRI study of fifty healthy subjects who each completed a psychometric test for grandiose narcissism. As predicted, the integrity of the neural fiber pathway connecting the nucleus accumbens to the medial prefrontal cortex was relatively deficient in the more narcissistic subjects.[47] The pathological exhibitionism of narcissists may simply be their way of compensating for a perpetual deficiency in self-affirmation.

As I emphasized in chapter 1, however, narcissistic leaders are only half the story of the cult phenomenon: their followers are the other half. These are people who are seduced by the narcissist's charisma, shower him with praise, fear his fits of rage, and make costly sacrifices to prove their loyalty. Their salient traits include a powerful need to belong, fear of ostracism, desire to sacrifice for a cause, and need for the comfort, certainty, or security that comes from being told what to do by a domineering leader. These traits, like the narcissistic ones, can be measured by psychometric tests and correlated with neural activity, gray matter volume, brain lesions, and so on. We encountered one such experiment in chapter 3, the neuroimaging study by Naomi Eisenberger and Matthew Lieberman that found similar activation in the anterior cingulate cortex both to physical pain and to the social pain of ostracism.[48]

In chapter 9 I discussed another more comprehensive study that subsumed essentially all of the defining traits of cult followers into two complex and related psychological dimensions: religious fundamentalism and authoritarianism.[49] This was the lesion mapping study, by Erik Asp and colleagues, of ten patients with damage to the ventromedial prefrontal cortex.[50] The authors hypothesized that this part of the brain, which is roughly equivalent to what I called the medial and orbital prefrontal networks in chapter 4, is important for detecting and rejecting false ideas and that patients with lesions there will be gullible, credulous, and susceptible to extreme and incoherent belief systems like those found in authoritarianism and religious fundamentalism. This was in fact the case: the patients were significantly more fundamentalist and authoritarian than the controls, and, in a related study, they were also more vulnerable to accepting false claims in deceptive advertisements.[51] This discovery is important, but the authors' interpretation of the ventromedial prefrontal cortex as the place where incorrect ideas get tagged as false may be too simplistic.

Based on the evidence I summarized in chapter 4 about the orbital and medial networks, skepticism does not stand out as a major function of these cortical areas. Neuroimaging studies of essential aspects of skepticism, like deductive reasoning and cognitive reflection, generally exclude the ventromedial prefrontal cortex from their rosters of activated brain regions.[52] So do studies that directly probe the rejection of statements as false.*[53] Instead, the orbital and medial networks appear to be the basis of reward valuation for reinforcement learning, repositories of innate knowledge about primary reinforcers, critical

---

* Two such studies implicated the ventromedial prefrontal cortex in accepting beliefs as true, in direct contradiction of the idea that this part of the brain does false tagging (Harris et al. 2008, 2009). A more recent study identified only the anterior cingulate cortex as being specifically activated when subjects judged statements as false (Howlett and Paulus 2015).

nodes in the neural networks of affiliative emotion, and part of the essential circuitry of moral intuition. One aspect of moral intuition, of course, is the uneasy gut feeling one gets when being accosted by an unscrupulous salesman; the absence of that feeling in Erik Asp's subjects might account for their apparent gullibility with deceptive advertisements. Another aspect is the tension between the moral virtues emphasized by the politically liberal (fairness; care for the weak and disadvantaged) and those emphasized by conservatives (loyalty; authority; sanctity). As psychologist Jonathan Haidt explains, this is a multidimensional gradient, not a strict dichotomy, but in practice it tends to collapse onto a one-dimensional political and moral spectrum.[54] The liberal and conservative ends of it map fairly well onto the neonatal and social roots of religion, respectively, with religious fundamentalism and right-wing authoritarianism being the part that survives the destruction of the medial prefrontal and orbitofrontal cortex in Asp's subjects. Exactly how this dimension of morality is mapped onto the brain is unknown, but the interplay between orbitofrontal cortex and amygdala may be part of it. Rather than obliterating a "false-tagging" circuit, lesions of the ventromedial prefrontal cortex may be boosting authoritarianism and fundamentalism by freeing the amygdala from inhibition by the orbitofrontal cortex.

Finally, there is one more aspect of the social root of religion that merits discussion here: group ritual. All human cultures speak, chant, sing, pray, or dance together in unison as part of their normal community life. Group rituals foster in-group trust and cooperation, from the exuberant dancing and singing of the early Shakers to the synchronous shouts of football fans who follow directions from cheerleaders. The neural basis of this was investigated in a neuroimaging study done by Sophie Scott and collaborators.[55] A subject in an fMRI scanner read a sentence aloud while they heard another person reading a sentence. In some trials, both read the same sentence, and so could read in synchrony, whereas in other trials they read different sentences, thus precluding synchrony. Another variable was the source of the other voice heard by the subject in the scanner: sometimes it was another person reading live, in which case both readers could contribute to their synchrony, but in other trials, and without the subject's knowledge, it was a recorded voice.

Normally when you speak, the activity in some parts of the auditory cortex is suppressed; in effect, the brain becomes less sensitive to the sound of your own voice while you are speaking. This is called speech-induced suppression, and although its function is unknown, analogous suppression in other sensory systems is thought to aid in distinguishing between self and other.[56] A surprising finding from this study was that the normally suppressed parts of the auditory cortex were released from suppression during synchronous speech, but *only* if the other speaker was live, not recorded. In this condition, the brain processes the sound of your own voice *as if it were the voice of another person*—a dissolution of the self-other distinction. Again, the reason for this is unknown, but the authors suggest it has something to do with the greater precision of synchrony that happens when both speakers contribute to the effort. Synchrony then is no longer controlled or directed by any one person; each person contributes equally, and the result is better than either could achieve with only a prerecorded voice as their partner. A group of chanting humans becomes like a huge flock of starlings that seems as if of a single mind—an ideal exercise for building group cohesion and trust. Religions, of course, exploit this to the fullest with their group prayers, chants, Scripture readings, liturgies, and hymns. This explains why fundamentalists complain, falsely, that prayer is banned in American public schools. What they really mean is that *group prayer* is banned; speaking to God silently, one-on-one, is just not adequate if your faith comes mostly from the social root of religion.

Epilogue

SO WHAT?

> In the end, we could learn that we are nothing more than
> nerve cells and molecules. But it is too early for believers
> to raise the white flag. . . . Perhaps our brains are reflect-
> ing an encounter with the divine—unseen, surely, but still
> real. Science can't referee that question.—Barbara Bradley
> Hagerty[1]

In this book and its predecessor, I have tried to explain an aspect of religious
experience that deserves more scientific attention than it has previously received.
In this volume especially, I have tried to make that explanation sufficiently
detailed and specific that it can be empirically tested, thus elevating it to the
status of hypothesis. Perhaps it will fail those tests and be discarded, but what if
it passes? Suppose we find that deep commonality in an infant's neural representation of its mother and an adult's neural representation of a mystical presence.
So what?

It was not neuroscience that made journalist Barbara Hagerty believe in
God. It was partly her upbringing in the Church of Christ, Scientist, but most of
all it was her personal experience of God's presence as described in the opening
pages of her book.[2] Why then did she devote a year of her life to interviewing
neurotheologists and writing a book about the brain and mystical experience?
Although not a scientist herself, she seems to appreciate, as most people vaguely
do, that science works. We are surrounded by and utterly dependent upon
what we call miracles of modern technology—things that would truly have
been miraculous in the prescientific age but that are, in fact, not miracles at all.
Hagerty and several of the neurotheologists she interviewed seem to want the
imprimatur of science bestowed upon their theistic beliefs, evidently not appre-
ciating that science does not work that way. It works by ruthlessly discarding

hypotheses that fail when tested and by refining, retesting, and building upon those that pass.

In the title of her book, Hagerty suggests that the patterns of neural activity seen by neurotheologists are "fingerprints of God," rather than mere physical manifestations of neural activity that can be completely understood in naturalistic terms. Science *can* referee this question. The supernatural contestant is immediately disqualified for lacking testable predictions.

If my hypothesis passes its tests, it will mean that we have a completely natural if tentative explanation for a widely reported aspect of human religious experience, an explanation grounded in evolutionary biology, ethology, and neuroscience. It will mean that the feeling of a divine presence can be understood as an aspect of mother-infant bonding, something we share with all mammals but that in humans has been greatly elaborated as the foundation of our highly social nature. It will mean that we better understand why we invent religions and long for that mystical *other*.

And if the hypothesis fails, it will not mean that God wins by default. With luck, the experiments that shoot it down will hint at better and equally testable hypotheses. With cleverness and hard work, we will eventually tease out the truth from among them. That is how science works.

But beyond the intellectual problem of understanding our spiritual longing, we humans face serious practical problems, largely of our own making. These include overpopulation, global warming, biodiversity loss, war, poverty, and disease—to name only the most obvious. The essence of our predicament is that only a small minority of us have transcended the limitations of intuitive thinking through cognitive reflection and the scientific method. Collectively we are like a toddler playing with a loaded gun. The technological power unleashed by science is too often wielded thoughtlessly, in ignorance, on emotional impulse, or toward illusory and self-destructive ends because most of us view reality through the dark glass of intuition and magical thinking. We are not immortal souls temporarily trapped in physical bodies. We are not God's chosen people, locked in righteous struggle with the infidels. We do not live in a mindful and loving universe. We are not part of a universal consciousness, dreaming our lives away in a world of illusion. We are one species, a product of evolution, a newcomer on an ancient but fragile planet, and utterly dependent upon the integrity of its biosphere for our existence.

And, yes, each of our conscious minds is "nothing more than nerve cells and molecules," but that trite caricature trivializes the marvels of the human brain that neuroscience has discovered. The unity of mind and brain is not an assumption to which fearful and closed-minded scientists cling. It is neither a manifestation of dogmatic scientism nor a symptom of pathological nihilism. *It is reality*, confirmed by abundant empirical evidence, our innate intuitions to the contrary notwithstanding. Those intuitions are part of that reality, apprehensible in terms of their evolutionary origin and neural implementation. To understand them for what they are and to see past them is to take one more step toward the enlightenment we most need.

Acknowledgments

I thank the generous and helpful people at the libraries I used, notably the National Center for Biotechnology Information through its PubMed database, the San Diego Public Library, and the academic libraries of the University of California (San Diego and Los Angeles), San Diego State University, University of San Diego, and Westminster Seminary California. I am sincerely grateful to my editor at Prometheus Books, Jake Bonar, for his enthusiastic support of this project and for the substantial improvements to the structure and content of the book that he suggested. It was a pleasure to work with him and his colleagues at Prometheus, notably production editor Jessica McCleary and my talented and insightful copyeditor Erin McGarvey.

I am sincerely grateful to the many friends, family members, former colleagues, and scientific pen pals who read and commented on various parts of the manuscript. They include Susan Blackmore, Adam Lee, and Jack Pettigrew. Several others deserve special thanks for reading all or nearly all of it, in some cases multiple revisions. They are: Dan Barker, Mary Ann Buckles, Mike Huntley, Robert Lawrence Kuhn, Darrel Ray, J. Anderson Thomson, and David M. Wulff. All gave much-needed encouragement and criticism, and their questions and comments alerted me to errors and weak spots. What flaws remain are entirely my responsibility. I am especially indebted to Sarah Hrdy, Darrel Ray, Andy Thomson, and David Wulff for their kindness, generosity, and enthusiastic appreciation of this book and its predecessor. Words cannot convey how much this helped.

No one, however, is more deserving of my thanks and praise than my beloved wife, Mary Ann Buckles, who encouraged and supported me in this project through good times and bad, who read more manuscript pages than anyone else, and who, like me, is delighted it's done.

Acknowledgments

Appendix A

GETTING ORIENTED IN THE BRAIN

> The explosively expanding branch of knowledge called
> "neuroscience," still in its relative youth, already claims to
> embrace a wider spectrum of complexity, a greater range
> of levels of explanation, than any other science.—neu-
> roethologist Theodore H. Bullock[1]

The human brain is a massively parallel computational machine of staggering
complexity, comprising about ninety billion neurons,[2] each of which is itself an
intricate and complex information processing device. Neuroscientists struggle to
understand how it all works by studying the effects of localized damage to it, ana-
lyzing the structure, function, and genetic control of its component parts, following its embryonic development, reconstructing its evolutionary history, and, most
recently, imaging its physiological activity in living, thinking human subjects.

In this brief overview I can convey only the bare essentials of the technical
concepts and words used in this book. You can find more details in the resources
listed at the end of this appendix. To understand this book, you need to know the
basics of three subjects: what a neuron is and how it works, how to find your way
around inside a brain, and what the methods of neuroscience can and cannot tell us.

Neurons

Most of what the brain does is done by neurons, tiny cells that generate and
propagate electrical signals.*[3] These signals may represent brightness, color, or

---

* For a more comprehensive treatment than I present here, see Martin et al. (2020). The other major
category of brain cells, the glia, may also have important if more subtle roles in information processing
in the brain; see Fields (2009).

contrast in a visual scene, temperature of the skin, a sound of a specific pitch, a unique odor or taste, or a command to contract a specific muscle. The meaning of the signal is determined mainly by where in the nervous system it occurs. For neurons deep in the brain, far removed both from sense organs and muscles, the meaning is more complex and abstract.

A typical neuron sends its output signals along a fibrous outgrowth called an axon and receives most of its input signals from other neurons on its dendrites—other outgrowths that branch like the limbs of a tree. These input signals arrive at points of close contact between an axon terminal of the transmitting neuron and the dendrite of the receiving one. Each such contact is a synapse.

The transmission of the signal at the synapse is most often a chemical process in which the transmitting or presynaptic neuron releases into the synapse molecules called, appropriately enough, neurotransmitters. Examples include acetylcholine, glutamate, glycine, dopamine, serotonin, GABA, and various peptides like oxytocin, vasopressin, and the endogenous opiates. These are relatively small molecules that diffuse across the synapse and bind to large protein molecules called receptors, which are embedded in the membrane of the receiving or postsynaptic neuron. The binding of the neurotransmitter affects the activity of the receiving neuron in a way that depends upon the type of receptor molecule.

Some receptors have a channel that opens when the neurotransmitter binds, allowing electrically charged atoms (ions) to flow across the membrane of the receiving neuron. This flow of electric current may either excite or inhibit the electrical activity of the neuron, depending on the charge and direction of flow of the ions in the channel.

Another variety of receptor does not directly open a channel, but instead triggers a cascade of biochemical reactions within the neuron. The result is a chemical signal that can propagate inside the postsynaptic cell causing various effects, including the opening or closing of channels elsewhere, the regulation of the number or sensitivity of receptor molecules at the synapse, or changes in gene expression. These changes tend to be modulatory in nature: slower and longer in duration than the effects of a receptor that simply opens a channel, and neurotransmitters that act in this way are sometimes called neuromodulators. How a neurotransmitter affects a cell, however, is mainly determined by the kind of receptor that binds and responds to it, so a given neurotransmitter, like acetylcholine, may function as a modulator at some synapses but not at others.

One of the most important roles for neuromodulators is their effect on the strength of synaptic connections: these changes are the substance of memory. The ability of a synapse to be modified in this way is called synaptic plasticity, and some are more modifiable than others. A prominent and abundant variety of highly plastic synapses are the ones that occur on dendritic spines. A spine is a small protuberance on a dendritic branch, rather like a thorn on a rosebush, only blunt and enlarged at the tip. Changes in the size, shape, and number of spines can occur rapidly during learning.[4]

When a neuron is sufficiently excited by activity at its synapses, it generates an action potential, also commonly called a spike, which is a brief and drastic change in the electrical potential (voltage) across the cellular membrane. At any one place in a neuron, an action potential lasts only a few milliseconds, but it never affects only one place. Instead it travels along the axon as a wave, propagating down every branch and triggering the release of neurotransmitter molecules at every synaptic contact made by those branches.[5] The precise time of arrival of the action potential at a synapse can be extremely important. If an action potential arrives at the presynaptic side of a synapse almost in synchrony with the firing of an action potential in the postsynaptic neuron, the strength of that synapse may be decreased or increased depending on whether the presynaptic spike occurs slightly before or slightly after the postsynaptic spike.[6] This is called spike-timing-dependent plasticity, and it is one of several mechanisms that contribute to the formation of networks of neurons, widely distributed in the brain, that fire rhythmically and in synchrony.[7] Synchronous oscillations of this kind appear to be critically important in conscious perception, selective attention, and the formation and retrieval of memories.[8]

Neuroanatomy

At a larger scale, neurons in the brain tend to be laid out in orderly patterns, with their cell bodies gathered together into clumps or layers, called gray matter, and their axons coalescing into bundles known as white matter. These patterns of organization arise as the brain develops in the embryo. In the early stages of this process, the entire nervous system is a hollow tube of cells filled with fluid. Under the direction of genetic instructions, the dividing cells proliferate and specialize more in some places than in others, causing large bulges and folds of neural tissue in the walls of the tube. Meanwhile, axons grow throughout the brain, following chemical signals, to form roughly the right connections with other parts of the brain—connections that will ultimately be made more precise and specific through subsequent pruning.

From front to back, the neural tube differentiates into regions known as forebrain, midbrain, hindbrain, and spinal cord. The most prominent bulges in the growing human brain are those in the roof of the forebrain that ultimately become the highly convoluted cerebral cortex. The floor of the forebrain develops into much smaller but extremely important clusters of neurons that together are called the basal forebrain. This region includes several groups of modulatory neurons, important for memory and the focusing of attention, that innervate the entire cortex and secrete acetylcholine as their neurotransmitter. Also in this neighborhood lie the basal ganglia, part of which—nucleus accumbens—figures prominently in this book.

Most of the book, however, concerns cerebral cortex. Like the rest of the brain, the cortex has bilateral symmetry, meaning that its left and right halves are essentially mirror images of one another. On close inspection, however, some areas of cortex differ in size or shape on the left and right sides, and there are interesting differences in function as well. These left-right differences in special-ization are called lateralization, the subject of chapter 12.

There are special terms for describing spatial directions in the brain, analo-gous to points of the compass in navigation, but in three dimensions. Imagine a line that passes through the two ears. Moving along this line from one ear toward the midpoint between the ears is movement in the medial direction. Moving from this midpoint toward either ear is movement in the lateral direction (left lateral or right lateral). Likewise, any movement parallel to this line, anywhere in the brain, is movement along a medial to lateral direction. The medial pre-frontal cortex is, therefore, the region where the two frontal lobes of opposite hemispheres face each other in the middle of the brain. Movement toward the forehead is movement in the rostral or anterior direction; movement toward the back of the head is in the caudal or posterior direction. Movement toward the top of the head is in the dorsal or superior direction; toward the bottom is ventral or inferior.*[9]

Just as diagonal directions can be described by combining directions of the compass (northwest, for example), so can we describe diagonal relationships in the brain with combined terms. The front part of the cingulate cortex is the anterior cingulate cortex, the most rostral and ventral part of which is the ros-troventral part.

The cortex is subdivided into lobes, each of which is named after the bone of the skull that covers it. As its name suggests, the frontal lobe is the most anterior of them, and it contains several important regions discussed in this book. The part of the frontal lobe anterior to the motor cortex is often called prefrontal cor-tex. At the most posterior end of the brain is the occipital lobe, which is mainly concerned with vision. Rostral and dorsal to this lies the parietal lobe, and ven-tral to that, on the left and right sides of the brain, are the temporal lobes. The transition zone between the temporal and parietal lobes, the temporoparietal junction, turns out to be especially relevant in several chapters, as is the insula,

---

* My apologies for the seemingly superfluous synonyms. The confusion only gets worse in the parts of the brain caudal to the thalamus, where the meanings of the synonyms change—essentially because humans walk upright on two legs. Luckily this book deals mainly with forebrain structures, so we can avoid those subtleties of nomenclature. For the full details, see chapter 2 of Blumenfeld (2002).

a deeply infolded cortical region buried and surrounded by the frontal, parietal, and temporal lobes.

The hills and valleys of the highly convoluted cortex have special names as well. An outward bulging hill is called a gyrus and an inward-going valley is a sulcus. The fusiform gyrus on the ventral surface of the temporal lobe contains a region specialized for the visual recognition of faces. More dorsally on the temporal lobe lies the superior temporal sulcus, which is specialized for the visual perception of biological motion, including facial expressions.[10]

At an even finer level of detail lies an aspect of cortical anatomy that is of fundamental importance. When seen in cross section, the cerebral cortex is clearly organized into characteristic layers of neuronal cell bodies and their fibers. This pattern of layering, the types of neurons in each layer, and the pattern of local connections among them are essentially the same in all areas of cortex, whether visual, auditory, motor, or somatosensory in function. As neurophysiologist Vernon Mountcastle noted long ago, this regularity of microanatomy implies that, whatever it does and however it works, the cortex is applying the same computational algorithm to all of these modalities.[11] Jeff Hawkins explores and builds upon this deep insight in his book *On Intelligence*, where he argues that the essential function of the cortex is the melding of memory, expectation, and perception, and that behavior and perception rely on a process of continual cortical prediction based on expectation from past experience.[12] The same theme recurs in this book, especially in chapter 10.

The Methods of Neuroscience

There are several different techniques for brain imaging, but functional magnetic resonance imaging (fMRI) now dominates the field as the least invasive and most widely applicable. Colorful and impressive images from fMRI experiments are now commonplace in popular media. The rate of progress in neuroimaging is so stunning that it is easy to lose sight of just how crude these methods really are in relation to the thing they study.

The finest spatial detail that can be resolved in a typical fMRI experiment—a piece of neural tissue about the size and shape of a grain of rice—contains tens or hundreds of thousands of neurons, each of which has connections to thousands of other neurons. They communicate with each other through electrical and chemical signals that occur in a few milliseconds. The activity of this intricate circuitry, over several seconds or tens of seconds, is grossly averaged and smeared out in space and time to yield a single tiny spot of color (a voxel) in the final fMRI image. The activity value encoded by that color is not even a direct measure of neural activity, but is instead a measure of a change in the oxygen content of the blood in that region, a change that is known to be correlated with neural activity.*[13]

Using fMRI to understand the brain is a bit like trying to understand the software that controls a robot by carefully monitoring temperature changes in the components on its circuit board while the robot does interesting tasks. If the robot is controlled by a single powerful microprocessor chip—the kind of design used in most personal computers—then this "thermal imaging" approach might tell us which tasks are more or less intellectually difficult for the robot but little more. It could never tell us what we really want to know: the detailed logic in the thousands or millions of lines of programming instructions that constitute its software.

If, however, the robot is controlled by many separate microprocessor chips that work in parallel and if each controls only a limited aspect of the robot's sensory input, decision making, or movement, then the thermal imaging technique could tell us something truly meaningful about how the thing really works. The intricate logic of the software would still be out of reach, but at least we could learn which processors are specifically involved in vision, hearing, touch, memory, or specific behaviors. We might also discover that some of these processors cooperate in unexpected and interesting ways.

Fortunately, brains are more like this multiprocessor design than like a PC. A great strength of fMRI is that it can monitor activity in the entire brain in a noninvasive way. When this is done in carefully designed and properly controlled experiments, it can reveal unexpected cooperation among widely separated brain regions during some complex task or when a stimulus evokes an emotional reaction. It can find those parts of the brain specifically involved in some higher cognitive function, like language comprehension, by means of a "contrast" or "difference image": an image of brain activity in response to speech, from which has been subtracted the image of activity in response to a similar but nonlinguistic control sound.[14] It can reveal correlations between changes in neural activity and changes in state of mind.[15] Although its resolution is crude relative to the details of neural circuitry, it has the best spatial resolution of all currently available methods for imaging neural activity.

Despite these advantages, fMRI has some serious practical limitations, especially for the imaging of neural activity in the neonatal brain. An fMRI machine is large, confining, and noisy, and the subject's head must remain motionless for at least several minutes or tens of minutes to acquire meaningful data. These are

---

* For a review of the evidence that the signal measured in fMRI is related to neural activity, see Casey et al. (2002). For an overview of the data processing and statistical analyses commonly used to distinguish meaningful signals from random noise in fMRI experiments, see Smith (2004). There are many good tutorial videos on the web, such as Ward (2020). For a more complete introduction, see Huettel et al. (2009).

serious problems for some of the experiments we would most like to do—like imaging neural activity in a neonate while it cries for its mother or while it is nursing at her breast. Other techniques may be more useful for such experiments, though practitioners of fMRI have recently made great strides in adapting it to the study of alert infants.[16]

The most widely used method for studying neural activity in small infants is electroencephalography (EEG), in which many electrodes on the scalp monitor electrical activity of the brain as reflected in weak currents that pass through the skull and other tissue. Most of the signals powerful enough to be detected in this way come from the synchronous activity of many neurons. Such synchronous activity occurs naturally in some normal brain states and during epileptic seizures, but it can also be elicited by an abrupt sensory stimulus, like a flash of light. Although EEG has much better temporal resolution than fMRI, its spatial resolution is far worse, and it is nearly blind to interesting but unsynchronized neural activity.

Another neuroimaging method suitable for use with infants is magnetoencephalography (MEG). This is similar to EEG in that it mainly detects the synchronous activity of many neurons, but rather than measuring electric fields, it measures the extremely weak magnetic fields induced by the flow of electric current around active neurons. This requires elaborate and expensive equipment, but it has the advantage that magnetic fields, unlike the electric ones measured by EEG, are not significantly distorted by the fluid, bone, and other tissue that surrounds the brain. For this reason MEG can even be used to measure neural responses in an unborn fetus.[17]

Both EEG and MEG are often used to measure neural responses to specific and precisely timed stimulus events, such as a briefly flashed visual image or a brief sound. The response to such a stimulus is called an evoked potential (EP) in the case of EEG or evoked field (EF) in the case of MEG. If the stimulus is given repeatedly at regular intervals, the responses can be averaged to reduce the effects of random noise. If, during such series of repeated stimuli, one of them is different enough from the others that it stands out as an oddball, then a unique response to the oddball may occur. This is called an event-related potential (ERP) in the case of EEG or event-related field (ERF) in the case of MEG. This kind of experiment can yield deep insight into the perceptual abilities and cortical competence of human infants.[18]

A promising new technique for use with infants is near-infrared spectroscopy (NIRS). This method uses infrared light, which can pass through biological tissue better than visible light, to measure local changes in blood oxygenation in the brain. Its temporal resolution is about the same as fMRI, but its spatial resolution is much worse, though potentially better than EEG. Its greatest advantage is that it uses fairly unobtrusive sensors and emitters attached to the scalp, and it can be used even when the subject's head is moving. Its greatest limitation is that it cannot image structures deep in the brain, though this is less of a problem with the smaller brains of infants. There is hope that the poor spatial resolution of NIRS, EEG, and MEG can be improved by simultaneously using combinations of these methods[19] or by processing the data in conjunction with high-resolution structural information from MRI scans of the same subject.[20]

The same physical principle at work in MEG can be used in reverse to stimulate neurons in a noninvasive way. This is transcranial magnetic stimulation (TMS), in which a brief, intense, and localized magnetic field is applied to the head, thus inducing the flow of electric current in the part of the brain nearest the electromagnetic coil. The effects are stronger if the pulse is repeatedly applied over a period of seconds or minutes, in which case the method is called repetitive TMS (rTMS). The pattern of neural activity induced by rTMS is highly abnormal, and its main effect is temporarily to disrupt the normal functioning of the part of the brain to which it is applied. For this reason, it is sometimes described as a virtual lesion: temporary brain damage that can be used safely with human subjects.[21] When applied to the motor cortex, it causes muscle twitching in the part of the body represented by the stimulated area. In the visual cortex it produces illusory spots of light in one part of the visual field and temporarily blocks normal vision there. When applied to the speech cortex, it temporarily renders the subject unable to speak. When applied for tens of minutes, subtle effects persist for tens of minutes after the stimulus is removed, during which time the effect on neural function can be tested.*[22] The rTMS technique is extremely valuable for demonstrating that the normal functioning of some part of the brain is necessary for a specific behavior or cognitive task. Just to give one especially fascinating example, when blind subjects are reading braille, applying rTMS to the cortex that handles the sense of touch in their fingertips interferes with their reading, as you might expect. But for subjects who were blind since early infancy, applying rTMS to the primary visual cortex interferes *even more*—stunning proof that their visual cortex has been rewired for handling other senses and tasks, like reading braille, and that a window of extraordinary neural plasticity is closed beyond early childhood in some parts of the brain.[23]

Unfortunately, no existing neuroimaging method can tell us what we most want to know: the detailed logic of how the brain does its information processing. Other more invasive techniques can reveal some of those details but not in the whole brain at once and normally not in human subjects. Part of the brain can be temporarily anesthetized or permanently destroyed to test the contribution of that brain region to some behavior or sensation. A small part of the brain,

---

* This so-called offline protocol for using rTMS opens up many new experimental possibilities, though one variant of it has recently proved less reliable than originally thought (Boucher et al. 2021). I discuss this issue in chapter 9.

or even a single neuron, can be electrically stimulated with a microelectrode to mimic sensation or to elicit behavior. The chemical signals between neurons can by mimicked or blocked by drugs or neurotoxins. The electrical activity of a single neuron can be monitored with a microelectrode. Some experiments can even monitor the opening and closing of a single ion channel in the membrane of a neuron—the molecular basis of its electrical excitability. Arguably the most powerful new technique, optogenetics, uses genetic manipulations to induce specific neurons of interest to express in their membranes artificial channel proteins that open in response to flashes of light. In this way, scientists can activate or inhibit exactly the neurons they are trying to understand in, for example, a freely behaving mouse.[24] These and a host of other methods are routinely used to get at the fine details of brain function, but the details examined are so fine and the amount of neural tissue involved so tiny that these methods are no more able than fMRI to reveal the computational logic of the thinking, feeling brain.

It might seem that a super-high-resolution imaging method—one that records every action potential of every neuron, every channel opening, and every change in strength of every synaptic connection in the entire brain—would solve the problem of how the brain works. Even if such a thing were possible, the result would just be an overwhelming deluge of incomprehensible data. What neuroscience most needs are specific theories of higher brain function that make predictions we can test with the methods we have.[25] The field is moving rapidly but still has far to go.

Resources

If this terse introduction to the language and concepts of neuroscience seems more confusing than helpful, other resources are available. If you encounter a mysterious neuroanatomical term and want an explanation, search for it in Wikipedia. You'll likely get some nice images showing where it is in the brain along with a simple description of its function. For greater depth and lucid prose, I highly recommend the fascinating book *Brainscapes* by Rebecca Schwarzlose.[26] There are several helpful introductions to neuroscience on the web,[*27] where you can also find David Eagleman's engaging documentary series, *The Brain*.[28]

---

* Good websites devoted to general neuroscience include: Dingman (2021), Dubuc (2021), Krebs and Fejtek (2021), and Rose and Kandel (2017). The paper by Chudler and Bergsman (2014) offers more suggestions.

Appendix B

MEASURING RELIGIOUSNESS

> Religion consists in a set of things which the average
> man thinks he believes and wishes he was certain.—Mark
> Twain[1]

Psychologists have devised many psychometric tests for measuring various
aspects of religiousness or spirituality.[2] I cite such research throughout this book,
but chapter 9 in particular illustrates the importance of using the right test when
trying to work out the contribution to religiousness of some specific part of
the brain. In the first four parts of this appendix, I list items from four tests I
discussed in chapter 9. In the last part, I present a rough draft of my proposed
test for measuring the contributions of the social and neonatal roots of religion
to a person's beliefs.

Persinger's Post-Helmet Poll

After a session in a sensory deprivation chamber wearing the God helmet,
Michael Persinger's subjects were asked to choose from a list of twenty sensations
those most closely matching their experience. Persinger and Healey list eighteen
of the twenty sensations in their publication, noting that two of them were not
related to the purposes of that study.[3] The two omitted were probably related to
the illusory feeling of alien abduction.[4] Here is the published list:

1. Dizzy or odd
2. Sensed presence
3. Tingling sensations
4. Visual images

5. Vibrations through body
6. Detachment or out of body
7. Anger
8. Sadness
9. Thoughts "not from own mind"
10. Ticking sounds
11. Odd smells
12. Fear or terror
13. Odd tastes
14. Felt as if somewhere else
15. Memories from childhood
16. Same thought kept repeating
17. Spinning or vortical experiences
18. Remembered a dream

Urgesi's Implicit Association Test

These are the religious and nonreligious words used in Urgesi's implicit association test. The Italian equivalents (in parentheses) are the ones actually used with the Italian subjects. For more details, see Crescentini et al. (2014).[5]

Religious/spiritual	*Nonreligious/nonspiritual*
Soul (*Anima*)	Agnostic (*Agnostico*)
Believer (*Credente*)	Atheist (*Ateo*)
God (*Dio*)	Carnal (*Carnale*)
Divine (*Divino*)	Cynical (*Cinico*)
Eternal (*Eterno*)	Concreteness (*Concretezza*)
Faith (*Fede*)	Body (*Corpo*)
Angelic (*Angelico*)	Physical (*Fisico*)
Inner (*Interiore*)	Irreligious (*Irreligioso*)
Meditation (*Meditazione*)	Limited (*Limitato*)
Omnipotent (*Onnipotente*)	Logical (*Logico*)
Religious (*Religioso*)	Material (*Materiale*)
Sacred (*Sacro*)	Objective (*Oggettivo*)
Saint (*Santo*)	Profane (*Profano*)
Supernatural (*Soprannaturale*)	Tangible (*Tangibile*)
Spirit (*Spirito*)	Earthly (*Terreno*)

Grafman's Personal Relationship with God Scale

Here is the complete psychometric test used by Grafman's group to quantify a personal relationship with God.[6] Subjects responded to items on a Likert scale.

God does not notice me.*
God lifts me up.
I am never really sure that God is really listening to me.*
God doesn't feel very personal to me.*
I can talk to God on an intimate basis.
I get no feeling of closeness to God, even in prayer.*
I feel that God knows me by name.
God never reached out to me.*
I feel warm inside when I pray.
God does not answer when I call.*
Prayer is very meaningful to me.
I prefer to face my problems without prayer.*
God tells me what he wants from me.
I do not think about God very often.*
In my religious experience, I felt that God or a higher power communicated with me.
I find a precise meaning was communicated to me through my religious experience.
My religious experience filled me with strong emotion.

*Items reverse-scored

Sample of Items from Spiritual Acceptance Test

The Self-Transcendence test is a subscale of a much larger personality test, the Temperament and Character Inventory.[7] To convey what the "spiritual acceptance" part of it is testing, here are five of its items:

I seem to have a "sixth sense" that sometimes allows me to know what is going to happen.
I sometimes feel a spiritual connection to other people that I cannot explain in words.

Sometimes I have felt my life being guided by a spiritual force greater than any human being.

I have had personal experiences in which I felt in contact with a divine and wonderful spiritual power.

I am fascinated by the many things in life that cannot be scientifically explained.

Social and Neonatal Dimensions of Religion

This is a rough draft of a proposed psychometric test, the purpose of which is to measure the relative extent to which a person's spiritual or religious beliefs derive from the two biological roots discussed in chapter 1 of this book and in chapter 7 of *The Illusion of God's Presence*, wherein an even rougher version appeared as an appendix. In this version, I have corrected a few ambiguities and added more items, but the test still needs validation and refinement.

There are two subscales that separately represent the neonatal and social roots of religion. For each item the subject is asked to indicate on a Likert scale his or her level of agreement with the statement. Statements that imply belief in the existence of God should have, in addition to the Likert scale, an alternative response labeled "Not applicable or don't believe in God," so that atheists and agnostics can take the test in a meaningful way without confusion; that response is scored the same as maximal disagreement for items written in the positive sense. As written here, all items are worded in a positive sense. In the final version, half the items in each subscale should be reworded in a negative sense and reverse scored, to eliminate acquiescence bias.

In the neonatal subscale, the dominant themes are the loving and dependable nature of God; God as a source of nurturance, safety, and protection; God's strength and the believer's weakness; God as parent and the believer as child; God's responsiveness to prayer; and the certainty of God's existence that comes from the feeling of his presence. In the social subscale, the dominant themes are the need to belong; the need for sacrifice and other public demonstrations of belief; affinity for members of the in-group and distrust of out-groups; obedience and submission to God's will; the appeal of group ritual; the equation of moral virtue with cleanliness and sin with contamination; and images of a fearsome and judgmental God.

I offer this test as an empirical prediction. If what I call the social and neonatal roots of religion really are the dominant sources of religious emotion, then the social and neonatal items should cohere as two distinct factors in a factor analysis when the test is given to many subjects of diverse religious backgrounds.

For each subject, the raw sums of the Likert scores could be normalized by the highest possible scores to yield two numbers, N and S, that vary from

0.0 (no belief) to 1.0 (maximal belief) for the neonatal and social dimensions, respectively. These two values map a person into a two-dimensional space of religiousness.

For some experimental questions, the relative contribution of the neonatal and social roots to a subject's beliefs may be of greater interest than the scores themselves. In that case, the normalized ratio $(N-S)/(N+S)$ might be useful, where N and S are the normalized neonatal and social scores, respectively. This normalized ratio varies from -1.0, for someone whose religiousness is purely social, to $+1.0$, for someone whose religiousness is purely neonatal. I predict, for example, that this ratio will be positive and significantly closer to 1.0 for elite religious scientists than for the general population.

NEONATAL ROOT

1. God loves all of his earthly children unconditionally.
2. God provides for all of my needs.
3. I know that God loves me.
4. I believe in the power of prayer.
5. I sometimes think of God more as a loving mother than as a heavenly father.
6. God always answers my prayers.
7. In times of hardship I turn to God in prayer.
8. I know God is real, because I have felt his presence.
9. God looks after and protects my loved ones.
10. God lives in my heart.
11. God knows what is best for us.
12. God will never abandon me.
13. We are all God's children.
14. God is love.
15. I could not get by in life without God's help.
16. I feel safe in God's arms.
17. There are many paths to God.
18. We are all part of a loving spirit.
19. God gives me strength in adversity.
20. God is everywhere in nature.
21. I would feel lost without God.
22. Nothing could be worse than separation from God.

SOCIAL ROOT

1. God expects us to make sacrifices when we follow the life he has chosen for us.
2. We should all obey God and submit to his will.
3. God rewards the righteous and punishes the wicked.
4. I enjoy fellowship with others who worship God as I do.
5. Hell is a real place where sinners and unbelievers are punished.
6. Holy Scripture is the ultimate source of truth.
7. I would rather my child die than lose his/her faith in God.
8. At times I have been fearful of God's wrath.
9. Teachers should lead prayer in public schools.
10. Anyone who does not believe in God cannot be trusted.
11. I feel a strong sense of belonging in my religious community.
12. Atheists should not be allowed to teach in public schools.
13. There is only one path to religious salvation.
14. Attending worship services is important to me.
15. People who don't contribute money at religious services should stop coming.
16. So-called natural disasters are really punishments from God.
17. God hates sin.
18. I would gladly give my life to defend my faith.
19. God washes away your sins if you repent.
20. Sin is filthy.
21. You can always trust godly people.
22. It feels good to praise God.

Notes

Epigraphs

1. Barker D (2008) *Godless: How an Evangelical Preacher Became One of America's Leading Atheists.* Ulysses: Berkeley, CA, p 38.
2. Campbell J (1959) [1991] *The Masks of God: Primitive Mythology.* Penguin/Arkana: New York, p 42.

Prologue

1. Segalowitz SJ, Davies PL (2004) Charting the Maturation of the Frontal Lobe: An Electrophysiological Strategy. *Brain Cogn* 55(1):116–33. Johnson MH, Griffin R, Csibra G, Halit H, Farroni T, de Haan M, Tucker LA, Baron-Cohen S, Richards J (2005) The Emergence of the Social Brain Network: Evidence from Typical and Atypical Development. *Dev Psychopathol* 17(3):599–619.
2. Ainsworth MD (1985) Attachments across the Life Span. *Bull N Y Acad Med* 61(9):792–812. Fillion TJ, Blass EM (1986) Infantile Experience with Suckling Odors Determines Adult Sexual Behavior in Male Rats. *Science* 231(4739):729–31. Melo AI, Lovic V, Gonzalez A, Madden M, Sinopoli K, Fleming AS (2006) Maternal and Littermate Deprivation Disrupts Maternal Behavior and Social-Learning of Food Preference in Adulthood: Tactile Stimulation, Nest Odor, and Social Rearing Prevent These Effects. *Dev Psychobiol* 48(3):209–19. Stevens HE, Leckman JF, Coplan JD, Suomi SJ (2009) Risk and Resilience: Early Manipulation of Macaque Social Experience and Persistent Behavioral and Neurophysiological Outcomes. *J Am Acad Child Adolesc Psychiatry* 48(2):114–27. Weaver A, de Waal FB (2003) The Mother-Offspring Relationship as a Template in Social Development: Reconciliation in Captive Brown Capuchins *(Cebus apella). J Comp Psychol* 117(1):101–10. Weaver A, Richardson R, Worlein J, De Waal F, Laudenslager M (2004) Response to Social Challenge in Young Bonnet *(Macaca radiata)*

and Pigtail *(Macaca nemestrina)* Macaques Is Related to Early Maternal Experiences. *Am J Primatol* 62(4):243–59.

3. Lee A (2011) The Apologist's Handbook. www.patheos.com/blogs/daylightathe ism/essays/the-apologists-handbook/ (accessed 27 June 2021).

4. Kuhn RL (2014) If God Knows the Future, What Is Free Will? www.closertotruth .com/series/if-god-knows-the-future-what-free-will (accessed 27 June 2021).

5. Wathey JC (2019) Electoral Reform Redux. www.watheyresearch.com/electoral -reform-redux/ (accessed 9 September 2021).

Chapter 1

1. Carlin G (1999) *You Are All Diseased.* Performance at the Beacon Theater, New York, 6 February 1999. MPI Home Video: Chicago, IL.

2. Yeh TD (2006) The Way to Peace: A Buddhist Perspective. *Int J Peace Stud* 11(1):91–112.

3. Beech H (2013) The Face of Buddhist Terror. *Time* [Jul 1] http://content.time .com/time/subscriber/article/0,33009,2146000-3,00.html (accessed 10 June 2021).

4. Koran Surah 15:49–50.

5. Koenig LB, McGue M, Krueger RF, Bouchard TJ Jr (2005) Genetic and Environmental Influences on Religiousness: Findings for Retrospective and Current Religiousness Ratings. *J Pers* 73(2):471–88.

6. Dawkins R (1976) *The Selfish Gene.* Oxford University Press: Oxford.

7. Ratnieks FL, Helanterä (2009) The Evolution of Extreme Altruism and Inequality in Insect Societies. *Philos Trans R Soc Lond B Biol Sci* 364(1533):3169–79.

8. Schino G, Aureli F (2008) Grooming Reciprocation among Female Primates: A Meta-Analysis. *Biol Lett* 4(1):9–11.

9. Dawkins (1976) chapter 10; Trivers RL (1971) The Evolution of Reciprocal Altruism. *Q Rev Biol* 46:35–57.

10. Haidt J (2012) *The Righteous Mind: Why Good People Are Divided by Politics and Religion.* Pantheon Books: New York.

11. Bloom P (2004) *Descartes' Baby: How the Science of Child Development Explains What Makes Us Human.* Basic Books: New York, chapter 4.

12. Garcia HA (2015) *Alpha God: The Psychology of Religious Violence and Oppression.* Prometheus Books: Amherst, NY. Irons W (2001) Religion as a Hard-to-Fake Sign of Commitment. In: Nesse RM [ed] *Evolution and the Capacity for Commitment.* Russell Sage: New York, pp 292–309. Norenzayan A (2013) *Big Gods: How Religion Transformed Cooperation and Conflict.* Princeton University Press: Princeton, NJ.

13. Shaw D (2014) *Traumatic Narcissism: Relational Systems of Subjugation.* Routledge: New York, chapter 3.

14. Russ E, Shedler J, Bradley R, Westen D (2008) Refining the Construct of Narcissistic Personality Disorder: Diagnostic Criteria and Subtypes. *Am J Psychiatry* 165(11):1473–81.

15. Livesley WJ, Jang KL, Jackson DN, Vemon PA (1993) Genetic and Environmental Contributions to Dimensions of Personality Disorder. *Am J Psychiatry* 150:1826–31.

16. Stinson FS, Dawson DA, Goldstein RB, Chou SP, Huang B, Smith SM, Ruan WJ, Pulay AJ, Saha TD, Pickering RP, Grant BF (2008) Prevalence, Correlates, Disability, and Comorbidity of DSM-IV Narcissistic Personality Disorder: Results from the Wave 2 National Epidemiologic Survey on Alcohol and Related Conditions. *J Clin Psychiatry* 69(7):1033–45.

17. Saha S, Chant D, Welham J, McGrath J (2005) A Systematic Review of the Prevalence of Schizophrenia. *PLoS Med* 2(5):e141.

18. For the classic exposition of frequency-dependent selection, see: Maynard Smith J (1982) *Evolution and the Theory of Games.* Cambridge University Press: Cambridge, UK. The hypothesis I suggest here for the evolutionary forces behind narcissistic personality disorder is similar to one proposed to explain sociopathic behavior in humans: Mealey L (1995) The Sociobiology of Sociopathy: An Integrated Evolutionary Model. *Behav Brain Sci* 18(3):523–99. Other evolutionary dynamics such as sexual selection may also have shaped the narcissistic personality: Baselice KF, Thomson JA (2018) An Evolutionary Hypothesis on Arrogance. In: Akhtar S, Smolen A [eds] *Arrogance: Developmental, Cultural, and Clinical Realms.* Routledge: New York, pp 25–45.

19. Dawkins R (1980) *Good Strategy or Evolutionarily Stable Strategy?* In: Barlow GW, Silverberg J [eds] *Sociobiology: Beyond Nature/Nurture? Reports, Definitions and Debate.* Westview Press: Boulder, CO, pp 331–67. Wathey JC (2016) *The Illusion of God's Presence: The Biological Origins of Spiritual Longing.* Prometheus Books: Amherst, New York, pp 168–69.

20. Need to belong: Boomsma DI, Willemsen G, Dolan CV, Hawkley LC, Cacioppo JT (2005) Genetic and Environmental Contributions to Loneliness in Adults: The Netherlands Twin Register Study. *Behav Genet* 35(6):745–52. Fear of ostracism: Stein MB, Jang KL, Livesley WJ (2002) Heritability of Social Anxiety-Related Concerns and Personality Characteristics: A Twin Study. *J NervMent Dis* 190(4):219–24. Willingness to sacrifice: Huml AM, Thornton JD, Figueroa M, Cain K, Dolata J, Scott K, Sullivan C, Sehgal AR (2019) Concordance of Organ Donation and Other Altruistic Behaviors among Twins. *Prog Transpl* 29(3):225–29.

21. Van Vugt M, Hogan R, Kaiser RB (2008) Leadership, Followership, and Evolution: Some Lessons from the Past. *Am Psychol* 63(3):182–96.

22. Pruetz JD, Ontl KB, Cleaveland E, Lindshield S, Marshack J, Erin G, Wessling EG (2017) Intragroup Lethal Aggression in West African Chimpanzees (*Pan troglodytes* Verus): Inferred Killing of a Former Alpha Male at Fongoli, Senegal. *Int J Primatol* 38:31–57.

23. Loomis RN (1999) Testimony of Ronald N. Loomis to the Maryland Cult Task Force in Support of House Joint Resolution 22. https://web.archive.org/web/20040818214228/http://religiousmovements.lib.virginia.edu/cultsect/mdtaskforce/loomis_testimony.htm (accessed 25 June 2021).

24. Moore R (2009) *Understanding Jonestown and Peoples Temple.* Praeger: Westport, CT, pp 16–17, 87–102.

25. Larson EJ, Witham L (1998) Leading Scientists Still Reject God. *Nature* 394:313. Stirrat M, Cornwell RE (2013) Eminent Scientists Reject the Supernatural: A Survey of the Fellows of the Royal Society. *Evol Educ Outreach* 6:33, doi:10.1186/1936-6434-6-33.

26. See minutes 12–15 of: Tyson ND (2006) *The Perimeter of Ignorance.* Beyond Belief: Science, Reason, Religion & Survival. Salk Institute, La Jolla, CA [5 November] www.youtube.com/watch?v=N7rR8stuQfk (accessed 11 September 2021).

27. Hay D (2013) Zoology and Religion: The Work of Alister Hardy. *Metanexus* [3 June]. https://metanexus.net/zoology-and-religion-work-alister-hardy/ (accessed 15 June 2021).

28. For example, see: Hardy AC (1966) *The Divine Flame: An Essay Towards a Natural History of Religion.* Collins: London, p 218.

29. Hardy AC (1985) The Significance of Religious Experience (Templeton Prize Acceptance Speech). *RERC Second Ser Occas Paper 12* https://core.ac.uk/download/pdf/96773534.pdf (accessed 11 September 2021).

30. Hardy AC (1979) *The Spiritual Nature of Man: A Study of Contemporary Religious Experience.* Clarendon: Oxford, pp 26–29.

31. Beardsworth T (1977) *A Sense of Presence: The Phenomenology of Certain Kinds of Visionary and Ecstatic Experience, Based on a Thousand Contemporary First-Hand Accounts.* Religious Experience Research Unit: Oxford, UK, p 121.

32. Hagerty BB (2009) *Fingerprints of God: The Search for the Science of Spirituality.* Riverhead: New York, p 12.

33. Collins FS (2006) *The Language of God: A Scientist Presents Evidence for Belief.* Free Press: New York, p 38. Pascal B (1669) [1958] *Pascal's Pensées.* Dutton: New York, 7.425.

34. Wathey (2016) pp 21–23.

35. Scott J (1989) *The Great Migration.* Rodale Press: Emmaus, PA, pp 15–19.

36. Lutz P, Musick J [eds] (1996) *The Biology of Sea Turtles.* CRC Press: Boca Raton, FL, chapters 3 and 5.

37. Tinbergen N, Perdeck AC (1950) On the Stimulus Situation Releasing the Begging Response in the Newly Hatched Herring Gull Chick (*Larus argentatus argentatus* Pont.). *Behaviour* 3:1–39.

38. Gould SJ, Lewontin R (1979) The Spandrels of San Marco and the Panglossian Paradigm: A Critique of the Adaptationist Program. *Proc R Soc Lond B Biol Sci* 205:581–98. Sosis R (2009) The Adaptationist-Byproduct Debate on the Evolution of Religion: Five Misunderstandings of the Adaptationist Program. *J Cogn Culture* 9:315–32.

39. John 3:3, Mark 10:15, Matt 18:3–4.

40. Kirkpatrick LA (2005) *Attachment, Evolution, and the Psychology of Religion.* Guilford Press: New York, p 60.

41. McDaniel J (2004) *Offering Flowers, Feeding Skulls: Popular Goddess Worship in West Bengal.* Oxford University Press: New York.

42. For more examples and additional interpretations of such role reversals, see Wathey (2016) pp 73–74, 101–3, 125–26, 152, 156–57, 183, 257–58.

43. Fadiman J, Frager R [eds] (1997) *Essential Sufism.* Harper San Francisco: San Francisco, CA, p 99.

44. Eck DL (1993) *Encountering God: A Spiritual Journey from Bozeman to Banaras.* Beacon Press: Boston, pp 104–5.

45. Ball A, Hinojosa D (2006) *Holy Infant Jesus: Stories, Devotions, and Pictures of the Holy Child around the World.* Crossroad Publishing: New York. Pope Benedict

XVI (2008) Solemnity of the Nativity of the Lord: Homily of His Holiness Benedict XVI. Libreria Editrice Vaticana: Rome [25 December] http://w2.vatican.va/content/benedict-xvi/en/homilies/2008/documents/hf_ben-xvi_hom_20081224_christmas.html (accessed 13 December 2021).

46. Hardy (1966) p 9.

47. Hardy (1966) p 52.

48. Dawkins R (1986) *The Blind Watchmaker*. WW Norton: New York.

49. Hay (2013).

50. Pylyshyn Z (1999) Is Vision Continuous with Cognition? The Case for Cognitive Impenetrability of Visual Perception. *Behav Brain Sci* 22:341–65; discussion 366–423.

51. Wathey JC (2018) The Mystery of Elite Religious Scientists. *Skeptic* 23(3):10–13.

52. Stirrat and Cornwell (2013).

53. Wathey (2016) p 63.

54. Freud S (1913) [1950] *Totem and Taboo: Some Points of Agreement between the Mental Lives of Savages and Neurotics*. Norton: New York, pp 146–55. Freud S (1927) [1964] *The Future of an Illusion*. Anchor Books: Garden City, NY, chapter 4.

55. Wathey (2016) pp 121–22.

56. Campbell J (1959) [1991] *The Masks of God: Primitive Mythology*. Penguin/Arkana: New York, chapter 1. Wathey (2016) chapter 4.

57. Granqvist P (2020) *Attachment in Religion and Spirituality: A Wider View*. Guilford Press: New York. Granqvist P, Kirkpatrick LA (2008) Attachment and Religious Representations and Behavior. In: Cassidy IJ, Shaver PR [eds] *Handbook of Attachment: Theory, Research, and Clinical Applications (Second Edition)*. Guilford: New York, pp 906–33. Kirkpatrick (2005).

58. Ainsworth MD, Blehar MC, Waters E, Wall S (1978) *Patterns of Attachment: A Psychological Study of the Strange Situation*. Erlbaum: Hillsdale, NJ. Bowlby J (1969) *Attachment and Loss*. Basic Books: New York.

59. Bartholomew K, Shaver PR (1998) Methods of Assessing Adult Attachment: Do They Converge? In: Simpson JA, Rholes WS [eds] *Attachment Theory and Close Relationships*. Guilford: New York, pp 25–45.

60. Kirkpatrick (2005) chapters 3 and 4.

61. Bokhorst CL, Bakermans-Kranenburg MJ, Fearon RM, van IJzendoorn MH, Fonagy P, Schuengel C (2003) The Importance of Shared Environment in Mother-Infant Attachment Security: A Behavioral Genetic Study. *Child Dev* 74(6):1769–82. Grossmann KE, Grossmann K, Waters E [eds] (2005) *Attachment from Infancy to Adulthood: The Major Longitudinal Studies*. Guilford: New York.

62. After fig. 5.4 of Kirkpatrick (2005) p 113 (annotations added). The table qualitatively summarizes statistical data from: Kirkpatrick LA, Shaver PR (1990) Attachment Theory and Religion: Childhood Attachments, Religious Beliefs, and Conversion. *J Sci Study Relig* 29(3):315–34. The pattern shown in the table has been reproduced in several subsequent studies (Granqvist and Kirkpatrick 2008).

63. This relationship between attachment security and religiosity accounts for individual differences in religious belief only to a small or moderate degree, though the effects are statistically significant and have now been reproduced in several independent studies and in a variety of cultural settings. See: Granqvist P, Kirkpatrick LA (2004) Religious

Conversion and Perceived Childhood Attachment: A Meta-Analysis. *Int J Psychol Relig* 14:223–50, table 1. Kirkpatrick (2005) pp 112–14. Granqvist and Kirkpatrick (2008) pp 917–18.

64. Granqvist and Kirkpatrick (2004) tables 2 and 3.

65. Granqvist and Kirkpatrick (2004) table 4.

66. Kirkpatrick (2005) chapter 5.

67. Kirkpatrick (2005) chapter 6.

68. Ainsworth MD (1985) Attachments across the Life Span. *Bull N Y Acad Med* 61(9):792–812. Passman RH, Halonen JS (1979) A Developmental Survey of Young Children's Attachments to Inanimate Objects. *J Genet Psychol* 134:165–78.

69. James W (1902) [1922] *The Varieties of Religious Experience: A Study in Human Nature.* Longmans, Green: New York, lectures 9 and 10. Starbuck ED (1911) *The Psychology of Religion: An Empirical Study of the Growth of Religious Consciousness.* Walter Scott: London, chapter 8.

70. Ullman C (1982) Cognitive and Emotional Antecedents of Religious Conversion. *J Pers Soc Psychol* 43:183–92.

71. Kirkpatrick (2005) chapters 4, 6, and 8.

72. See also Wathey (2016) chapter 8.

73. Kirkpatrick (2005) p 211, original emphasis.

74. See chapter 6 of Wathey (2016) for a more detailed discussion of how my hypothesis meshes with the attachment theory of religion.

75. Oberholtzer G (2014) Out of the Light and into the Darkness: Managing the Impacts of Artificial Light on Sea Turtles. www.oceanhealthindex.org/news/Out_Of_The_Light_And_Into_The_Darkness (accessed 17 June 2021).

76. Craig W (1917) Appetites and Aversions as Constituents of Instincts. *Proc Natl Acad Sci USA* 3(12):685–88. For recent and more specialized reviews, see: Balthazart J, Ball GF (2007) Topography in the Preoptic Region: Differential Regulation of Appetitive and Consummatory Male Sexual Behaviors. *Front Neuroendocrinol* 28(4):161–78. Berridge KC (2009) "Liking" and "Wanting" Food Rewards: Brain Substrates and Roles in Eating Disorders. *Physiol Behav* 97(5):537–50.

77. Levine S, Wiener SG (1988) Psychoendocrine Aspects of Mother-Infant Relationships in Nonhuman Primates. *Psychoneuroendocrinology* 13(1–2):143–54.

78. Hennessy MB, Miller EE, Shair HN (2006) Brief Exposure to the Biological Mother "Potentiates" the Isolation Behavior of Precocial Guinea Pig Pups. *Dev Psychobiol* 48(8):653–59. Moles A, Kieffer BL, D'Amato FR (2004) Deficit in Attachment Behavior in Mice Lacking the Mu-Opioid Receptor Gene. *Science* 304(5679):1983–86. Robison WT, Myers MM, Hofer MA, Shair HN, Welch MG (2016) Prairie Vole Pups Show Potentiated Isolation-Induced Vocalizations Following Isolation from Their Mother, but Not Their Father. *Dev Psychobiol* 58(6):687–99. Shair HN (2007) Acquisition and Expression of a Socially Mediated Separation Response. *Behav Brain Res* 182(2):180–92.

79. Permission from the Canadian Psychological Association Inc. to reprint image from "Figure 1. 40 closure faces arranged with the most difficult 5 at upper left, next most difficult 5 at upper right, and, so on down to easiest 5 at lower right." *Canadian*

Journal of Psychology, Volume 11, No. 4, 1 image from figure 1, p 220, Copyright © 1957 by the Canadian Psychological Association Inc.

80. Resnick B (2020) "Reality" Is Constructed by Your Brain. Here's What That Means, and Why It Matters. *Vox* [22 June] www.vox.com/science-and-health/20978285/optical-illusion-science-humility-reality-polarization (accessed 18 June 2021).

81. Mooney CM (1957) Age in the Development of Closure Ability in Children. *Can J Psychol* 11(4):219–26.

82. Zhu Q, Song Y, Hu S, Li X, Tian M, Zhen Z, Dong Q, Kanwisher N, Liu J (2010) Heritability of the Specific Cognitive Ability of Face Perception. *Curr Biol* 20(2):137–42.

83. Slater A, Kirby R (1998) Innate and Learned Perceptual Abilities in the Newborn Infant. *Exp Brain Res* 123(1–2):90–94.

84. Cocker KD, Moseley MJ, Bissenden JG, Fielder AR (1994) Visual Acuity and Pupillary Responses to Spatial Structure in Infants. *Invest Ophthalmol Vis Sci* 35(5):2620–25.

85. Brown AM, Lindsey DT (2009) Contrast Insensitivity: The Critical Immaturity in Infant Visual Performance. *Optom Vis Sci* 86(6):572–76.

86. Adams RJ, Courage ML (1998) Human Newborn Color Vision: Measurement with Chromatic Stimuli Varying in Excitation Purity. *J Exp Child Psychol* 68(1):22–34.

87. Held R, Birch E, Gwiazda J (1980) Stereoacuity of Human Infants. *Proc Natl Acad Sci USA* 77(9):5572–74.

88. Hrdy SB (2009) *Mothers and Others: The Evolutionary Origins of Mutual Understanding.* Harvard University Press: Cambridge, MA, chapter 3.

89. Konner M (2005) Hunter-Gatherer Infancy and Childhood: The !Kung and Others. In: Hewlett BS, Lamb ME [eds] *Hunter-Gatherer Childhoods: Evolutionary, Developmental and Cultural Perpectives.* Aldine/Transaction: New Brunswick, NJ, pp 19–64.

90. Hrdy (2009) chapter 9.

91. Hrdy SB (1999) *Mother Nature: A History of Mothers, Infants, and Natural Selection.* Pantheon: New York, chapters 12 and 14.

92. Hrdy (1999) chapter 21. Soltis J (2004) The Signal Functions of Early Infant Crying. *Behav Brain Sci* 27(4):443–58.

93. Falk D (2004) Prelinguistic Evolution in Early Hominins: Whence Motherese? *Behav Brain Sci* 27(4):491–503; discussion 503–83. Hrdy (2009) chapter 4. Lavelli M, Fogel A (2002) Developmental Changes in Mother-Infant Face-to-Face Communication: Birth to 3 Months. *Dev Psychol* 38(2):288–305.

94. Hrdy (2009).

Chapter 2

1. Campbell J (1959) [1991] *The Masks of God: Primitive Mythology.* Penguin/Arkana: New York, p 35.

2. Hrdy SB (2009) *Mothers and Others: The Evolutionary Origins of Mutual Understanding.* Harvard University Press: Cambridge, MA, p 60.

3. Pinker S (2002) *The Blank Slate: The Modern Denial of Human Nature.* Penguin Books: New York.

4. For a more detailed summary, see chapter 11 in Wathey JC (2016) *The Illusion of God's Presence: The Biological Origins of Spiritual Longing.* Prometheus Books: Amherst, NY.

5. Lai CH, Chan YS (2002) Development of the Vestibular System. *Neuroembryology Aging* 1(2):61–71.

6. Clark DL, Cordero L, Goss KC, Manos D (1989) Effects of Rocking on Neuromuscular Development in the Premature. *Biol Neonate* 56(6):306–14.

7. Cioni G, Favilla M, Ghelarducci B, La Noce A (1984) Development of the Dynamic Characteristics of the Horizontal Vestibulo-Ocular Reflex in Infancy. *Neuropediatrics* 15(3):125–30. Clark DL, Kreutzberg JR, Chee FK (1977) Vestibular Stimulation Influence on Motor Development in Infants. *Science* 196(4295):1228–29.

8. Howard CR, Lanphear N, Lanphear BP, Eberly S, Lawrence RA (2006) Parental Responses to Infant Crying and Colic: The Effect on Breastfeeding Duration. *Breastfeed Med* 1(3):146–55.

9. Korner AF, Thoman EB (1972) The Relative Efficacy of Contact and Vestibular-Proprioceptive Stimulation in Soothing Neonates. *Child Dev* 43(2):443–53.

10. Beckett C, Bredenkamp D, Castle J, Groothues C, O'Connor TG, Rutter M, English and Romanian Adoptees (ERA) Study Team (2002) Behavior Patterns Associated with Institutional Deprivation: A Study of Children Adopted from Romania. *J Dev Behav Pediatr* 23(5):297–303. Harlow HF, Zimmermann RR (1959) Affectional Responses in the Infant Monkey; Orphaned Baby Monkeys Develop a Strong and Persistent Attachment to Inanimate Surrogate Mothers. *Science* 130(3373):421–32.

11. Mason WA, Berkson G (1975) Effects of Maternal Mobility on the Development of Rocking and Other Behaviors in Rhesus Monkeys: A Study with Artificial Mothers. *Dev Psychobiol* 8(3):197–211.

12. Gregg CL, Haffner ME, Korner AF (1976) The Relative Efficacy of Vestibular-Proprioceptive Stimulation and the Upright Position in Enhancing Visual Pursuit in Neonates. *Child Dev* 47(2):309–14.

13. Field TM, Woodson R, Greenberg R, Cohen D (1982) Discrimination and Imitation of Facial Expression by Neonates. *Science* 218:180–81.

14. Varendi H, Porter RH, Winberg J (1994) Does the Newborn Baby Find the Nipple by Smell? *Lancet* 344(8928):989–90.

15. Varendi H, Porter RH (2001) Breast Odour as the Only Maternal Stimulus Elicits Crawling towards the Odour Source. *Acta Paediatr* 90(4):372–75.

16. Doucet S, Soussignan R, Sagot P, Schaal B (2009) The Secretion of Areolar (Montgomery's) Glands from Lactating Women Elicits Selective, Unconditional Responses in Neonates. *PLoS One* 4(10):e7579.

17. Marlier L, Schaal B (2005) Human Newborns Prefer Human Milk: Conspecific Milk Odor Is Attractive without Postnatal Exposure. *Child Dev* 76(1):155–68.

18. Romantshik O, Porter RH, Tillmann V, Varendi H (2007) Preliminary Evidence of a Sensitive Period for Olfactory Learning by Human Newborns. *Acta Paediatr* 96(3):372–76.

19. Varendi H, Porter RH, Winberg J (2002) The Effect of Labor on Olfactory Exposure Learning within the First Postnatal Hour. *Behav Neurosci* 116(2):206–11.

20. Delaunay-El Allam M, Marlier L, Schaal B (2006) Learning at the Breast: Preference Formation for an Artificial Scent and Its Attraction against the Odor of Maternal Milk. *Infant Behav Dev* 29(3):308–21.

21. Delaunay-El Allam M, Soussignan R, Patris B, Marlier L, Schaal B (2010) Long-Lasting Memory for an Odor Acquired at the Mother's Breast. *Dev Sci* 13(6):849–63.

22. Sullivan RM, Taborsky-Barba S, Mendoza R, Itano A, Leon M, Cotman CW, Payne TF, Lott I (1991) Olfactory Classical Conditioning in Neonates. *Pediatrics* 87(4):511–18.

23. Korner AF, Thoman EB (1972) The Relative Efficacy of Contact and Vestibular-Proprioceptive Stimulation in Soothing Neonates. *Child Dev* 43(2):443–53.

24. Im H, Kim E (2009) Effect of Yakson and Gentle Human Touch versus Usual Care on Urine Stress Hormones and Behaviors in Preterm Infants: A Quasi-Experimental Study. *Int J Nurs Stud* 46(4):450–58.

25. Shibata M, Fuchino Y, Naoi N, Kohno S, Kawai M, Okanoya K, Myowa-Yamakoshi M (2012) Broad Cortical Activation in Response to Tactile Stimulation in Newborns. *Neuroreport* 23(6):373–77.

26. Fernald A (1991) Prosody in Speech to Children: Prelinguistic and Linguistic Functions. *Ann Child Dev* 8:43–80.

27. Luef EM, Liebal K (2012) Infant-Directed Communication in Lowland Gorillas *(Gorilla gorilla)*: Do Older Animals Scaffold Communicative Competence in Infants? *Am J Primatol* 74(9):841–52. Rendall D (2003) Acoustic Correlates of Caller Identity and Affect Intensity in the Vowel-Like Grunt Vocalizations of Baboons. *J Acoust Soc Am* 113(6):3390–402. Whitham JC, Gerald MS, Maestripieri D (2007) Intended Receivers and Functional Significance of Grunt and Girney Vocalizations in Free-Ranging Female Rhesus Macaques. *Ethology* 113:862–74.

28. Bryant GA, Barrett HC (2007) Recognizing Intentions in Infant-Directed Speech: Evidence for Universals. *Psychol Sci* 18(8):746–51. Grieser DL, Kuhl PK (1988) Maternal Speech to Infants in a Tonal Language: Support for Universal Prosodic Features in Motherese. *Dev Psychol* 24(1):14–20.

29. Masataka N (1998) Perception of Motherese in Japanese Sign Language by 6-Month-Old Hearing Infants. *Dev Psychol* 34(2):241–46.

30. Gordon I, Zagoory-Sharon O, Leckman JF, Feldman R (2010) Oxytocin and the Development of Parenting in Humans. *Biol Psychiatry* 68(4):377–82.

31. Cooper RP, Aslin RN (1990) Preference for Infant-Directed Speech in the First Month after Birth. *Child Dev* 61(5):1584–95.

32. DeCasper AJ, Fifer WP (1980) Of Human Bonding: Newborns Prefer Their Mothers' Voices. *Science* 208:1174–76. Kisilevsky BS, Hains SM, Lee K, Xie X, Huang H, Ye HH, Zhang K, Wang Z (2003) Effects of Experience on Fetal Voice Recognition. *Psychol Sci* 14(3):220–24. Moon CM, Fifer WP (2000) Evidence of Transnatal Auditory Learning. *J Perinatol* 20(8 Pt 2):S36–43.

33. Cooper RP, Aslin RN (1994) Developmental Differences in Infant Attention to the Spectral Properties of Infant-Directed Speech. *Child Dev* 65(6):1663–77.

34. Dehaene-Lambertz G, Pena M (2001) Electrophysiological Evidence for Automatic Phonetic Processing in Neonates. *Neuroreport* 12(14):3155–58. Perani D, Saccuman MC, Scifo P, Anwander A, Spada D, Baldoli C, Poloniato A, Lohmann G, Friederici AD (2011) Neural Language Networks at Birth. *Proc Natl Acad Sci USA* 108(38):16056–61. Reeb-Sutherland BC, Fifer WP, Byrd DL, Hammock EA, Levitt P, Fox NA (2011) One-Month-Old Human Infants Learn about the Social World While They Sleep. *Dev Sci* 14(5):1134–41.

35. Morton J, Johnson MH (1991) CONSPEC and CONLERN: A Two-Process Theory of Infant Face Recognition. *Psychol Rev* 98(2):164–81. Simion F, Di Giorgio E, Leo I, Bardi L (2011) The Processing of Social Stimuli in Early Infancy: From Faces to Biological Motion Perception. *Prog Brain Res* 189:173–93.

36. Yin RK (1969) Looking at Upside-Down Faces. *J Exp Psychol* 81(1):141–45.

37. Mondloch CJ, Le Grand R, Maurer D (2002) Configural Face Processing Develops More Slowly Than Featural Face Processing. *Perception* 31(5):553–66.

38. Thompson P (1980) Margaret Thatcher: A New Illusion. *Perception* 9(4):483–84.

39. De Heering A, Turati C, Rossion B, Bulf H, Goffaux V, Simion F (2008) Newborns' Face Recognition Is Based on Spatial Frequencies below 0.5 Cycles per Degree. *Cognition* 106(1):444–54.

40. Morton and Johnson (1991).

41. Johnson MH (2005) Subcortical Face Processing. *Nat Rev Neurosci* 6(10):766–74.

42. Turati C, Bulf H, Simion F (2008) Newborns' Face Recognition over Changes in Viewpoint. *Cognition* 106(3):1300–321.

43. Leo I, Simion F (2009) Newborns' Mooney-Face Perception. *Infancy* 14(6):641–53.

44. Leo I, Simion F (2009) Face Processing at Birth: A Thatcher Illusion Study. *Dev Sci* 12(3):492–98.

45. Farroni T, Csibra G, Simion F, Johnson MH (2002) Eye Contact Detection in Humans from Birth. *Proc Natl Acad Sci USA* 99(14):9602–5.

46. Farroni T, Massaccesi S, Pividori D, Johnson M (2004) Gaze Following in Newborns. *Infancy* 5:39–60.

47. Adamson LB, Frick JE (2003) The Still-Face: A History of a Shared Experimental Paradigm. *Infancy* 4:451–73.

48. Bertin E, Striano T (2006) The Still-Face Response in Newborn, 1.5-, and 3-Month-Old Infants. *Infant Behav Dev* 29(2):294–97.

49. Nagy E (2008) Innate Intersubjectivity: Newborns' Sensitivity to Communication Disturbance. *Dev Psychol* 44(6):1779–84.

50. Field et al. (1982); Gregg et al. (1976).

51. Johansson G (1973) Visual Perception of Biological Motion and a Model of Its Analysis. *Percept Psychophys* 14:202–11.

52. Blakemore SJ, Decety J (2001) From the Perception of Action to the Understanding of Intention. *Nat Rev Neurosci* 2(8):561–67.

53. Allison T, Puce A, McCarthy G (2000) Social Perception from Visual Cues: Role of the STS Region. *Trends Cogn Sci* 4(7):267–78.

54. Bardi L, Regolin L, Simion F (2011) Biological Motion Preference in Humans at Birth: Role of Dynamic and Configural Properties. *Dev Sci* 14(2):353–59. Simion F, Regolin L, Bulf H (2008) A Predisposition for Biological Motion in the Newborn Baby. *Proc Natl Acad Sci USA* 105(2):809–13.

55. Kuhlmeier VA, Troje NF, Lee V (2010) Young Infants Detect the Direction of Biological Motion in Point-Light Displays. *Infancy* 15(1):83–93.

56. Von Hofsten C (2004) An Action Perspective on Motor Development. *Trends Cogn Sci* 8(6):266–72.

57. Rochat P, Hespos SJ (1997) Differential Rooting Response by Neonates: Evidence for an Early Sense of Self. *Early Dev Parent* 6(3–4):105–12.

58. Craig CM, Lee DN (1999) Neonatal Control of Nutritive Sucking Pressure: Evidence for an Intrinsic Tau-Guide. *Exp Brain Res* 124(3):371–82.

59. DeCasper and Fifer (1980).

60. Meltzoff AN, Moore MK (1977) Imitation of Facial and Manual Gestures by Human Neonates. *Science* 198:75–78. Meltzoff AN, Moore MK (1997) Explaining Facial Imitation: A Theoretical Model. *Early Dev Parent* 6:179–92.

61. Field et al. (1982).

62. Ekman P (1989) The Argument and Evidence about Universals in Facial Expression of Emotions. In: Wagner H, Manstead A [eds] *Handbook of Social Psychophysiology*. Wiley: Chichester, pp 143–64. Izard CE (1994) Innate and Universal Facial Expressions: Evidence from Developmental and Cross-Cultural Research. *Psychol Bull* 115(2):288–99. Tracy JL, Matsumoto D (2008) The Spontaneous Expression of Pride and Shame: Evidence for Biologically Innate Nonverbal Displays. *Proc Natl Acad Sci USA* 105(33):11655–60.

63. Gregg et al. (1976).

64. Meltzoff AN, Moore MK (1994) Imitation, Memory, and the Representation of Persons. *Infant Behav Dev* 17:83–99. Meltzoff and Moore (1997). Nagy E, Compagne H, Orvos H, Pal A, Molnar P, Janszky I, Loveland KA, Bardos G (2005) Index Finger Movement Imitation by Human Neonates: Motivation, Learning, and Left-Hand Preference. *Pediatr Res* 58(4):749–53.

65. Nagy et al. (2005).

66. Meltzoff and Moore (1997).

67. Iacoboni M, Molnar-Szakacs I, Gallese V, Buccino G, Mazziotta JC, Rizzolatti G (2005) Grasping the Intentions of Others with One's Own Mirror Neuron System. *PLoS Biol* 3:e79. Mukamel R, Ekstrom AD, Kaplan J, Iacoboni M, Fried I (2010) Single-Neuron Responses in Humans during Execution and Observation of Actions. *Curr Biol* 20(8):750–56. Rizzolatti G, Fogassi L, Gallese V (2001) Neurophysiological Mechanisms Underlying the Understanding and Imitation of Action. *Nat Rev Neurosci* 2:661–70.

68. Meltzoff and Moore (1997).

69. Reissland N (1988) Neonatal Imitation in the First Hour of Life: Observations in Rural Nepal. *Dev Psychol* 24(4):464–69.

70. Bard KA (2007) Neonatal Imitation in Chimpanzees *(Pan troglodytes)* Tested with Two Paradigms. *Anim Cogn* 10(2):233–42. Ferrari PF, Bonini L, Fogassi L (2009)

From Monkey Mirror Neurons to Primate Behaviours: Possible "Direct" and "Indirect" Pathways. *Philos Trans R Soc Lond B Biol Sci* 364(1528):2311–23.

71. Jones SS (2009) The Development of Imitation in Infancy. *Philos Trans R Soc Lond B Biol Sci* 364(1528):2325–35. Meltzoff AN, Murray L, Simpson E, Heimann M, Nagy E, Nadel J, Pedersen EJ, Brooks R, Messinger DS, Pascalis L, Subiaul F, Paukner A, Ferrari PF (2018) Re-examination of Oostenbroek et al. (2016): Evidence for Neonatal Imitation of Tongue Protrusion. *Dev Sci* 21(4):e12609. Oostenbroek J, Suddendorf T, Nielsen M, Redshaw J, Kennedy-Constantini S, Davis J, Clark S, Slaughter V (2016) Comprehensive Longitudinal Study Challenges the Existence of Neonatal Imitation in Humans. *Curr Biol* 26(10):1334–38.

72. Heimann M, Nelson KE, Schaller J (1989) Neonatal Imitation of Tongue Protrusion and Mouth Opening: Methodological Aspects and Evidence of Early Individual Differences. *Scand J Psychol* 30(2):90–101.

73. Ferrari et al. (2009).

74. Kuhl PK, Meltzoff AN (1982) The Bimodal Perception of Speech in Infancy. *Science* 218(4577):1138–41. Kuhl PK, Meltzoff AN (1996) Infant Vocalizations in Response to Speech: Vocal Imitation and Developmental Change. *J Acoust Soc Am* 100(4 Pt 1):2425–38.

75. Chen X, Striano T, Rakoczy H (2004) Auditory-Oral Matching Behavior in Newborns. *Dev Sci* 7(1):42–47.

76. Kuhl PK, Meltzoff AN (1984) The Intermodal Representation of Speech in Infants. *Infant Behav Dev* 7(3):361–81.

77. Kuhl PK, Conboy BT, Coffey-Corina S, Padden D, Rivera-Gaxiola M, Nelson T (2008) Phonetic Learning as a Pathway to Language: New Data and Native Language Magnet Theory Expanded (NLM-E). *Philos Trans R Soc Lond B Biol Sci* 363(1493):979–1000.

78. Kuhl PK (2003) Human Speech and Birdsong: Communication and the Social Brain. *Proc Natl Acad Sci USA* 100(17):9645–46.

79. Nagy E, Molnar P (2004) Homo Imitans or Homo Provocans? The Phenomenon of Neonatal Initiation. *Infant Behav Dev* 27:57–63.

80. Jennings JR, van der Molen MW, Brock K (1990) Forearm, Chest, and Skin Vascular Changes during Simple Performance Tasks. *Biol Psychol* 31(1):23–45. Richards JE (2008) Attention in Young Infants: A Developmental Psychophysiological Perspective. In: Nelson CA, Luciana M [eds] *Handbook of Developmental Cognitive Neuroscience*. MIT Press: Cambridge, MA, pp 321–38.

81. Nagy and Molnar (2004). Nagy E (2006) From Imitation to Conversation: The First Dialogues with Human Neonates. *Infant Child Dev* 15(3):223–32.

82. Thanukos A (2004) Understanding Evolution: How Your Eye Works. University of California Museum of Paleontology. https://evolution.berkeley.edu/evolibrary/article/side_0_0/eyeworks_01 (accessed 21 June 2021). See also appendix 1 of Wathey (2016).

83. Jacob F (1977) Evolution and Tinkering. *Science* 196(4295):1161–66.

84. Bowlby J (1969) *Attachment and Loss*. Basic Books: New York, p 233.

85. Kirkpatrick LA (2005) *Attachment, Evolution, and the Psychology of Religion*. Guilford Press: New York. pp 200–206. See also: Zeifman D, Hazan C (1997) Attachment: The Bond in Pair-Bonds. In: Simpson JA, Kendrick DT [eds] *Evolutionary Social*

Psychology. Erlbaum: Mahwah, NJ, pp 237–63. Frank RH (1988) *Passions within Reason: The Strategic Role of the Emotions.* Norton: New York.

86. For a more complete discussion of these puzzles of religion that can be explained by cross talk in underlying neural circuitry, see chapter 8 of Wathey (2016).

87. Garrett C (1987) *Spirit Possession and Popular Religion: From the Camisards to the Shakers.* Johns Hopkins University Press: Baltimore, chapter 9.

88. Wathey (2016) pp 178–80.

89. Ray D (2012) *Sex & God: How Religion Distorts Sexuality.* IPC Press: Bonner Springs, KS.

90. Teresa of Avila (1957) *The Life of Saint Teresa of Avila by Herself: Translated with an Introduction by JM Cohen.* Penguin: New York, p 210.

91. Lieberman D, Hatfield E (2006) Lust and Disgust: Crosscultural and Evolutionary Perspectives. In: Sternberg RJ, Weis K [eds] *The Psychology of Love (Second Edition).* Yale University Press: New Haven, CT, pp 274–97.

92. Blackmon RA (1984) *The Hazards of Ministry.* Doctoral Dissertation, Fuller Theological Seminary, Graduate School of Psychology: Pasadena, CA, chapter 8. Hart AD (1988) Special Report: How Common Is Pastoral Indiscretion? *Leadership Journal* 9(1):12–13.

93. Holroyd JC, Brodsky AM (1977) Psychologist's Attitudes and Practices Regarding Erotic and Nonerotic Contact with Patients. *Am Psychol* 32:843–49.

94. Kardener SH, Fuller M, Mensh IN (1973) A Survey of Physicians Attitudes and Practices Regarding Erotic and Nonerotic Contact with Patients. *Am J Psychiatry* 130(10):1077–81. Gartrell NK, Milliken N, Goodson WH III, Thiemann S, Lo B (1992) Physician-Patient Sexual Contact. Prevalence and Problems. *West J Med* 157(2):139–43.

95. Blackmon (1984) pp 121–22.

96. Blackmon (1984) pp 125–26; Hart (1988).

97. Anonymous (1982) The War Within: An Anatomy of Lust. *Leadership Journal* 3(3):30–48. Anonymous (1988) The War Within Continues: An Update on a Christian Leader's Struggle with Lust. *Leadership Journal* 9(1):24–41.

98. John Jay College of Criminal Justice (2004) *The Nature and Scope of Sexual Abuse of Minors by Catholic Priests and Deacons in the United States, 1950–2002.* United States Conference of Catholic Bishops: Washington, DC.

99. O'Donohue W, Cirlugea O, Benuto L (2012) Some Key Misunderstandings Regarding the Child Sexual Abuse Scandal and the Catholic Church. www.catholicleague.org/some-key-misunderstandings-regarding-the-child-sexual-abuse-scandal-and-the-catholic-church/ (accessed 14 December 2021).

100. Parkinson P (2014) Child Sexual Abuse and the Churches: A Story of Moral Failure? *Current Issues in Criminal Justice* 26(1):119–38. https://papers.ssrn.com/sol3/papers.cfm?abstract_id=2348413 (accessed 14 December 2021), p 5.

101. Baran M (2014) Number of Alleged Sex Abusers Greater than Archdiocese Has Revealed. Minnesota Public Radio [19 February] http://minnesota.publicradio.org/collections/catholic-church/2014/02/19/investigation-more-priests-accused-of-sexual-abuse-in-twin-cities-catholic-church/ (accessed 14 December 2021). Boston Globe (2002) *Betrayal: The Crisis in the Catholic Church.* Little, Brown: Boston.

102. See p 54 of: Kafka MP (2004) Sexual Molesters of Adolescents, Ephebophilia, and Catholic Clergy: A Review and Synthesis. In: Hanson RK, Pfäfflin F, Lütz M [eds] *Sexual Abuse in the Catholic Church: Scientific and Legal Perspectives.* Libreria Editrico Vaticana: Rome, pp 51–62.

103. Parkinson (2014) pp 5–7.

104. Kafka (2004), Parkinson (2014).

105. Wolf JG (1989) *Gay Priests.* Harper & Row: San Francisco, pp 59–63. Sipe AW (1990) *A Secret World: Sexuality and the Search for Celibacy.* Brunner/Mazel: New York.

106. Kinsey Institute (2011) Diversity of Sexual Orientation. https://kinseyinstitute .org/research/publications/historical-report-diversity-of-sexual-orientation.php (accessed 14 December 2021).

107. Sipe (1990).

108. Thomas A (2014) Secrets of the Vatican. *Frontline* WGBH/PBS: Boston [25 February] www.pbs.org/wgbh/frontline/film/secrets-of-the-vatican (accessed 9 February 2022).

109. Foster L (1981) *Religion and Sexuality: The Shakers, the Mormons, and the Oneida Community.* Oxford University Press: New York.

110. Palmer SJ (2004) *Aliens Adored: Raël's UFO Religion.* Rutgers University Press: New Brunswick, NJ.

111. Ayres BD Jr (1997) Families Learning of 39 Cultists Who Died Willingly. *New York Times* [29 Mar] www.nytimes.com/1997/03/29/us/families-learning-of-39-cultists -who-died-willingly.html (accessed 15 September 2021). Balch RW (1995) Waiting for the Ships: Disillusionment and Revitalization of Faith in Bo and Peep's UFO Cult. In: Lewis JR [ed] *The Gods Have Landed: New Religions from Other Worlds.* State University of New York Press: New York, pp 137–66.

112. Williams M (1998) *Heaven's Harlots: My Fifteen Years as a Sacred Prostitute in the Children of God Cult.* Eagle Brook: New York.

Chapter 3

1. Chinmoy S (1970) *Meditations: Food for the Soul.* Harper & Row: New York [online version: www.srichinmoylibrary.com/mfs (accessed 9 February 2022)]. Tamm J (2009) *Cartwheels in a Sari: A Memoir of Growing Up Cult.* Harmony Books: New York.

2. Hagerty BB (2009) *Fingerprints of God: The Search for the Science of Spirituality.* Riverhead: New York, p 12. Wathey JC (2016) *The Illusion of God's Presence: The Biological Origins of Spiritual Longing.* Prometheus Books: Amherst, NY, chapter 4.

3. Kamiya K, Fumoto M, Seki Y, Sato-Suzuki I, Arita H (2008) Activation of the Medial Prefrontal Cortex Prior to Crying. *Auton Neurosci* 144(1–2):90. Laureys S, Goldman S (2004) Imagine Imaging Neural Activity in Crying Infants and in Their Caring Parents. *Behav Brain Sci* 27(4):465–67. Sato-Suzuki I, Fumoto M, Seki Y, Kamiya K, Yu X, Kikuchi H, Sekiyama T, Arita H (2007) Activation of the Medial Prefrontal Cortex during Crying with Emotional Tears: Near-Infrared Spectroscopy Study. *Auton Neurosci* 135(1–2):128–37.

4. Bass AH, Gilland EH, Baker R (2008) Evolutionary Origins for Social Vocalization in a Vertebrate Hindbrain-Spinal Compartment. *Science* 321(5887):417–21.

5. Branchi I, Santucci D, Alleva E (2001) Ultrasonic Vocalisation Emitted by Infant Rodents: A Tool for Assessment of Neurobehavioural Development. *Behav Brain Res* 125(1–2):49–56. Pettijohn TF (1979) Attachment and Separation Distress in the Infant Guinea Pig. *Dev Psychobiol* 12(1):73–81. Shair HN (2007) Acquisition and Expression of a Socially Mediated Separation Response. *Behav Brain Res* 182(2):180–92.

6. MacLean PD, Newman JD (1988) Role of Midline Frontolimbic Cortex in Production of the Isolation Call of Squirrel Monkeys. *Brain Res* 450(1–2):111–23. Overall KL, Dunham AE, Frank D (2001) Frequency of Nonspecific Clinical Signs in Dogs with Separation Anxiety, Thunderstorm Phobia, and Noise Phobia, Alone or in Combination. *J Am Vet Med Assoc* 219(4):467–73.

7. Newman JD (2007) Neural Circuits Underlying Crying and Cry Responding in Mammals. *Behav Brain Res* 182(2):155–65. Soltis J (2004) The Signal Functions of Early Infant Crying. *Behav Brain Sci* 27(4):443–58.

8. Walter C (2006) Why Do We Cry? *Sci Am Mind* 17(6):44–51.

9. Darwin C (1874) [1998] *The Descent of Man.* Prometheus Books: Amherst, NY, p 103. Hardy AC (1966) *The Divine Flame: An Essay Towards a Natural History of Religion.* Collins: London, pp 173–75. Leuba J (1925) *The Psychology of Religious Mysticism.* Harcourt, Brace: New York, p 280. Pettijohn TF, Wong TW, Ebert PD, Scott JP (1977) Alleviation of Separation Distress in 3 Breeds of Young Dogs. *Dev Psychobiol* 10(4):373–81.

10. Brand S, Furlano R, Sidler M, Schulz J, Holsboer-Trachsler E (2011) "Oh, Baby, Please Don't Cry!": In Infants Suffering from Infantile Colic Hypothalamic-Pituitary-Adrenocortical Axis Activity Is Related to Poor Sleep and Increased Crying Intensity. *Neuropsychobiology* 64(1):15–23. Gunnar MR, Connors J, Isensee J, Wall L (1988) Adrenocortical Activity and Behavioral Distress in Human Newborns. *Dev Psychobiol* 21(4):297–310. Ludington-Hoe SM, Cong X, Hashemi F (2002) Infant Crying: Nature, Physiologic Consequences, and Select Interventions. *Neonatal Netw* 21(2):29–36.

11. Newman (2007).

12. Nielsen JM, Sedgwick RP (1949) Instincts and Emotions in an Anencephalic Monster. *J Nerv Ment Dis* 110(5):387–94. Barnet A, Bazelon M, Zappella M (1966) Visual and Auditory Function in an Hydranencephalic Infant. *Brain Res* 2(4):351–60.

13. Hofer MA, Shair HN (1992) Ultrasonic Vocalization by Rat Pups during Recovery from Deep Hypothermia. *Dev Psychobiol* 25(7):511–28.

14. Jürgens U, Ploog D (1970) Cerebral Representation of Vocalization in the Squirrel Monkey. *Exp Brain Res* 10(5):532–54. Kaada BR (1951) Somato-Motor, Autonomic and Electrocorticographic Responses to Electrical Stimulation of Rhinencephalic and Other Structures in Primates, Cat, and Dog; A Study of Responses from the Limbic, Subcallosal, Orbito-Insular, Piriform and Temporal Cortex, Hippocampus-Fornix and Amygdala. *Acta Physiol Scand Suppl* 24(83):1–262. Robinson BW (1967) Vocalization Evoked from the Forebrain in *Macaca mulatta. Physiol Behav* 2:345–54. Smith WK (1945) The Functional Significance of the Rostral Cingular Cortex as Revealed by Its Responses to Electrical Excitation. *J Neurophysiol* 8:241–55.

15. MacLean and Newman (1988).

16. Lorberbaum JP, Newman JD, Horwitz AR, Dubno JR, Lydiard RB, Hamner MB, Bohning DE, George MS (2002) A Potential Role for Thalamocingulate Circuitry in Human Maternal Behavior. *Biol Psychiatry* 51:431–45. Strathearn L, Li J, Fonagy P, Montague PR (2008) What's in a Smile? Maternal Brain Responses to Infant Facial Cues. *Pediatrics* 122(1):40–51.

17. Murphy MR, MacLean PD, Hamilton SC (1981) Species-Typical Behavior of Hamsters Deprived from Birth of the Neocortex. *Science* 213(4506):459–61. Stamm JS (1955) The Function of the Median Cerebral Cortex in Maternal Behavior of Rats. *J Comp Physiol Psychol* 48(4):347–56. Slotnick BM (1967) Disturbances of Maternal Behavior in the Rat following Lesions of the Cingulate Cortex. *Behaviour* 29(2):204–36.

18. Jensen AR, Rohwer WD Jr (1966) The Stroop Color-Word Test: A Review. *Acta Psychol (Amst)* 25(1):36–93.

19. Bush G, Luu P, Posner MI (2000) Cognitive and Emotional Influences in Anterior Cingulate Cortex. *Trends Cogn Sci* 4(6):215–22.

20. Bunge SA, Ochsner KN, Desmond JE, Glover GH, Gabrieli JD (2001) Prefrontal Regions Involved in Keeping Information in and out of Mind. *Brain* 124(Pt 10):2074–86. Critchley HD, Mathias CJ, Josephs O, O'Doherty J, Zanini S, Dewar BK, Cipolotti L, Shallice T, Dolan RJ (2003) Human Cingulate Cortex and Autonomic Control: Converging Neuroimaging and Clinical Evidence. *Brain* 126(Pt 10):2139–52.

21. Bennett PJG, Maier E, Brecht M (2019) Involvement of Rat Posterior Prelimbic and Cingulate Area 2 in Vocalization Control. *Eur J Neurosci* 50(7):3164–80. Green DB, Shackleton TM, Grimsley JMS, Zobay O, Palmer AR, Wallace MN (2018) Communication Calls Produced by Electrical Stimulation of Four Structures in the Guinea Pig Brain. *PLoS One* 13(3):e0194091.

22. Von Cramon D, Jürgens U (1983) The Anterior Cingulate Cortex and the Phonatory Control in Monkey and Man. *Neurosci Biobehav Rev* 7(3):423–25.

23. Hornak J, Bramham J, Rolls ET, Morris RG, O'Doherty J, Bullock PR, Polkey CE (2003) Changes in Emotion after Circumscribed Surgical Lesions of the Orbitofrontal and Cingulate Cortices. *Brain* 126(Pt 7):1691–712.

24. Peyron R, Laurent B, Garcia-Larrea L (2000) Functional Imaging of Brain Responses to Pain. A Review and Meta-Analysis (2000). *Neurophysiol Clin* 30(5):263–88.

25. Eisenberger NI, Lieberman MD (2004) Why Rejection Hurts: A Common Neural Alarm System for Physical and Social Pain. *Trends Cogn Sci* 8(7):294–300.

26. Morrison I, Downing PE (2007) Organization of Felt and Seen Pain Responses in Anterior Cingulate Cortex. *Neuroimage* 37(2):642–51.

27. Fowler CJ, Griffiths DJ (2010) A Decade of Functional Brain Imaging Applied to Bladder Control. *Neurourol Urodyn* 29(1):49–55.

28. Price JL, Drevets WC (2010) Neurocircuitry of Mood Disorders. *Neuropsychopharmacology* 35(1):192–216.

29. Mayberg HS, Liotti M, Brannan SK, McGinnis S, Mahurin RK, Jerabek PA, Silva JA, Tekell JL, Martin CC, Lancaster JL, Fox PT (1999) Reciprocal Limbic-Cortical Function and Negative Mood: Converging PET Findings in Depression and Normal Sadness. *Am J Psychiatry* 156(5):675–82.

30. Varnas K, Halldin C, Hall H (2004) Autoradiographic Distribution of Serotonin Transporters and Receptor Subtypes in Human Brain. *Hum Brain Mapp* 22(3):246–60.

31. Mayberg HS (2009) Targeted Electrode-Based Modulation of Neural Circuits for Depression. *J Clin Invest* 119(4):717–25.

32. Hadland KA, Rushworth MF, Gaffan D, Passingham RE (2003) The Effect of Cingulate Lesions on Social Behaviour and Emotion. *Neuropsychologia* 41(8):919–31.

33. Hornak et al. (2003).

34. Di Martino A, Ross K, Uddin LQ, Sklar AB, Castellanos FX, Milham MP (2009) Functional Brain Correlates of Social and Nonsocial Processes in Autism Spectrum Disorders: An Activation Likelihood Estimation Meta-Analysis. *Biol Psychiatry* 65(1):63–74.

35. Bush et al. (2000).

36. Beckmann M, Johansen-Berg H, Rushworth MF (2009) Connectivity-Based Parcellation of Human Cingulate Cortex and Its Relation to Functional Specialization. *J Neurosci* 29(4):1175–90. Pandya DN, Van Hoesen GW, Mesulam MM (1981) Efferent Connections of the Cingulate Gyrus in the Rhesus Monkey. *Exp Brain Res* 42(3–4):319–30. Vogt BA, Nimchinsky EA, Vogt LJ, Hof PR (1995) Human Cingulate Cortex: Surface Features, Flat Maps, and Cytoarchitecture. *J Comp Neurol* 359(3):490–506. Zilles K, Schlaug G, Matelli M, Luppino G, Schleicher A, Qu M, Dabringhaus A, Seitz R, Roland PE (1995) Mapping of Human and Macaque Sensorimotor Areas by Integrating Architectonic, Transmitter Receptor, MRI and PET Data. *J Anat* 187(Pt 3):515–37.

37. Allman JM, Hakeem A, Erwin JM, Nimchinsky E, Hof P (2001) The Anterior Cingulate Cortex. The Evolution of an Interface between Emotion and Cognition. *Ann N Y Acad Sci* 935:107–17. Hayden BY, Platt ML (2010) Neurons in Anterior Cingulate Cortex Multiplex Information about Reward and Action. *J Neurosci* 30(9):3339–46. Holroyd CB, Verguts T (2021) The Best Laid Plans: Computational Principles of Anterior Cingulate Cortex. *Trends Cogn Sci* 25(4):316–29.

38. Oades RD, Halliday GM (1987) Ventral Tegmental (A10) System: Neurobiology. 1. Anatomy and Connectivity. *Brain Res* 434(2):117–65.

39. Beckmann et al. (2009). Hatanaka N, Tokuno H, Hamada I, Inase M, Ito Y, Imanishi M, Hasegawa N, Akazawa T, Nambu A, Takada M (2003) Thalamocortical and Intracortical Connections of Monkey Cingulate Motor Areas. *J Comp Neurol* 462(1):121–38.

40. Beckmann et al. (2009). Freedman LJ, Insel TR, Smith Y (2000) Subcortical Projections of Area 25 (Subgenual Cortex) of the Macaque Monkey. *J Comp Neurol* 421(2):172–88. Öngür D, An X, Price JL (1998) Prefrontal Cortical Projections to the Hypothalamus in Macaque Monkeys. *J Comp Neurol* 401(4):480–505.

41. Dampney RA (2011) The Hypothalamus and Autonomic Regulation: An Overview. In: Llewellyn-Smith IJ, Verberne AJ [eds] *Central Regulation of Autonomic Functions*. Oxford University Press: New York, chapter 3.

42. Freedman et al. (2000). Newman JD, MacLean PD (1982) Effects of Tegmental Lesions on the Isolation Call of Squirrel Monkeys. *Brain Res* 232(2):317–30. Price JL (2005) Free Will versus Survival: Brain Systems That Underlie Intrinsic Constraints on Behavior. *J Comp Neurol* 493(1):132–39.

43. Eisenberger and Lieberman (2004).

44. On this point they are restating the insights of: Panksepp J, Nelson E, Bekkedal M (1997) Brain Systems for the Mediation of Social Separation-Distress and

Social-Reward. Evolutionary Antecedents and Neuropeptide Intermediaries. *Ann N Y Acad Sci* 807:78–100.

45. Newman (2007).

46. Panksepp J, Nelson E, Siviy S (1994) Brain Opioids and Mother-Infant Social Motivation. *Acta Paediatr Suppl* 397:40–46.

47. Martel FL, Nevison CM, Rayment FD, Simpson MJ, Keverne EB (1993) Opioid Receptor Blockade Reduces Maternal Affect and Social Grooming in Rhesus Monkeys. *Psychoneuroendocrinology* 18(4):307–21.

48. Vogt BA, Wiley RG, Jensen EL (1995) Localization of Mu and Delta Opioid Receptors to Anterior Cingulate Afferents and Projection Neurons and Input/Output Model of Mu Regulation. *Exp Neurol* 135(2):83–92.

49. Zubieta JK, Ketter TA, Bueller JA, Xu Y, Kilbourn MR, Young EA, Koeppe RA (2003) Regulation of Human Affective Responses by Anterior Cingulate and Limbic Mu-Opioid Neurotransmission. *Arch Gen Psychiatry* 60(11):1145–53.

Chapter 4

1. Darwin C, Barlow N (1958) *The Autobiography of Charles Darwin: 1809–1882 with Original Omissions Restored.* Harcourt, Brace: New York, p 96.

2. Carmichael ST, Price JL (1996) Connectional Networks within the Orbital and Medial Prefrontal Cortex of Macaque Monkeys. *J Comp Neurol* 371(2):179–207.

3. Carmichael and Price (1996). Cavada C, Company T, Tejedor J, Cruz-Rizzolo RJ, Reinoso-Suarez F (2000) The Anatomical Connections of the Macaque Monkey Orbitofrontal Cortex. A Review. *Cereb Cortex* 10(3):220–42. Price JL (2005) Free Will versus Survival: Brain Systems That Underlie Intrinsic Constraints on Behavior. *J Comp Neurol* 493(1):132–39. Price JL, Drevets WC (2010) Neurocircuitry of Mood Disorders. *Neuropsychopharmacology* 35(1):192–216.

4. Carmichael ST, Price JL (1994) Architectonic Subdivision of the Orbital and Medial Prefrontal Cortex in the Macaque Monkey. *J Comp Neurol* 346(3):366–402. Hof PR, Mufson EJ, Morrison JH (1995) Human Orbitofrontal Cortex: Cytoarchitecture and Quantitative Immunohistochemical Parcellation. *J Comp Neurol* 359(1):48–68. Öngür D, Price JL (2000) The Organization of Networks within the Orbital and Medial Prefrontal Cortex of Rats, Monkeys and Humans. *Cereb Cortex* 10(3):206–19.

5. Rolls ET (2004) The Functions of the Orbitofrontal Cortex. *Brain Cogn* 55(1):11–29.

6. Thorpe SJ, Rolls ET, Maddison S (1983) The Orbitofrontal Cortex: Neuronal Activity in the Behaving Monkey. *Exp Brain Res* 49(1):93–115.

7. Rolls ET, Treves A (1998) *Neural Networks and Brain Function.* Oxford University Press: Oxford, chapter 7.

8. Rolls ET, Critchley HD, Browning AS, Inoue K (2006) Face-Selective and Auditory Neurons in the Primate Orbitofrontal Cortex. *Exp Brain Res* 170(1):74–87.

4555

9. Kringelbach ML, Lehtonen A, Squire S, Harvey AG, Craske MG, Holliday IE, Green AL, Aziz TZ, Hansen PC, Cornelissen PL, Stein A (2008) A Specific and Rapid Neural Signature for Parental Instinct. *PLoS One* 3(2):e1664.

10. Rolls ET (2010) The Affective and Cognitive Processing of Touch, Oral Texture, and Temperature in the Brain. *Neurosci Biobehav Rev* 34(2):237–45.

11. Rolls et al. (2006). Romanski LM, Tian B, Fritz J, Mishkin M, Goldman-Rakic PS, Rauschecker JP (1999) Dual Streams of Auditory Afferents Target Multiple Domains in the Primate Prefrontal Cortex. *Nat Neurosci* 2(12):1131–36. Romanski LM, Goldman-Rakic PS (2002) An Auditory Domain in Primate Prefrontal Cortex. *Nat Neurosci* 5(1):15–16.

12. Ebata S, Sugiuchi Y, Izawa Y, Shinomiya K, Shinoda Y (2004) Vestibular Projection to the Periarcuate Cortex in the Monkey. *Neurosci Res* 49(1):55–68. Fasold O, von Brevern M, Kuhberg M, Ploner CJ, Villringer A, Lempert T, Wenzel R (2002) Human Vestibular Cortex as Identified with Caloric Stimulation in Functional Magnetic Resonance Imaging. *Neuroimage* 17(3):1384–93. Stephan T, Hufner K, Brandt T (2009) Stimulus Profile and Modeling of Continuous Galvanic Vestibular Stimulation in Functional Magnetic Resonance Imaging. *Ann N Y Acad Sci* 1164:472–75.

13. Cavada et al. (2000); Fasold et al. (2002).

14. Liu J, Harris A, Kanwisher N (2010) Perception of Face Parts and Face Configurations: An FMRI Study. *J Cogn Neurosci* 22(1):203–11.

15. Johnson MH (2005) Subcortical Face Processing. *Nat Rev Neurosci* 6(10):766–74. Long KD, Kennedy G, Balaban E (2001) Transferring an Inborn Auditory Perceptual Predisposition with Interspecies Brain Transplants. *Proc Natl Acad Sci USA* 98(10):5862–67.

16. LeDoux J (2003) The Emotional Brain, Fear, and the Amygdala. *Cell Mol Neurobiol* 23(4–5):727–38.

17. Adolphs R (2008) Fear, Faces, and the Human Amygdala. *Curr Opin Neurobiol* 18(2):166–72.

18. Murray EA, Wise SP (2010) Interactions between Orbital Prefrontal Cortex and Amygdala: Advanced Cognition, Learned Responses and Instinctive Behaviors. *Curr Opin Neurobiol* 20(2):212–20.

19. Baxter MG, Parker A, Lindner CC, Izquierdo AD, Murray EA (2000) Control of Response Selection by Reinforcer Value Requires Interaction of Amygdala and Orbital Prefrontal Cortex. *J Neurosci* 20(11):4311–19.

20. Nelson EE, Shelton SE, Kalin NH (2003) Individual Differences in the Responses of Naive Rhesus Monkeys to Snakes. *Emotion* 3(1):3–11.

21. Machado CJ, Kazama AM, Bachevalier J (2009) Impact of Amygdala, Orbital Frontal, or Hippocampal Lesions on Threat Avoidance and Emotional Reactivity in Nonhuman Primates. *Emotion* 9(2):147–63. Prather MD, Lavenex P, Mauldin-Jourdain ML, Mason WA, Capitanio JP, Mendoza SP, Amaral DG (2001) Increased Social Fear and Decreased Fear of Objects in Monkeys with Neonatal Amygdala Lesions. *Neuroscience* 106(4):653–58.

22. Gothard KM, Battaglia FP, Erickson CA, Spitler KM, Amaral DG (2007) Neural Responses to Facial Expression and Face Identity in the Monkey Amygdala. *J Neurophysiol* 97(2):1671–83.

23. Kondo H, Saleem KS, Price JL (2003) Differential Connections of the Temporal Pole with the Orbital and Medial Prefrontal Networks in Macaque Monkeys. *J Comp Neurol* 465(4):499–523.

24. Quiroga RQ, Reddy L, Kreiman G, Koch C, Fried I (2005) Invariant Visual Representation by Single Neurons in the Human Brain. *Nature* 435(7045):1102–7.

25. Bauman MD, Lavenex P, Mason WA, Capitanio JP, Amaral DG (2004) The Development of Mother-Infant Interactions after Neonatal Amygdala Lesions in Rhesus Monkeys. *J Neurosci* 24(3):711–21.

26. Kennedy DP, Glascher J, Tyszka JM, Adolphs R (2009) Personal Space Regulation by the Human Amygdala. *Nat Neurosci* 12(10):1226–27.

27. Emery NJ, Capitanio JP, Mason WA, Machado CJ, Mendoza SP, Amaral DG (2001) The Effects of Bilateral Lesions of the Amygdala on Dyadic Social Interactions in Rhesus Monkeys *(Macaca mulatta)*. *Behav Neurosci* 115(3):515–44.

28. Machado CJ, Bachevalier J (2006) The Impact of Selective Amygdala, Orbital Frontal Cortex, or Hippocampal Formation Lesions on Established Social Relationships in Rhesus Monkeys *(Macaca mulatta)*. *Behav Neurosci* 120(4):761–86.

29. Moriceau S, Wilson DA, Levine S, Sullivan RM (2006) Dual Circuitry for Odor-Shock Conditioning during Infancy: Corticosterone Switches between Fear and Attraction via Amygdala. *J Neurosci* 26(25):6737–48. Roth TL, Moriceau S, Sullivan RM (2006) Opioid Modulation of Fos Protein Expression and Olfactory Circuitry Plays a Pivotal Role in What Neonates Remember. *Learn Mem* 13(5):590–98.

30. Moriceau S, Sullivan RM (2006) Maternal Presence Serves as a Switch between Learning Fear and Attraction in Infancy. *Nat Neurosci* 9(8):1004–6. Roth et al. (2006).

31. Domes G, Heinrichs M, Glascher J, Buchel C, Braus DF, Herpertz SC (2007) Oxytocin Attenuates Amygdala Responses to Emotional Faces Regardless of Valence. *Biol Psychiatry* 62(10):1187–90. Kosfeld M, Heinrichs M, Zak PJ, Fischbacher U, Fehr E (2005) Oxytocin Increases Trust in Humans. *Nature* 435(7042):673–76. Petrovic P, Kalisch R, Singer T, Dolan RJ (2008) Oxytocin Attenuates Affective Evaluations of Conditioned Faces and Amygdala Activity. *J Neurosci* 28(26):6607–15.

32. Damasio AR, Tranel D, Damasio H (1990) Individuals with Sociopathic Behavior Caused by Frontal Damage Fail to Respond Autonomically to Social Stimuli. *Behav Brain Res* 41(2):81–94.

33. Bechara A, Tranel D, Damasio H (2000) Characterization of the Decision-Making Deficit of Patients with Ventromedial Prefrontal Cortex Lesions. *Brain* 123(11):2189–202. Ciaramelli E, Muccioli M, Ladavas E, di Pellegrino G (2007) Selective Deficit in Personal Moral Judgment Following Damage to Ventromedial Prefrontal Cortex. *Soc Cogn Affect Neurosci* 2(2):84–92. Schneider B, Koenigs M (2017) Human Lesion Studies of Ventromedial Prefrontal Cortex. *Neuropsychologia* 107:84–93. Shamay-Tsoory SG, Tomer R, Berger BD, Aharon-Peretz J (2003) Characterization of Empathy Deficits Following Prefrontal Brain Damage: The Role of the Right Ventromedial Prefrontal Cortex. *J Cogn Neurosci* 15(3):324–37.

34. Koenigs M, Young L, Adolphs R, Tranel D, Cushman F, Hauser M, Damasio A (2007) Damage to the Prefrontal Cortex Increases Utilitarian Moral Judgements. *Nature* 446:908–11.

35. Eslinger PJ, Damasio AR (1985) Severe Disturbance of Higher Cognition after Bilateral Frontal Lobe Ablation: Patient EVR. *Neurology* 35(12):1731–41.

36. Kertesz A (2006) Progress in Clinical Neurosciences: Frontotemporal Dementia-Pick's Disease. *Can J Neurol Sci* 33(2):141–48.

37. Michels S (2008) Researchers Explore New Technologies, Treatments for Dementia Patients. *PBS Newshour* [12 November] www.pbs.org/newshour/show/researchers-explore-new-technologies-treatments-for-dementia-patients (accessed 11 January 2022).

38. Haidt J (2007) The New Synthesis in Moral Psychology. *Science* 316(5827): 998–1002.

39. Anderson SW, Damasio H, Tranel D, Damasio AR (2000) Long-Term Sequelae of Prefrontal Cortex Damage Acquired in Early Childhood. *Dev Neuropsychol* 18(3):281–96.

40. Vargha-Khadem F, O'Gorman AM, Watters GV (1985) Aphasia and Handedness in Relation to Hemispheric Side, Age at Injury and Severity of Cerebral Lesion during Childhood. *Brain* 108(Pt 3):677–96.

41. Young L, Saxe R (2009) An fMRI Investigation of Spontaneous Mental State Inference for Moral Judgment. *J Cogn Neurosci* 21(7):1396–405.

42. Mitchell JP, Macrae CN, Banaji MR (2006) Dissociable Medial Prefrontal Contributions to Judgments of Similar and Dissimilar Others. *Neuron* 50(4):655–63. St Jacques PL, Conway MA, Lowder MW, Cabeza R (2011) Watching My Mind Unfold versus Yours: An fMRI Study Using a Novel Camera Technology to Examine Neural Differences in Self-Projection of Self versus Other Perspectives. *J Cogn Neurosci* 23(6):1275–84.

43. Camille N, Coricelli G, Sallet J, Pradat-Diehl P, Duhamel JR, Sirigu A (2004) The Involvement of the Orbitofrontal Cortex in the Experience of Regret. *Science* 304(5674):1167–70.

44. Nitschke JB, Nelson EE, Rusch BD, Fox AS, Oakes TR, Davidson RJ (2004) Orbitofrontal Cortex Tracks Positive Mood in Mothers Viewing Pictures of Their Newborn Infants. *Neuroimage* 21(2):583–92. Noriuchi M, Kikuchi Y, Senoo A (2008) The Functional Neuroanatomy of Maternal Love: Mother's Response to Infant's Attachment Behaviors. *Biol Psychiatry* 63(4):415–23.

45. Noriuchi M, Kikuchi Y, Mori K, Kamio Y (2019) The Orbitofrontal Cortex Modulates Parenting Stress in the Maternal Brain. *Sci Rep* 9(1):1658.

46. Provenzi L, Lindstedt J, De Coen K, Gasparini L, Peruzzo D, Grumi S, Arrigoni F, Ahlqvist-Björkroth S (2021) The Paternal Brain in Action: A Review of Human Fathers' fMRI Brain Responses to Child-Related Stimuli. *Brain Sci* 11(6):816 [Epub only].

47. Minagawa-Kawai Y, Matsuoka S, Dan I, Naoi N, Nakamura K, Kojima S (2009) Prefrontal Activation Associated with Social Attachment: Facial-Emotion Recognition in Mothers and Infants. *Cereb Cortex* 19(2): 284–92.

48. Lauronen L, Nevalainen P, Pihko E (2012) Magnetoencephalography in Neonatology. *Neurophysiol Clin* 42(1–2):27–34. Martynova O, Kirjavainen J, Cheour M (2003) Mismatch Negativity and Late Discriminative Negativity in Sleeping Human Newborns. *Neurosci Lett* 340(2):75–78.

49. Saito Y, Aoyama S, Kondo T, Fukumoto R, Konishi N, Nakamura K, Kobayashi M, Toshima T (2007) Frontal Cerebral Blood Flow Change Associated with Infant-Directed Speech. *Arch Dis Child Fetal Neonatal Ed* 92(2):F113–16.

50. Aoyama S, Toshima T, Saito Y, Konishi N, Motoshige K, Ishikawa N, Nakamura K, Kobayashi M (2010) Maternal Breast Milk Odour Induces Frontal Lobe Activation in Neonates: A NIRS Study. *Early Hum Dev* 86(9):541–45.

51. Bartocci M, Winberg J, Ruggiero C, Bergqvist LL, Serra G, Lagercrantz H (2000) Activation of Olfactory Cortex in Newborn Infants after Odor Stimulation: A Functional Near-Infrared Spectroscopy Study. *Pediatr Res* 48(1):18–23.

52. Ellis CT, Skalaban LJ, Yates TS, Bejjanki VR, Córdova NI, Turk-Browne NB (2020) Re-imagining fMRI for Awake Behaving Infants. *Nat Commun* 11(1):4523.

Chapter 5

1. Miller DP (1995) *The Book of Practical Faith. A Path to Useful Spirituality.* Henry Holt: New York, p 42.

2. Smith H, Kernochan S, Gortner M, Silk L, Palevsky M (1972) [2008] *Marjoe.* Docurama: New York.

3. Shair HN (2007) Acquisition and Expression of a Socially Mediated Separation Response. *Behav Brain Res* 182(2):180–92.

4. Hennessy MB, Miller EE, Shair HN (2006) Brief Exposure to the Biological Mother "Potentiates" the Isolation Behavior of Precocial Guinea Pig Pups. *Dev Psychobiol* 48(8):653–59. Levine S, Wiener SG (1988) Psychoendocrine Aspects of Mother-Infant Relationships in Nonhuman Primates. *Psychoneuroendocrinology* 13(1–2):143–54. Moles A, Kieffer BL, D'Amato FR (2004) Deficit in Attachment Behavior in Mice Lacking the Mu-Opioid Receptor Gene. *Science* 304(5679):1983–86. Robison WT, Myers MM, Hofer MA, Shair HN, Welch MG (2016) Prairie Vole Pups Show Potentiated Isolation-Induced Vocalizations following Isolation from Their Mother, but Not Their Father. *Dev Psychobiol* 58(6):687–99.

5. Muller JM, Brunelli SA, Moore H, Myers MM, Shair HN (2005) Maternally Modulated Infant Separation Responses Are Regulated by D2-Family Dopamine Receptors. *Behav Neurosci* 119(5):1384–88.

6. Muller JM, Moore H, Myers MM, Shair HN (2008) Ventral Striatum Dopamine D2 Receptor Activity Inhibits Rat Pups' Vocalization Response to Loss of Maternal Contact. *Behav Neurosci* 122(1):119–28.

7. Smeets WJ, Marin O, Gonzalez A (2000) Evolution of the Basal Ganglia: New Perspectives through a Comparative Approach. *J Anat* 196 (Pt 4):501–17.

8. Strausfeld NJ, Hirth F (2013) Deep Homology of Arthropod Central Complex and Vertebrate Basal Ganglia. *Science* 340(6129):157–61.

9. DeLong MR, Wichmann T (2007) Circuits and Circuit Disorders of the Basal Ganglia. *Arch Neurol* 64(1):20–24. Kelly RM, Strick PL (2004) Macro-Architecture of Basal Ganglia Loops with the Cerebral Cortex: Use of Rabies Virus to Reveal Multisynaptic Circuits. *Prog Brain Res* 143:449–59.

10. Jankovic J (2008) Parkinson's Disease: Clinical Features and Diagnosis. *J Neurol Neurosurg Psychiatry* 79(4):368–76.

11. Cotzias GC, Papavasiliou PS, Gellene R (1969) Modification of Parkinsonism—Chronic Treatment with L-Dopa. *N Engl J Med* 280(7):337–45. Fahn S (1999) Parkinson Disease, the Effect of Levodopa, and the ELLDOPA Trial. Earlier Vs Later L-DOPA. *Arch Neurol* 56(5):529–35.

12. Doupe AJ, Perkel DJ, Reiner A, Stern EA (2005) Birdbrains Could Teach Basal Ganglia Research a New Song. *Trends Neurosci* 28(7):353–63. Fee MS, Scharff C (2010) The Songbird as a Model for the Generation and Learning of Complex Sequential Behaviors. *ILAR J* 51(4):362–77.

13. Doupe et al. (2005).

14. Adret P (2004) In Search of the Song Template. *Ann N Y Acad Sci* 1016:303–24.

15. Adret (2004); Doupe et al. (2005).

16. Scharff C, Nottebohm F (1991) A Comparative Study of the Behavioral Deficits Following Lesions of Various Parts of the Zebra Finch Song System: Implications for Vocal Learning. *J Neurosci* 11(9):2896–913.

17. Kao MH, Doupe AJ, Brainard MS (2005) Contributions of an Avian Basal Ganglia-Forebrain Circuit to Real-Time Modulation of Song. *Nature* 433(7026):638–43.

18. Andalman AS, Fee MS (2009) A Basal Ganglia-Forebrain Circuit in the Songbird Biases Motor Output to Avoid Vocal Errors. *Proc Natl Acad Sci USA* 106(30):12518–23.

19. Ding L, Perkel DJ (2004) Long-Term Potentiation in an Avian Basal Ganglia Nucleus Essential for Vocal Learning. *J Neurosci* 24(2):488–94. Soha JA, Shimizu T, Doupe AJ (1996) Development of the Catecholaminergic Innervation of the Song System of the Male Zebra Finch. *J Neurobiol* 29(4):473–89.

20. Dave AS, Margoliash D (2000) Song Replay during Sleep and Computational Rules for Sensorimotor Vocal Learning. *Science* 290(5492):812–16. Doupe AJ (1997) Song- and Order-Selective Neurons in the Songbird Anterior Forebrain and Their Emergence during Vocal Development. *J Neurosci* 17(3):1147–67.

21. Hessler NA, Doupe AJ (1999) Singing-Related Neural Activity in a Dorsal Forebrain-Basal Ganglia Circuit of Adult Zebra Finches. *J Neurosci* 19(23):10461–81.

22. Scharff and Nottebohm (1991).

23. Brainard MS, Doupe AJ (2000) Interruption of a Basal Ganglia-Forebrain Circuit Prevents Plasticity of Learned Vocalizations. *Nature* 404(6779):762–66. Kao et al. (2005).

24. Samuel M, Caputo E, Brooks DJ, Schrag A, Scaravilli T, Branston NM, Rothwell JC, Marsden CD, Thomas DG, Lees AJ, Quinn NP (1998) A Study of Medial Pallidotomy for Parkinson's Disease: Clinical Outcome, MRI Location and Complications. *Brain* 121 (Pt 1):59–75.

25. Haber SN, Knutson B (2010) The Reward Circuit: Linking Primate Anatomy and Human Imaging. *Neuropsychopharmacology* 35(1):4–26.

26. Shair (2007).

27. Getz LL, Carter CS, Gavish L (1981) The Mating System of the Prairie Vole, *Microtus ochrogaster*: Field and Laboratory Evidence for Pair-Bonding. *Behav Ecol Sociobiol* 8(3):189–94.

28. Williams JR, Catania KC, Carter CS (1992) Development of Partner Preferences in Female Prairie Voles *(Microtus ochrogaster)*: The Role of Social and Sexual Experience. *Horm Behav* 26(3):339–49.

29. Getz LL, Carter CS, Gavish L (1981) The Mating System of the Prairie Vole, *Microtus ochrogaster*: Field and Laboratory Evidence for Pair-Bonding. *Behav Ecol Sociobiol* 8(3):189–94. Ophir AG, Phelps SM, Sorin AB, Wolff JO (2008) Social but Not Genetic Monogamy Is Associated with Greater Breeding Success in Prairie Voles. *Anim Behav* 75(3):1143–54. Wang Z, Hulihan TJ, Insel TR (1997) Sexual and Social Experience Is Associated with Different Patterns of Behavior and Neural Activation in Male Prairie Voles. *Brain Res* 767(2):321–32.

30. Jannett FJ (1980) Social Dynamics of the Montane Vole *Microtus montanus*, as a Paradigm. *Biologist* 62:3–19.

31. Shapiro LE, Insel TR (1990) Infant's Response to Social Separation Reflects Adult Differences in Affiliative Behavior: A Comparative Developmental Study in Prairie and Montane Voles. *Dev Psychobiol* 23(5):375–93.

32. Negi CS (2009) *Introduction to Endocrinology*. PHI Learning: New Delhi, chapter 6.

33. Goodson JL (2008) Nonapeptides and the Evolutionary Patterning of Sociality. *Prog Brain Res* 170:3–15.

34. Insel TR (2010) The Challenge of Translation in Social Neuroscience: A Review of Oxytocin, Vasopressin, and Affiliative Behavior. *Neuron* 65(6):768–79. Young KA, Liu Y, Wang Z (2008) The Neurobiology of Social Attachment: A Comparative Approach to Behavioral, Neuroanatomical, and Neurochemical Studies. *Comp Biochem Physiol C Toxicol Pharmacol* 148(4):401–10.

35. Lim MM, Young LJ (2004) Vasopressin-Dependent Neural Circuits Underlying Pair Bond Formation in the Monogamous Prairie Vole. *Neuroscience* 125(1):35–45.

36. Liu Y, Wang ZX (2003) Nucleus Accumbens Oxytocin and Dopamine Interact to Regulate Pair Bond Formation in Female Prairie Voles. *Neuroscience* 121(3):537–44. Young LJ, Lim MM, Gingrich B, Insel TR (2001) Cellular Mechanisms of Social Attachment. *Horm Behav* 40(2):133–38.

37. Young KA, Gobrogge KL, Liu Y, Wang Z (2011) The Neurobiology of Pair Bonding: Insights from a Socially Monogamous Rodent. *Front Neuroendocrinol* 32(1):53–69.

38. Aragona BJ, Liu Y, Curtis JT, Stephan FK, Wang Z (2003) A Critical Role for Nucleus Accumbens Dopamine in Partner-Preference Formation in Male Prairie Voles. *J Neurosci* 23(8):3483–90. Gingrich B, Liu Y, Cascio C, Wang Z, Insel TR (2000) Dopamine D2 Receptors in the Nucleus Accumbens Are Important for Social Attachment in Female Prairie Voles *(Microtus ochrogaster)*. *Behav Neurosci* 114(1):173–83.

39. Aragona BJ, Liu Y, Yu YJ, Curtis JT, Detwiler JM, Insel TR, Wang Z (2006) Nucleus Accumbens Dopamine Differentially Mediates the Formation and Maintenance of Monogamous Pair Bonds. *Nat Neurosci* 9(1):133–39. Gingrich et al. (2000).

40. Aragona et al. (2006).

41. Insel TR, Winslow JT (1991) Central Administration of Oxytocin Modulates the Infant Rat's Response to Social Isolation. *Eur J Pharmacol* 203(1):149–52.

42. Nelson E, Panksepp J (1996) Oxytocin Mediates Acquisition of Maternally Associated Odor Preferences in Preweanling Rat Pups. *Behav Neurosci* 110(3):583–92.

43. Brischoux F, Chakraborty S, Brierley DI, Ungless MA (2009) Phasic Excitation of Dopamine Neurons in Ventral VTA by Noxious Stimuli. *Proc Natl Acad Sci USA* 106(12):4894–99. Faure A, Richard JM, Berridge KC (2010) Desire and Dread from the Nucleus Accumbens: Cortical Glutamate and Subcortical GABA Differentially Generate Motivation and Hedonic Impact in the Rat. *PLoS One* 5(6):e11223. Salamone JD, Correa M, Mingote SM, Weber SM (2005) Beyond the Reward Hypothesis: Alternative Functions of Nucleus Accumbens Dopamine. *Curr Opin Pharmacol* 5(1):34–41. Schultz W (2010) Dopamine Signals for Reward Value and Risk: Basic and Recent Data. *Behav Brain Funct* 6:24.

44. Surmeier DJ, Ding J, Day M, Wang Z, Shen W (2007) D1 and D2 Dopamine-Receptor Modulation of Striatal Glutamatergic Signaling in Striatal Medium Spiny Neurons. *Trends Neurosci* 30(5):228–35.

45. Neve KA, Seamans JK, Trantham-Davidson H (2004) Dopamine Receptor Signaling. *J Recept Signal Transduct Res* 24(3):165–205.

46. Ding L, Perkel DJ (2002) Dopamine Modulates Excitability of Spiny Neurons in the Avian Basal Ganglia. *J Neurosci* 22(12):5210–18. Mizuno T, Schmauss C, Rayport S (2007) Distinct Roles of Presynaptic Dopamine Receptors in the Differential Modulation of the Intrinsic Synapses of Medium-Spiny Neurons in the Nucleus Accumbens. *BMC Neurosci* 8:8.

47. Schultz W (2007) Multiple Dopamine Functions at Different Time Courses. *Annu Rev Neurosci* 30:259–88.

48. Zweifel LS, Parker JG, Lobb CJ, Rainwater A, Wall VZ, Fadok JP, Darvas M, Kim MJ, Mizumori SJ, Paladini CA, Phillips PE, Palmiter RD (2009) Disruption of NMDAR-Dependent Burst Firing by Dopamine Neurons Provides Selective Assessment of Phasic Dopamine-Dependent Behavior. *Proc Natl Acad Sci USA* 106(18):7281–88.

49. Floresco SB, St Onge JR, Ghods-Sharifi S, Winstanley CA (2008) Cortico-Limbic-Striatal Circuits Subserving Different Forms of Cost-Benefit Decision Making. *Cogn Affect Behav Neurosci* 8(4):375–89.

50. Palmiter RD (2008) Dopamine Signaling in the Dorsal Striatum Is Essential for Motivated Behaviors: Lessons from Dopamine-Deficient Mice. *Ann N Y Acad Sci* 1129:35–46.

51. Beninger RJ, Miller R (1998) Dopamine D1-Like Receptors and Reward-Related Incentive Learning. *Neurosci Biobehav Rev* 22(2):335–45. Ding and Perkel (2004). Eyny YS, Horvitz JC (2003) Opposing Roles of D1 and D2 Receptors in Appetitive Conditioning. *J Neurosci* 23(5):1584–87.

52. Aragona et al. (2006).

53. Shair HN, Muller JM, Moore H (2009) Dopamine's Role in Social Modulation of Infant Isolation-Induced Vocalization: I. Reunion Responses to the Dam, but Not Littermates, Are Dopamine Dependent. *Dev Psychobiol* 51(2):131–46.

54. Levine and Wiener (1988).

55. Beardsworth T (1977) *A Sense of Presence: The Phenomenology of Certain Kinds of Visionary and Ecstatic Experience, Based on a Thousand Contemporary First-Hand Accounts.* Religious Experience Research Unit: Oxford, UK.

56. Champagne FA, Chretien P, Stevenson CW, Zhang TY, Gratton A, Meaney MJ (2004) Variations in Nucleus Accumbens Dopamine Associated with Individual Differences in Maternal Behavior in the Rat. *J Neurosci* 24(17):4113–23.

57. Shahrokh DK, Zhang TY, Diorio J, Gratton A, Meaney MJ (2010) Oxytocin-Dopamine Interactions Mediate Variations in Maternal Behavior in the Rat. *Endocrinology* 151(5):2276–86.

58. Numan M, Numan MJ, Pliakou N, Stolzenberg DS, Mullins OJ, Murphy JM, Smith CD (2005) The Effects of D1 or D2 Dopamine Receptor Antagonism in the Medial Preoptic Area, Ventral Pallidum, or Nucleus Accumbens on the Maternal Retrieval Response and Other Aspects of Maternal Behavior in Rats. *Behav Neurosci* 119(6):1588–604.

59. Mattson BJ, Williams S, Rosenblatt JS, Morrell JI (2001) Comparison of Two Positive Reinforcing Stimuli: Pups and Cocaine throughout the Postpartum Period. *Behav Neurosci* 115(3):683–94.

60. Kalivas PW, Volkow ND (2005) The Neural Basis of Addiction: A Pathology of Motivation and Choice. *Am J Psychiatry* 162(8):1403–13. Wanat MJ, Willuhn I, Clark JJ, Phillips PE (2009) Phasic Dopamine Release in Appetitive Behaviors and Drug Addiction. *Curr Drug Abuse Rev* 2(2):195–213.

61. Crombag HS, Bossert JM, Koya E, Shaham Y (2008) Review. Context-Induced Relapse to Drug Seeking: A Review. *Philos Trans R Soc Lond B Biol Sci* 363(1507):3233–43.

62. Davey CG, Allen NB, Harrison BJ, Dwyer DB, Yucel M (2010) Being Liked Activates Primary Reward and Midline Self-Related Brain Regions. *Hum Brain Mapp* 31(4):660–68. Depue RA, Morrone-Strupinsky JV (2005) A Neurobehavioral Model of Affiliative Bonding: Implications for Conceptualizing a Human Trait of Affiliation. *Behav Brain Sci* 28(3):313–50; discussion 350–95. Insel TR (2003) Is Social Attachment an Addictive Disorder? *Physiol Behav* 79(3):351–57.

63. Cavada C, Goldman-Rakic PS (1989) Posterior Parietal Cortex in Rhesus Monkey: II. Evidence for Segregated Corticocortical Networks Linking Sensory and Limbic Areas with the Frontal Lobe. *J Comp Neurol* 287(4):422–45. Pandya DN, Van Hoesen GW, Mesulam MM (1981) Efferent Connections of the Cingulate Gyrus in the Rhesus Monkey. *Exp Brain Res* 42(3–4):319–30.

Chapter 6

1. Critchley M (1955) The Idea of a Presence. *Acta Psychiatr Neurol Scand* 30(1–2):155–68, p 159.

2. Piaget J, Inhelder B (1969) *The Psychology of the Child.* Basic Books: New York, pp 22–24.

3. Gallese V (2007) Before and Below "Theory of Mind": Embodied Simulation and the Neural Correlates of Social Cognition. *Philos Trans R Soc Lond B Biol Sci* 362(1480):659–69.

4. Niedenthal PM, Mermillod M, Maringer M, Hess U (2010) The Simulation of Smiles (SIMS) Model: Embodied Simulation and the Meaning of Facial Expression. *Behav Brain Sci* 33(6):417–33; discussion 433–80.

5. Blackmore S (2017) The New Science of Out-of-Body Experiences. In: Almqvist K, Haag A, [eds] *The Return of Consciousness: A New Science on Old Questions.* Axess Publishing AB: Stockholm, Sweden.

6. Bahnemann M, Dziobek I, Prehn K, Wolf I, Heekeren HR (2010) Sociotopy in the Temporoparietal Cortex: Common versus Distinct Processes. *Soc Cogn Affect Neurosci* 5(1):48–58.

7. Bundick T Jr, Spinella M (2000) Subjective Experience, Involuntary Movement, and Posterior Alien Hand Syndrome. *J Neurol Neurosurg Psychiatry* 68(1):83–85. Leiguarda R, Starkstein S, Nogues M, Berthier M, Arbelaiz R (1993) Paroxysmal Alien Hand Syndrome. *J Neurol Neurosurg Psychiatry* 56(7):788–92.

8. Banks G, Short P, Martinez J, Latchaw R, Ratcliff G, Boller F (1989) The Alien Hand Syndrome. Clinical and Postmortem Findings. *Arch Neurol* 46(4):456–59. Kubrick S, George P, Southern T (1964) *Dr Strangelove, or How I Learned to Stop Worrying and Love the Bomb.* Columbia Pictures.

9. Mesulam MM (1999) Spatial Attention and Neglect: Parietal, Frontal and Cingulate Contributions to the Mental Representation and Attentional Targeting of Salient Extrapersonal Events. *Philos Trans R Soc Lond B Biol Sci* 354(1387):1325–46. Vallar G (2007) Spatial Neglect, Balint-Homes' and Gerstmann's Syndrome, and Other Spatial Disorders. *CNS Spectr* 12(7):527–36. Vallar G, Ronchi R (2009) Somatoparaphrenia: A Body Delusion. A Review of the Neuropsychological Literature. *Exp Brain Res* 192(3):533–51.

10. Lenggenhager B, Smith ST, Blanke O (2006) Functional and Neural Mechanisms of Embodiment: Importance of the Vestibular System and the Temporal Parietal Junction. *Rev Neurosci* 17(6):643–57.

11. French CC (2001) Dying to Know the Truth: Visions of a Dying Brain, or False Memories? *Lancet* 358:2010–11.

12. Wathey JC (2016) *The Illusion of God's Presence: The Biological Origins of Spiritual Longing.* Prometheus Books: Amherst, NY, pp 284–86. Woerlee G (2005) *Mortal Minds: The Biology of Near-Death Experiences.* Prometheus Books: Amherst, NY.

13. Husain M, Nachev P (2007) Space and the Parietal Cortex. *Trends Cogn Sci* 11(1):30–36.

14. Yamakawa Y, Kanai R, Matsumura M, Naito E (2009) Social Distance Evaluation in Human Parietal Cortex. *PLoS One* 4(2):e4360.

15. Downar J, Crawley AP, Mikulis DJ, Davis KD (2001) The Effect of Task Relevance on the Cortical Response to Changes in Visual and Auditory Stimuli: An Event-Related fMRI Study. *Neuroimage* 14(6):1256–67.

16. Brandt T, Dieterich M (1999) The Vestibular Cortex. Its Locations, Functions, and Disorders. *Ann N Y Acad Sci* 871:293–312. Lewis JW, Van Essen DC (2000) Corticocortical Connections of Visual, Sensorimotor, and Multimodal Processing Areas in the Parietal Lobe of the Macaque Monkey. *J Comp Neurol* 428(1):112–37. Matsuhashi M, Ikeda A, Ohara S, Matsumoto R, Yamamoto J, Takayama M, Satow T, Begum T, Usui K, Nagamine T, Mikuni N, Takahashi J, Miyamoto S, Fukuyama H, Shibasaki H

(2004) Multisensory Convergence at Human Temporo-Parietal Junction—Epicortical Recording of Evoked Responses. *Clin Neurophysiol* 115(5):1145–60.

17. Allison T, Puce A, McCarthy G (2000) Social Perception from Visual Cues: Role of the STS Region. *Trends Cogn Sci* 4(7):267–78. Puce A, Perrett D (2003) Electrophysiology and Brain Imaging of Biological Motion. *Philos Trans R Soc Lond B Biol Sci* 358(1431):435–45.

18. Grossman ED, Blake R (2001) Brain Activity Evoked by Inverted and Imagined Biological Motion. *Vision Res* 41(10–11):1475–82.

19. Grossman ED, Battelli L, Pascual-Leone A (2005) Repetitive TMS over Posterior STS Disrupts Perception of Biological Motion. *Vision Res* 45(22):2847–53.

20. Kontaris I, Wiggett AJ, Downing PE (2009) Dissociation of Extrastriate Body and Biological-Motion Selective Areas by Manipulation of Visual-Motor Congruency. *Neuropsychologia* 47(14):3118–24. Leube DT, Knoblich G, Erb M, Grodd W, Bartels M, Kircher TT (2003) The Neural Correlates of Perceiving One's Own Movements. *Neuroimage* 20(4):2084–90.

21. Farrer C, Frith CD (2002) Experiencing Oneself vs Another Person as Being the Cause of an Action: The Neural Correlates of the Experience of Agency. *Neuroimage* 15(3):596–603.

22. Decety J, Chaminade T, Grezes J, Meltzoff AN (2002) A PET Exploration of the Neural Mechanisms Involved in Reciprocal Imitation. *Neuroimage* 15(1):265–72.

23. Uddin LQ, Kaplan JT, Molnar-Szakacs I, Zaidel E, Iacoboni M (2005) Self-Face Recognition Activates a Frontoparietal "Mirror" Network in the Right Hemisphere: An Event-Related fMRI Study. *Neuroimage* 25(3):926–35.

24. Uddin LQ, Molnar-Szakacs I, Zaidel E, Iacoboni M (2006) rTMS to the Right Inferior Parietal Lobule Disrupts Self-Other Discrimination. *Soc Cogn Affect Neurosci* 1(1):65–71.

25. Vander Wyk BC, Hudac CM, Carter EJ, Sobel DM, Pelphrey KA (2009) Action Understanding in the Superior Temporal Sulcus Region. *Psychol Sci* 20(6):771–77.

26. Saxe R, Kanwisher N (2003) People Thinking about Thinking People. The Role of the Temporo-Parietal Junction in "Theory of Mind." *Neuroimage* 19(4):1835–42, p 1841.

27. Saxe and Kanwisher (2003).

28. Young L, Camprodon JA, Hauser M, Pascual-Leone A, Saxe R (2010) Disruption of the Right Temporoparietal Junction with Transcranial Magnetic Stimulation Reduces the Role of Beliefs in Moral Judgments. *Proc Natl Acad Sci USA* 107(15):6753–58.

29. Lenggenhager B, Smith ST, Blanke O (2006) Functional and Neural Mechanisms of Embodiment: Importance of the Vestibular System and the Temporal Parietal Junction. *Rev Neurosci* 17(6):643–57.

30. Rorsman I, Magnusson M, Johansson BB (1999) Reduction of Visuo-Spatial Neglect with Vestibular Galvanic Stimulation. *Scand J Rehabil Med* 31(2):117–24.

31. Grossmann T, Johnson MH, Lloyd-Fox S, Blasi A, Deligianni F, Elwell C, Csibra G (2008) Early Cortical Specialization for Face-to-Face Communication in Human Infants. *Proc R Soc Lond B Biol Sci* 275(1653):2803–11.

32. Senju A, Johnson MH, Csibra G (2006) The Development and Neural Basis of Referential Gaze Perception. *Soc Neurosci* 1(3–4):220–34.

33. Lloyd-Fox S, Blasi A, Volein A, Everdell N, Elwell CE, Johnson MH (2009) Social Perception in Infancy: A Near Infrared Spectroscopy Study. *Child Dev* 80(4):986–99.

34. Nakato E, Otsuka Y, Kanazawa S, Yamaguchi MK, Watanabe S, Kakigi R (2009) When Do Infants Differentiate Profile Face from Frontal Face? A Near-Infrared Spectroscopic Study. *Hum Brain Mapp* 30(2):462–72.

35. Otsuka Y, Nakato E, Kanazawa S, Yamaguchi MK, Watanabe S, Kakigi R (2007) Neural Activation to Upright and Inverted Faces in Infants Measured by Near Infrared Spectroscopy. *Neuroimage* 34(1):399–406.

36. Honda Y, Nakato E, Otsuka Y, Kanazawa S, Kojima S, Yamaguchi MK, Kakigi R (2010) How Do Infants Perceive Scrambled Face? A Near-Infrared Spectroscopic Study. *Brain Res* 1308:137–46.

37. Carlsson J, Lagercrantz H, Olson L, Printz G, Bartocci M (2008) Activation of the Right Fronto-Temporal Cortex during Maternal Facial Recognition in Young Infants. *Acta Paediatr* 97(9):1221–25.

38. Critchley (1955).

39. Brugger P, Regard M, Landis T (1996) Unilaterally Felt "Presences": The Neuropsychiatry of One's Invisible Doppelganger. *Neuropsychiatry Neuropsychol Behav Neurol* 9(2):114–22.

40. Zijlmans M, van Eijsden P, Ferrier CH, Kho KH, van Rijen PC, Leijten FS (2009) Illusory Shadow Person Causing Paradoxical Gaze Deviations during Temporal Lobe Seizures. *J Neurol Neurosurg Psychiatry* 80(6):686–88.

41. Arzy S, Seeck M, Ortigue S, Spinelli L, Blanke O (2006) Induction of an Illusory Shadow Person. *Nature* 443(7109):287.

42. Spengler S, von Cramon DY, Brass M (2009) Was It Me or Was It You? How the Sense of Agency Originates from Ideomotor Learning Revealed by fMRI. *Neuroimage* 46(1):290–98.

43. Cavanna AE, Trimble MR (2006) The Precuneus: A Review of Its Functional Anatomy and Behavioural Correlates. *Brain* 129(Pt 3):564–83.

44. Astafiev SV, Stanley CM, Shulman GL, Corbetta M (2004) Extrastriate Body Area in Human Occipital Cortex Responds to the Performance of Motor Actions. *Nat Neurosci* 7(5):542–48.

45. Brandt T, Dieterich M (1999) The Vestibular Cortex. Its Locations, Functions, and Disorders. *Ann N Y Acad Sci* 871:293–312. Craig AD (2009) How Do You Feel—Now? The Anterior Insula and Human Awareness. *Nat Rev Neurosci* 10(1):59–70.

46. Farrer and Frith (2002); Karnath HO, Baier B (2010) Right Insula for Our Sense of Limb Ownership and Self-Awareness of Actions. *Brain Struct Funct* 214(5–6):411–17.

47. Small DM (2010) Taste Representation in the Human Insula. *Brain Struct Funct* 214(5–6):551–61.

48. Naqvi NH, Bechara A (2010) The Insula and Drug Addiction: An Interoceptive View of Pleasure, Urges, and Decision-Making. *Brain Struct Funct* 214(5–6):435–50.

49. Wiech K, Lin CS, Brodersen KH, Bingel U, Ploner M, Tracey I (2010) Anterior Insula Integrates Information about Salience into Perceptual Decisions about Pain. *J Neurosci* 30(48):16324–31.

50. Cavada C, Company T, Tejedor J, Cruz-Rizzolo RJ, Reinoso-Suarez F (2000) The Anatomical Connections of the Macaque Monkey Orbitofrontal Cortex. A Review.

Cereb Cortex 10(3):220–42. Medford N, Critchley HD (2010) Conjoint Activity of Anterior Insular and Anterior Cingulate Cortex: Awareness and Response. *Brain Struct Funct* 214(5–6):535–49.

51. Picard F, Craig AD (2009) Ecstatic Epileptic Seizures: A Potential Window on the Neural Basis for Human Self-Awareness. *Epilepsy Behav* 16(3):539–46.

52. Landtblom AM, Lindehammar H, Karlsson H, Craig AD (2011) Insular Cortex Activation in a Patient with "Sensed Presence"/Ecstatic Seizures. *Epilepsy Behav* 20(4):714–18.

Chapter 7

1. James W (1902) [1922] *The Varieties of Religious Experience: A Study in Human Nature.* Longmans, Green: New York, p 73.

2. Boorstin DJ (1983) *The Discoverers: A History of Man's Search to Know His World and Himself.* Random House: New York, p 92.

3. Penfield W, Perot P (1963) The Brain's Record of Auditory and Visual Experience: A Final Summary and Discussion. *Brain* 86:595–696.

4. O'Connor AR, Lever C, Moulin CJ (2010) Novel Insights into False Recollection: A Model of *Deja Vecu*. *Cogn Neuropsychiatry* 15(1):118–44.

5. Loftus E (2003) Our Changeable Memories: Legal and Practical Implications. *Nat Rev Neurosci* 4(3):231–34.

6. Burton R (2009) *On Being Certain: Believing You Are Right Even When You're Not.* St. Martin's Griffin: New York.

7. Kim H, Cabeza R (2007) Trusting Our Memories: Dissociating the Neural Correlates of Confidence in Veridical versus Illusory Memories. *J Neurosci* 27(45):12190–97.

8. Gallo DA (2010) False Memories and Fantastic Beliefs: 15 Years of the DRM Illusion. *Mem Cognit* 38(7):833–48. Straube B (2012) An Overview of the Neuro-Cognitive Processes Involved in the Encoding, Consolidation, and Retrieval of True and False Memories. *Behav Brain Funct* 8:35.

9. See table 4 and figure 5b of Kim and Cabeza (2007).

10. See table 2 and figure 3b of Kim and Cabeza (2007).

11. See table 4 and figure 5a of Kim and Cabeza (2007). A more recent study has found evidence that the anterior temporal lobe is also involved in the generation of high-confidence false memories in word-recall tests of this kind: Gallate J, Chi R, Ellwood S, Snyder A (2009) Reducing False Memories by Magnetic Pulse Stimulation. *Neurosci Lett* 449(3):151–54.

12. Howlett JR, Paulus MP (2015) The Neural Basis of Testable and Non-Testable Beliefs. *PLoS One* 10(5):e0124596.

13. Kahneman D (2011) *Thinking, Fast and Slow.* Farrar, Straus and Giroux: New York.

14. Howlett and Paulus (2015) p 6. Lindeman M, Svedholm AM, Riekki T, Raij T, Hari R (2013) Is It Just a Brick Wall or a Sign from the Universe? An fMRI Study of Supernatural Believers and Skeptics. *Soc Cogn Affect Neurosci* 8(8):943–49.

15. Howlett and Paulus (2015) pp 6–8.

16. Eliot L (1999) *What's Going on in There? How the Brain and Mind Develop in the First Five Years of Life*. Bantam: New York.

17. Sewards TV, Sewards MA (2002) Innate Visual Object Recognition in Vertebrates: Some Proposed Pathways and Mechanisms. *Comp Biochem Physiol A Mol Integr Physiol* 132(4):861–91.

18. Long KD, Kennedy G, Balaban E (2001) Transferring an Inborn Auditory Perceptual Predisposition with Interspecies Brain Transplants. *Proc Natl Acad Sci USA* 98(10):5862–67.

19. Dubowitz LM, Mushin J, De Vries L, Arden GB (1986) Visual Function in the Newborn Infant: Is It Cortically Mediated? *Lancet* 1(8490):1139–41.

20. Johnson MH (2005) Subcortical Face Processing. *Nat Rev Neurosci* 6(10):766–74. Vuilleumier P, Armony JL, Driver J, Dolan RJ (2003) Distinct Spatial Frequency Sensitivities for Processing Faces and Emotional Expressions. *Nat Neurosci* 6(6):624–31.

21. Rosander K, Nystrom P, Gredeback G, von Hofsten C (2007) Cortical Processing of Visual Motion in Young Infants. *Vision Res* 47(12):1614–23.

22. Csibra G, Tucker LA, Volein A, Johnson MH (2000) Cortical Development and Saccade Planning: The Ontogeny of the Spike Potential. *Neuroreport* 11(5):1069–73.

23. Csibra G, Davis G, Spratling MW, Johnson MH (2000) Gamma Oscillations and Object Processing in the Infant Brain. *Science* 290(5496):1582–85.

24. Nagy E, Molnar P (2004) Homo Imitans or Homo Provocans? The Phenomenon of Neonatal Initiation. *Infant Behav Dev* 27:57–63.

25. Buzsáki G, Draguhn A (2004) Neuronal Oscillations in Cortical Networks. *Science* 304(5679):1926–29. Engel AK, Fries P, Singer W (2001) Dynamic Predictions: Oscillations and Synchrony in Top-Down Processing. *Nat Rev Neurosci* 2(10):704–16. Senkowski D, Schneider TR, Foxe JJ, Engel AK (2008) Crossmodal Binding through Neural Coherence: Implications for Multisensory Processing. *Trends Neurosci* 31(8):401–9.

26. Permission from the Canadian Psychological Association Inc. to reprint image from "Figure 1. 40 closure faces arranged with the most difficult 5 at upper left, next most difficult 5 at upper right, and, so on down to easiest 5 at lower right." *Canadian Journal of Psychology*, Volume 11, No. 4, 1 image from Figure 1, p 220 Copyright © 1957 by the Canadian Psychological Association Inc.

27. Rodriguez E, George N, Lachaux JP, Martinerie J, Renault B, Varela FJ (1999) Perception's Shadow: Long-Distance Synchronization of Human Brain Activity. *Nature* 397(6718):430–33.

28. Womelsdorf T, Fries P (2007) The Role of Neuronal Synchronization in Selective Attention. *Curr Opin Neurobiol* 17(2):154–60. Baldauf D, Desimone R (2014) Neural Mechanisms of Object-Based Attention. *Science* 344(6182):424–27.

29. Kanayama N, Sato A, Ohira H (2009) The Role of Gamma Band Oscillations and Synchrony on Rubber Hand Illusion and Crossmodal Integration. *Brain Cogn* 69(1):19–29.

30. Summerfield C, Mangels JA (2005) Functional Coupling between Frontal and Parietal Lobes during Recognition Memory. *Neuroreport* 16(2):117–22.

246 THE PHANTOM GOD

31. Benasich AA, Gou Z, Choudhury N, Harris KD (2008) Early Cognitive and Language Skills Are Linked to Resting Frontal Gamma Power across the First 3 Years. *Behav Brain Res* 195(2):215–22. Uhlhaas PJ, Roux F, Rodriguez E, Rotarska-Jagiela A, Singer W (2010) Neural Synchrony and the Development of Cortical Networks. *Trends Cogn Sci* 14(2):72–80.

32. Imada T, Zhang Y, Cheour M, Taulu S, Ahonen A, Kuhl PK (2006) Infant Speech Perception Activates Broca's Area: A Developmental Magnetoencephalography Study. *Neuroreport* 17(10):957–62. Mahmoudzadeh M, Dehaene-Lambertz G, Fournier M, Kongolo G, Goudjil S, Dubois J, Grebe R, Wallois F (2013) Syllabic Discrimination in Premature Human Infants Prior to Complete Formation of Cortical Layers. *Proc Natl Acad Sci USA* 110(12):4846–51. Mangalathu-Arumana J, Beardsley SA, Liebenthal E (2012) Within-Subject Joint Independent Component Analysis of Simultaneous fMRI/ERP in an Auditory Oddball Paradigm. *Neuroimage* 60(4):2247–57.

33. Näätänen R, Paavilainen P, Rinne T, Alho K (2007) The Mismatch Negativity (MMN) in Basic Research of Central Auditory Processing: A Review. *Clin Neurophysiol* 118(12):2544–90.

34. Näätänen et al. (2007).

35. Atienza M, Cantero JL, Dominguez-Marin E (2002) Mismatch Negativity (MMN): An Objective Measure of Sensory Memory and Long-Lasting Memories during Sleep. *Int J Psychophysiol* 46(3):215–25.

36. Heinke W, Koelsch S (2005) The Effects of Anesthetics on Brain Activity and Cognitive Function. *Curr Opin Anaesthesiol* 18(6):625–31.

37. Atienza et al. (2002). Näätänen et al. (2007).

38. Edwards E, Soltani M, Deouell LY, Berger MS, Knight RT (2005) High Gamma Activity in Response to Deviant Auditory Stimuli Recorded Directly from Human Cortex. *J Neurophysiol* 94(6):4269–80.

39. Mashour GA (2004) Consciousness Unbound: Toward a Paradigm of General Anesthesia. *Anesthesiology* 100:428–33.

40. Lauronen L, Nevalainen P, Pihko E (2012) Magnetoencephalography in Neonatology. *Neurophysiol Clin* 42(1–2):27–34. Martynova O, Kirjavainen J, Cheour M (2003) Mismatch Negativity and Late Discriminative Negativity in Sleeping Human Newborns. *Neurosci Lett* 340(2):75–78.

41. Isler JR, Tarullo AR, Grieve PG, Housman E, Kaku M, Stark RI, Fifer WP (2012) Toward an Electrocortical Biomarker of Cognition for Newborn Infants. *Dev Sci* 15(2):260–71. Kushnerenko E, Ceponiene R, Balan P, Fellman V, Naatanen R (2002) Maturation of the Auditory Change Detection Response in Infants: A Longitudinal ERP Study. *Neuroreport* 13(15):1843–48.

42. Isler et al. (2012). Stefanics G, Haden G, Huotilainen M, Balazs L, Sziller I, Beke A, Fellman V, Winkler I (2007) Auditory Temporal Grouping in Newborn Infants. *Psychophysiology* 44(5):697–702.

43. Ruusuvirta T, Huotilainen M, Fellman V, Näätänen R (2003) The Newborn Human Brain Binds Sound Features Together. *Neuroreport* 14(16):2117–19.

44. Stefanics et al. (2007).

45. Dehaene-Lambertz G, Pena M (2001) Electrophysiological Evidence for Automatic Phonetic Processing in Neonates. *Neuroreport* 12(14):3155–58.

46. He C, Trainor LJ (2009) Finding the Pitch of the Missing Fundamental in Infants. *J Neurosci* 29(24):7718–8822.

47. Imada et al. (2006).

48. Miller DJ, Duka T, Stimpson CD, Schapiro SJ, Baze WB, McArthur MJ, Fobbs AJ, Sousa AM, Sestan N, Wildman DE, Lipovich L, Kuzawa CW, Hof PR, Sherwood CC (2012) Prolonged Myelination in Human Neocortical Evolution. *Proc Natl Acad Sci USA* 109(41):16480–85.

49. Mahmoudzadeh et al. (2013).

50. Benavides-Varela S, Gomez DM, Macagno F, Bion RA, Peretz I, Mehler J (2011) Memory in the Neonate Brain. *PLoS One* 6(11):e27497.

51. Benavides-Varela S, Hochmann JR, Macagno F, Nespor M, Mehler J (2012) Newborn's Brain Activity Signals the Origin of Word Memories. *Proc Natl Acad Sci USA* 109(44):17908–13.

52. Saito Y, Aoyama S, Kondo T, Fukumoto R, Konishi N, Nakamura K, Kobayashi M, Toshima T (2007) Frontal Cerebral Blood Flow Change Associated with Infant-Directed Speech. *Arch Dis Child Fetal Neonatal Ed* 92(2):F113–16.

53. Imada et al. (2006). Perani D, Saccuman MC, Scifo P, Anwander A, Spada D, Baldoli C, Poloniato A, Lohmann G, Friederici AD (2011) Neural Language Networks at Birth. *Proc Natl Acad Sci USA* 108(38):16056–61.

54. Aoyama S, Toshima T, Saito Y, Konishi N, Motoshige K, Ishikawa N, Nakamura K, Kobayashi M (2010) Maternal Breast Milk Odour Induces Frontal Lobe Activation in Neonates: A NIRS Study. *Early Hum Dev* 86(9):541–45. Bartocci M, Winberg J, Ruggiero C, Bergqvist LL, Serra G, Lagercrantz H (2000) Activation of Olfactory Cortex in Newborn Infants after Odor Stimulation: A Functional Near-Infrared Spectroscopy Study. *Pediatr Res* 48(1):18–23. Bartocci M, Winberg J, Papendieck G, Mustica T, Serra G, Lagercrantz H (2001) Cerebral Hemodynamic Response to Unpleasant Odors in the Preterm Newborn Measured by Near-Infrared Spectroscopy. *Pediatr Res* 50(3):324–30.

55. Bartocci M, Bergqvist LL, Lagercrantz H, Anand KJ (2006) Pain Activates Cortical Areas in the Preterm Newborn Brain. *Pain* 122(1–2):109–17.

56. Shibata M, Fuchino Y, Naoi N, Kohno S, Kawai M, Okanoya K, Myowa-Yamakoshi M (2012) Broad Cortical Activation in Response to Tactile Stimulation in Newborns. *Neuroreport* 23(6):373–77.

57. Lee MH, Hacker CD, Snyder AZ, Corbetta M, Zhang D, Leuthardt EC, Shimony JS (2012) Clustering of Resting State Networks. *PLoS One* 7(7):e40370.

58. Fransson P, Skiold B, Engstrom M, Hallberg B, Mosskin M, Aden U, Lagercrantz H, Blennow M (2009) Spontaneous Brain Activity in the Newborn Brain during Natural Sleep—An fMRI Study in Infants Born at Full Term. *Pediatr Res* 66(3):301–5.

59. For references describing these and related findings, see chapter 2 of this book and chapter 11 of Wathey JC (2016) *The Illusion of God's Presence: The Biological Origins of Spiritual Longing.* Prometheus Books: Amherst, NY.

60. Farah MJ, Rabinowitz C, Quinn GE, Liu GT (2000) Early Commitment of Neural Substrates for Face Recognition. *Cogn Neuropsychol* 17(1):117–23.

61. Anderson SW, Damasio H, Tranel D, Damasio AR (2000) Long-Term Sequelae of Prefrontal Cortex Damage Acquired in Early Childhood. *Dev Neuropsychol* 18(3):281–96.

62. Sugita Y (2008) Face Perception in Monkeys Reared with No Exposure to Faces. *Proc Natl Acad Sci USA* 105(1):394–98.

63. Ramachandran VS, Hirstein W (1998) The Perception of Phantom Limbs. The D. O. Hebb Lecture. *Brain* 121 (Pt 9):1603–30.

64. Parkes CM (1972) *Bereavement: Studies of Grief in Adult Life.* Tavistock: London, chapter 11.

65. Melzack R, Israel R, Lacroix R, Schultz G (1997) Phantom Limbs in People with Congenital Limb Deficiency or Amputation in Early Childhood. *Brain* 120 (Pt 9):1603–20.

66. McGeoch PD, Ramachandran VS (2012) The Appearance of New Phantom Fingers Post-Amputation in a Phocomelus. *Neurocase* 18(2):95–97.

67. Brugger P, Kollias SS, Müri RM, Crelier G, Hepp-Reymond MC, Regard M (2000) Beyond Re-Membering: Phantom Sensations of Congenitally Absent Limbs. *Proc Natl Acad Sci USA* 97(11):6167–72, figure 2.

68. Ersland L, Rosén G, Lundervold A, Smievoll AI, Tillung T, Sundberg H, Hugdahl K (1996) Phantom Limb Imaginary Fingertapping Causes Primary Motor Cortex Activation: An fMRI Study. *Neuroreport* 8(1):207–10.

69. Cohen LG, Bandinelli S, Findley TW, Hallett M (1991) Motor Reorganization after Upper Limb Amputation in Man. A Study with Focal Magnetic Stimulation. *Brain* 114(Pt 1B):615–27. Hess CW, Mills KR, Murray NM (1986) Magnetic Stimulation of the Human Brain: Facilitation of Motor Responses by Voluntary Contraction of Ipsilateral and Contralateral Muscles with Additional Observations on an Amputee. *Neurosci Lett* 71(2):235–40. Mercier C, Reilly KT, Vargas CD, Aballea A, Sirigu A (2006) Mapping Phantom Movement Representations in the Motor Cortex of Amputees. *Brain* 129(Pt 8):2202–10. Pascual-Leone A, Peris M, Tormos JM, Pascual AP, Catala MD (1996) Reorganization of Human Cortical Motor Output Maps Following Traumatic Forearm Amputation. *Neuroreport* 7(13):2068–70.

70. Brugger et al. (2000).

71. Brugger et al. (2000) p 6172.

72. Ramachandran VS, Brang D (2009) Sensations Evoked in Patients with Amputation from Watching an Individual Whose Corresponding Intact Limb Is Being Touched. *Arch Neurol* 66(10):1281–84.

73. Price EH (2006) A Critical Review of Congenital Phantom Limb Cases and a Developmental Theory for the Basis of Body Image. *Conscious Cogn* 15(2):310–22.

74. Ferrari PF, Bonini L, Fogassi L (2009) From Monkey Mirror Neurons to Primate Behaviours: Possible "Direct" and "Indirect" Pathways. *Philos Trans R Soc Lond B Biol Sci* 364(1528):2311–23. Nagy E (2006) From Imitation to Conversation: The First Dialogues with Human Neonates. *Infant Child Dev* 15(3):223–32.

75. Ramachandran VS, McGeoch PD (2007) Occurrence of Phantom Genitalia after Gender Reassignment Surgery. *Med Hypotheses* 69(5):1001–3.

76. Ramachandran VS, McGeoch PD (2008) Phantom Penises in Transsexuals: Evidence of an Innate Gender-Specific Body Image in the Brain. *J Conscious Stud* 15(1):5–16.

77. Hilti LM, Hanggi J, Vitacco DA, Kraemer B, Palla A, Luechinger R, Jancke L, Brugger P (2013) The Desire for Healthy Limb Amputation: Structural Brain Correlates and Clinical Features of Xenomelia. *Brain* 136(Pt 1):318–29. McGeoch PD, Brang D, Song T, Lee RR, Huang M, Ramachandran VS (2011) Xenomelia: A New Right Parietal Lobe Syndrome. *J Neurol Neurosurg Psychiatry* 82(12):1314–19.

Chapter 8

1. Coles A (2008) God, Theologian and Humble Neurologist. *Brain* 131:1953–59.

2. For other reviews of this literature, see: Cahn BR, Polich J (2006) Meditation States and Traits: EEG, ERP, and Neuroimaging Studies. *Psychol Bull* 132:180–211. Grafman J, Cristofori I, Zhong W, Bulbulia J (2020) The Neural Basis of Religious Cognition. *Curr Dir Psychol Sci* 29(2):126–33, doi:10.1177/0963721419898183. Lutz A, Dunne JD, Davidson RJ (2007) Meditation and the Neuroscience of Consciousness: An Introduction. In: Zelazo PD, Moscovitch M, Thompson E [eds] *The Cambridge Handbook of Consciousness*. Cambridge University Press: New York, pp 499–551. McNamara P, Durso R, Brown A, Harris E (2006) The Chemistry of Religiosity: Evidence from Patients with Parkinson's Disease. In: McNamara P [ed] *Where God and Science Meet: How Brain and Evolutionary Studies Alter Our Understanding of Religion. Vol 2: The Neurology of Religious Experience*. Praeger Perspectives: Westport, CT, pp 1–14. McNamara P (2009) *The Neuroscience of Religious Experience*. Cambridge University Press: Cambridge, UK.

3. Coles (2008). Dennett DC (2006) *Breaking the Spell: Religion as a Natural Phenomenon*. Viking: New York, p 316. Ratcliffe M (2006) Neurotheology: A Science of What? In: McNamara P [ed] *Where God and Science Meet: How Brain and Evolutionary Studies Alter Our Understanding of Religion. Vol 2: The Neurology of Religious Experience*. Praeger Perspectives: Westport, CT, pp 81–104.

4. Azari NP, Nickel J, Wunderlich G, Niedeggen M, Hefter H, Tellmann L, Herzog H, Stoerig P, Birnbacher D, Seitz RJ (2001) Neural Correlates of Religious Experience. *Eur J Neurosci* 13(8):1649–52.

5. Cavanna AE, Trimble MR (2006) The Precuneus: A Review of Its Functional Anatomy and Behavioural Correlates. *Brain* 129(Pt 3):564–83.

6. Harris S, Kaplan JT, Curiel A, Bookheimer SY, Iacoboni M, Cohen MS (2009) The Neural Correlates of Religious and Nonreligious Belief. *PLoS One* 4(10):e0007272.

7. Harris et al. (2009). Kim H, Cabeza R (2007) Trusting Our Memories: Dissociating the Neural Correlates of Confidence in Veridical versus Illusory Memories. *J Neurosci* 27(45):12190–97.

8. Compare figure 3 and table 5 of Harris et al. (2009) with figure 5b and table 4 of Kim and Cabeza (2007).

9. Poldrack RA (2006) Can Cognitive Processes Be Inferred from Neuroimaging Data? *Trends Cogn Sci* 10(2):59–63.

10. Pagnoni G, Cekic M, Guo Y (2008) "Thinking about Not-Thinking": Neural Correlates of Conceptual Processing during Zen Meditation. *PLoS One* 3(9):e3083.

11. See tables 3 and 4 in: Lutz A, Brefczynski-Lewis J, Johnstone T, Davidson RJ (2008) Regulation of the Neural Circuitry of Emotion by Compassion Meditation: Effects of Meditative Expertise. *PLoS One* 3(3):e1897.

12. See figure 3 and table 5 of Lutz et al. (2008).

13. Schjoedt U, Stodkilde-Jorgensen H, Geertz AW, Roepstorff A (2008) Rewarding Prayers. *Neurosci Lett* 443(3):165–68. Schjoedt U, Stødkilde-Jørgensen H, Geertz AW, Roepstorff A (2009) Highly Religious Participants Recruit Areas of Social Cognition in Personal Prayer. *Soc Cogn Affect Neurosci* 4(2):199–207.

14. Schjoedt et al. (2008).

15. Everitt BJ, Robbins TW (2005) Neural Systems of Reinforcement for Drug Addiction: From Actions to Habits to Compulsion. *Nat Neurosci* 8(11):1481–89. Gerdeman GL, Partridge JG, Lupica CR, Lovinger DM (2003) It Could Be Habit Forming: Drugs of Abuse and Striatal Synaptic Plasticity. *Trends Neurosci* 26(4):184–92.

16. See figure 1 and table 4 of Schjoedt et al. (2009).

17. See p 187 in: Beauregard M, Paquette V (2006) Neural Correlates of a Mystical Experience in Carmelite Nuns. *Neurosci Lett* 405(3):186–90.

18. See table 1 of Beauregard and Paquette (2006).

19. Ferguson MA, Nielsen JA, King JB, Dai L, Giangrasso DM, Holman R, Korenberg JR, Anderson JS (2018) Reward, Salience, and Attentional Networks Are Activated by Religious Experience in Devout Mormons. *Soc Neurosci* 13(1):104–16.

20. Wathey JC (2016) *The Illusion of God's Presence: The Biological Origins of Spiritual Longing.* Prometheus Books: Amherst, NY, chapter 12.

Chapter 9

1. Pascal B (1669) [1958] *Pascal's Pensées.* Dutton: New York, p x.

2. Persinger MA, Healey F (2002) Experimental Facilitation of the Sensed Presence: Possible Intercalation between the Hemispheres Induced by Complex Magnetic Fields. *J Nerv Ment Dis* 190(8):533–41.

3. Granqvist P, Fredrikson M, Unge P, Hagenfeldt A, Valind S, Larhammar D, Larsson M (2005) Sensed Presence and Mystical Experiences Are Predicted by Suggestibility, Not by the Application of Transcranial Weak Complex Magnetic Fields. *Neurosci Lett* 379(1):1–6.

4. St-Pierre LS, Persinger MA (2006) Experimental Facilitation of the Sensed Presence Is Predicted by the Specific Patterns of the Applied Magnetic Fields, Not by Suggestibility: Re-Analyses of 19 Experiments. *Int J Neurosci* 116(9):1079–96.

5. Leuba J (1925) *The Psychology of Religious Mysticism.* Harcourt, Brace: New York, pp 283–86.

6. Leuba (1925) p 286.

7. Bexton WH, Heron W, Scott RH (1954) Effects of Decreased Variation in the Sensory Environment. *Can J Psychol* 8(2):70–76.

8. Buzsáki G, Draguhn A (2004) Neuronal Oscillations in Cortical Networks. *Science* 304(5679):1926–29.

9. Richards PM, Koren SA, Persinger MA (1992) Experimental Stimulation by Burst-Firing Weak Magnetic Fields over the Right Temporal Lobe May Facilitate Apprehension in Women. *Percept Mot Skills* 75(2):667–70.

10. Blackmore S (1994) Alien Abduction: The Inside Story. *New Sci* [November 19]:29–31.

11. Persinger MA (2000) The UFO Experience: A Normal Correlate of Human Brain Function. In: Jacobs DM [ed] *UFOs and Abduction: Challenging the Borders of Knowledge.* University Press: Lawrence, KS, pp 262–302.

12. Persinger MA (1987) *Neuropsychological Bases of God Beliefs.* Praeger: New York.

13. For an irreverent critique of Persinger's ideas by a pseudonymous sleuth, see Clyde S (2018) Michael Persinger's Crank Magnetism. https://forbetterscience.com/2018/04/03/michael-persingers-crank-magnetism/ (accessed 1 November 2021). To browse the titles of Persinger's publications, see https://neurotree.org/beta/publications.php?pid=12461 (accessed 1 November 2021).

14. Pascual-Leone A, Bartres-Faz D, Keenan JP (1999) Transcranial Magnetic Stimulation: Studying the Brain-Behaviour Relationship by Induction of "Virtual Lesions." *Philos Trans R Soc Lond B Biol Sci* 354(1387):1229–38.

15. Crescentini C, Aglioti SM, Fabbro F, Urgesi C (2014) Virtual Lesions of the Inferior Parietal Cortex Induce Fast Changes of Implicit Religiousness/Spirituality. *Cortex* 54:1–15.

16. Lutz A, Brefczynski-Lewis J, Johnstone T, Davidson RJ (2008) Regulation of the Neural Circuitry of Emotion by Compassion Meditation: Effects of Meditative Expertise. *PLoS One* 3(3):e1897. Newberg A, Alavi A, Baime M, Pourdehnad M, Santanna J, d'Aquili E (2001) The Measurement of Regional Cerebral Blood Flow during the Complex Cognitive Task of Meditation: A Preliminary SPECT Study. *Psychiatry Res* 106(2):113–22. Urgesi C, Aglioti SM, Skrap M, Fabbro F (2010) The Spiritual Brain: Selective Cortical Lesions Modulate Human Self-Transcendence. *Neuron* 65(3):309–19.

17. Crescentini et al. (2014).

18. Holbrook C, Izuma K, Deblieck C, Fessler DM, Iacoboni M (2016) Neuromodulation of Group Prejudice and Religious Belief. *Soc Cogn Affect Neurosci* 11(3):387–94.

19. Burke BL, Martens A, Faucher EH (2010) Two Decades of Terror Management Theory: A Meta-Analysis of Mortality Salience Research. *Pers Soc Psychol Rev* 14(2):155–95.

20. Golec de Zavala A, Lantos D (2020) Collective Narcissism and Its Social Consequences: The Bad and the Ugly. *Curr Dir Psychol Sci* 29(3):273–78, doi:10.1177/0963721420917703.

21. Huang YZ, Edwards MJ, Rounis E, Bhatia KP, Rothwell JC (2005) Theta Burst Stimulation of the Human Motor Cortex. *Neuron* 45:201–6.

22. Holbrook et al. (2016).

23. Burke et al. (2010).

24. Wathey JC (2016) *The Illusion of God's Presence: The Biological Origins of Spiritual Longing.* Prometheus Books: Amherst, NY, chapter 7.

25. Brugger P, Regard M, Landis T (1996) Unilaterally Felt "Presences": The Neuropsychiatry of One's Invisible Doppelganger. *Neuropsychiatry Neuropsychol Behav Neurol* 9(2):114–22. Critchley M (1955) The Idea of a Presence. *Acta Psychiatr Neurol Scand* 30(1–2):155–68.

26. Min YS, Park JW, Jin SU, Jang KE, Lee BJ, Lee HJ, Lee J, Lee YS, Chang Y, Jung TD (2016) Neuromodulatory Effects of Offline Low-Frequency Repetitive Transcranial Magnetic Stimulation of the Motor Cortex: A Functional Magnetic Resonance Imaging Study. *Sci Rep* 6:36058.

27. Holbrook et al. (2016).

28. Huang et al. (2005).

29. Boucher PO, Ozdemir RA, Momi D, Burke MJ, Jannati A, Fried PJ, Pascual-Leone A, Shafi MM, Santarnecchi E (2021) Sham-Derived Effects and the Minimal Reliability of Theta Burst Stimulation. *Sci Rep* 11(1):21170.

30. Bergmann TO, Varatheeswaran R, Hanlon CA, Madsen KH, Thielscher A, Siebner HR (2021) Concurrent TMS-fMRI for Causal Network Perturbation and Proof of Target Engagement. *Neuroimage* 237:118093.

31. Michels S (2008) Researchers Explore New Technologies, Treatments for Dementia Patients. *PBS Newshour* [12 November] www.pbs.org/newshour/show/researchers -explore-new-technologies-treatments-for-dementia-patients (accessed 11 January 2022).

32. Onyike CU, Diehl-Schmid J (2013) The Epidemiology of Frontotemporal Dementia. *Int Rev Psychiatry* 25(2):130–37.

33. Chan D, Anderson V, Pijnenburg Y, Whitwell J, Barnes J, Scahill R, Stevens JM, Barkhof F, Scheltens P, Rossor MN, Fox NC (2009) The Clinical Profile of Right Temporal Lobe Atrophy. *Brain* 132(Pt 5):1287–98.

34. Miller BL, Seeley WW, Mychack P, Rosen HJ, Mena I, Boone K (2001) Neuroanatomy of the Self: Evidence from Patients with Frontotemporal Dementia. *Neurology* 57:817–21.

35. Karatzikou M, Afrantou T, Parissis D, Ioannidis P (2020) Hyper-Religiosity in Frontotemporal Dementia with Predominant Atrophy of the Right Temporal Lobe. *Pract Neurol* [Epub ahead of print].

36. Everhart DE, Watson EM, Bickel KL, Stephenson AJ (2015) Right Temporal Lobe Atrophy: A Case That Initially Presented as Excessive Piety. *Clin Neuropsychol* 29(7):1053–67.

37. Asp E, Ramchandran K, Tranel D (2012) Authoritarianism, Religious Fundamentalism, and the Human Prefrontal Cortex. *Neuropsychology* 26(4):414–21.

38. Asp et al. (2012).

39. Karnath HO, Sperber C, Rorden C (2018) Mapping Human Brain Lesions and Their Functional Consequences. *Neuroimage* 165:180–89.

40. Zhong W, Cristofori I, Bulbulia J, Krueger F, Grafman J (2017) Biological and Cognitive Underpinnings of Religious Fundamentalism. *Neuropsychologia* 100:18–25.

41. Cohen-Zimerman S, Cristofori I, Zhong W, Bulbulia J, Krueger F, Gordon B, Grafman J (2020) Neural Underpinning of a Personal Relationship with God and Sense of Control: A Lesion-Mapping Study. *Cogn Affect Behav Neurosci* 20(3):575–87.

42. Asp et al. (2012); Zhong et al. (2017).

43. Beauregard M, Paquette V (2006) Neural Correlates of a Mystical Experience in Carmelite Nuns. *Neurosci Lett* 405(3):186–90. Ferguson MA, Nielsen JA, King JB, Dai L, Giangrasso DM, Holman R, Korenberg JR, Anderson JS (2018) Reward, Salience, and Attentional Networks Are Activated by Religious Experience in Devout Mormons. *Soc Neurosci* 13(1):104–16.

44. Nitschke JB, Nelson EE, Rusch BD, Fox AS, Oakes TR, Davidson RJ (2004) Orbitofrontal Cortex Tracks Positive Mood in Mothers Viewing Pictures of Their Newborn Infants. *Neuroimage* 21(2):583–92. Noriuchi M, Kikuchi Y, Senoo A (2008) The Functional Neuroanatomy of Maternal Love: Mother's Response to Infant's Attachment Behaviors. *Biol Psychiatry* 63(4):415–23.

45. Aoyama S, Toshima T, Saito Y, Konishi N, Motoshige K, Ishikawa N, Nakamura K, Kobayashi M (2010) Maternal Breast Milk Odour Induces Frontal Lobe Activation in Neonates: A NIRS Study. *Early Hum Dev* 86(9):541–45. Bartocci M, Winberg J, Ruggiero C, Bergqvist LL, Serra G, Lagercrantz H (2000) Activation of Olfactory Cortex in Newborn Infants after Odor Stimulation: A Functional Near-Infrared Spectroscopy Study. *Pediatr Res* 48(1):18–23. Minagawa-Kawai Y, Matsuoka S, Dan I, Naoi N, Nakamura K, Kojima S (2009) Prefrontal Activation Associated with Social Attachment: Facial-Emotion Recognition in Mothers and Infants. *Cereb Cortex* 19(2):284–92. Saito Y, Aoyama S, Kondo T, Fukumoto R, Konishi N, Nakamura K, Kobayashi M, Toshima T (2007) Frontal Cerebral Blood Flow Change Associated with Infant-Directed Speech. *Arch Dis Child Fetal Neonatal Ed* 92(2):F113–16.

46. Beauregard and Paquette (2006); Ferguson et al. (2018).

47. Asp et al. (2012); Zhong et al. (2017).

48. Urgesi C, Aglioti SM, Skrap M, Fabbro F (2010) The Spiritual Brain: Selective Cortical Lesions Modulate Human Self-Transcendence. *Neuron* 65(3):309–19.

49. Urgesi et al. (2010).

50. Urgesi et al. (2010) p 313.

51. Toga AW, Clark KA, Thompson PM, Shattuck DW, Van Horn JD (2012) Mapping the Human Connectome. *Neurosurgery* 71(1):1–5.

52. Holmes AJ, Hollinshead MO, O'Keefe TM, Petrov VI, Fariello GR, Wald LL, Fischl B, Rosen BR, Mair RW, Roffman JL, Smoller JW, Buckner RL (2015) Brain Genomics Superstruct Project Initial Data Release with Structural, Functional, and Behavioral Measures. *Sci Data* 2:150031.

53. Boes AD, Prasad S, Liu H, Liu Q, Pascual-Leone A, Caviness VS Jr, Fox MD (2015) Network Localization of Neurological Symptoms from Focal Brain Lesions. *Brain* 138(Pt 10):3061–75.

54. Ferguson MA, Schaper FLWVJ, Cohen A, Siddiqi S, Merrill SM, Nielsen JA, Grafman J, Urgesi C, Fabbro F, Fox MD (2021) A Neural Circuit for Spirituality and Religiosity Derived from Patients with Brain Lesions. *Biol Psychiatry* [Epub ahead of print].

55. Urgesi et al. (2010).

56. Ferguson et al. (2021).

57. Olson I, Suryanarayana SM, Robertson B, Grillner S (2017) Griseum Centrale, a Homologue of the Periaqueductal Gray in the Lamprey. *IBRO Rep* 2:24–30.

58. Gross CT, Canteras NS (2012) The Many Paths to Fear. *Nat Rev Neurosci* 13(9):651–58.

59. Goodson JL (2005) The Vertebrate Social Behavior Network: Evolutionary Themes and Variations. *Horm Behav* 48(1):11–22. Motta SC, Carobrez AP, Canteras NS (2017) The Periaqueductal Gray and Primal Emotional Processing Critical to Influence Complex Defensive Responses, Fear Learning and Reward Seeking. *Neurosci Biobehav Rev* 76(Pt A):39–47.

60. Subramanian HH, Holstege G (2014) The Midbrain Periaqueductal Gray Changes the Eupneic Respiratory Rhythm into a Breathing Pattern Necessary for Survival of the Individual and of the Species. *Prog Brain Res* 212:351–84.

61. Bagley EE, Ingram SL (2020) Endogenous Opioid Peptides in the Descending Pain Modulatory Circuit. *Neuropharmacology* 173:108131.

62. Zare A, Jahanshahi A, Rahnama'i MS, Schipper S, van Koeveringe GA (2019) The Role of the Periaqueductal Gray Matter in Lower Urinary Tract Function. *Mol Neurobiol* 56(2):920–34.

63. Maney DL (2013) The Incentive Salience of Courtship Vocalizations: Hormone-Mediated "Wanting" in the Auditory System. *Hear Res* 305:19–30. Tschida K, Michael V, Han BX, Zhao S, Sakurai K, Mooney R, Wang F (2018) Identification of Midbrain Neurons Essential for Vocal Communication. *bioRχiv* 310250, doi: https://doi.org/10.1101/310250.

64. Beauregard M, Courtemanche J, Paquette V, St-Pierre EL (2009) The Neural Basis of Unconditional Love. *Psychiatry Res* 172(2):93–98.

65. Linnman C, Moulton EA, Barmettler G, Becerra L, Borsook D (2012) Neuroimaging of the Periaqueductal Gray: State of the Field. *Neuroimage* 60(1):505–22.

66. Noriuchi M, Kikuchi Y, Senoo A (2008) The Functional Neuroanatomy of Maternal Love: Mother's Response to Infant's Attachment Behaviors. *Biol Psychiatry* 63(4):415–23.

67. Parsons CE, Young KS, Joensson M, Brattico E, Hyam JA, Stein A, Green AL, Aziz TZ, Kringelbach ML (2014) Ready for Action: A Role for the Human Midbrain in Responding to Infant Vocalizations. *Soc Cogn Affect Neurosci* 9(7):977–84.

68. Lonstein JS, Simmons DA, Stern JM (1998) Functions of the Caudal Periaqueductal Gray in Lactating Rats: Kyphosis, Lordosis, Maternal Aggression, and Fearfulness. *Behav Neurosci* 112:1502–18.

69. Miranda-Paiva CM, Ribeiro-Barbosa ER, Canteras NS, Felicio LF (2003) A Role for the Periaqueductal Grey in Opioidergic Inhibition of Maternal Behaviour. *Eur J Neurosci* 18(3):667–74.

70. Wiedenmayer CP, Goodwin GA, Barr GA (2000) The Effect of Periaqueductal Gray Lesions on Responses to Age-Specific Threats in Infant Rats. *Brain Res Dev Brain Res* 120(2):191–98.

71. Liao PY, Chiu YM, Yu JH, Chen SK (2020) Mapping Central Projection of Oxytocin Neurons in Unmated Mice Using Cre and Alkaline Phosphatase Reporter. *Front Neuroanat* 14:559402.

72. Jenkins JS, Ang VT, Hawthorn J, Rossor MN, Iversen LL (1984) Vasopressin, Oxytocin and Neurophysins in the Human Brain and Spinal Cord. *Brain Res* 291(1):111–17.

73. Mantyh PW (1982) The Ascending Input to the Midbrain Periaqueductal Gray of the Primate. *J Comp Neurol* 211(1):50–64. Mantyh PW (1983) Connections of Midbrain Periaqueductal Gray in the Monkey. II. Descending Efferent Projections. *J Neurophysiol* 49(3):582–94.

74. Beitz AJ (1982) The Organization of Afferent Projections to the Midbrain Periaqueductal Gray of the Rat. *Neuroscience* 7(1):133–59. Mantyh PW (1982) Forebrain Projections to the Periaqueductal Gray in the Monkey, with Observations in the Cat and Rat. *J Comp Neurol* 206(2):146–58. Mantyh PW (1983) Connections of Midbrain Periaqueductal Gray in the Monkey. I. Ascending Efferent Projections. *J Neurophysiol* 49(3):567–81. Linnman C, Moulton EA, Barmettler G, Becerra L, Borsook D (2012) Neuroimaging of the Periaqueductal Gray: State of the Field. *Neuroimage* 60(1):505–22.

75. Beardsworth T (1977) *A Sense of Presence: The Phenomenology of Certain Kinds of Visionary and Ecstatic Experience, Based on a Thousand Contemporary First-Hand Accounts.* Religious Experience Research Unit: Oxford, UK. Chan D, Rossor MN (2002) "But Who Is That on the Other Side of You?" Extracampine Hallucinations Revisited. *Lancet* 360:2064–66. Cheyne JA, Girard TA (2007) Paranoid Delusions and Threatening Hallucinations: A Prospective Study of Sleep Paralysis Experiences. *Conscious Cogn* 16(4):959–74. Critchley M (1955) The Idea of a Presence. *Acta Psychiatr Neurol Scand* 30(1–2):155–68. Geiger J (2009) *The Third Man Factor: The Secret of Survival in Extreme Environments.* Weinstein Books: New York. Shermer M (2010) The Sensed-Presence Effect. *Sci Am* 302(4):34. Solomonova E, Frantova E, Nielsen T (2011) Felt Presence: The Uncanny Encounters with the Numinous Other. *AI Soc* 26:171–78. Suedfeld P, Mocellin JS (1987) The Sensed Presence in Unusual Environments. *Environ Psychol* 19:33–52.

76. Barnby JM, Bell V (2017) The Sensed Presence Questionnaire (SenPQ): Initial Psychometric Validation of a Measure of the "Sensed Presence" Experience. *PeerJ* 5:e3149.

77. Leuba (1925) pp 283–86.

78. Frantova E, Solomonova E, Sutton T (2011) Extra-Personal Awareness through the Media-Rich Environment. *AI Soc* 26(2):179–86.

79. Granqvist et al. (2005).

80. Andersen M, Schjoedt U, Nielbo K, Sorensen J (2014) Mystical Experience in the Lab. *Method Theory Study Relig* 26:217–45, pp 227–29.

81. Andersen et al. (2014) figure 4.

Chapter 10

1. Wasson RG (1957) Seeking the Magic Mushroom. *Life* 42(19):100–120.

2. Wasson (1957).

3. González-Maeso J, Weisstaub NV, Zhou M, Chan P, Ivic L, Ang R, Lira A, Bradley-Moore M, Ge Y, Zhou Q, Sealfon SC, Gingrich JA (2007) Hallucinogens Recruit Specific Cortical 5-HT(2A) Receptor-Mediated Signaling Pathways to Affect Behavior. *Neuron* 53(3):439–52.

4. Muetzelfeldt L, Kamboj SK, Rees H, Taylor J, Morgan CJ, Curran HV (2008) Journey through the K-Hole: Phenomenological Aspects of Ketamine Use. *Drug Alcohol Depend* 95(3):219–29. Pomarol-Clotet E, Honey GD, Murray GK, Corlett PR, Absalom AR, Lee M, McKenna PJ, Bullmore ET, Fletcher PC (2006) Psychological Effects of Ketamine in Healthy Volunteers. Phenomenological Study. *Br J Psychiatry* 189:173–79. Wilkins LK, Girard TA, Cheyne JA (2011) Ketamine as a Primary Predictor of Out-of-Body Experiences Associated with Multiple Substance Use. *Conscious Cogn* 20(3):943–50.

5. Johnson MW, MacLean KA, Reissig CJ, Prisinzano TE, Griffiths RR (2011) Human Psychopharmacology and Dose-Effects of Salvinorin A, a Kappa Opioid Agonist Hallucinogen Present in the Plant *Salvia divinorum*. *Drug Alcohol Depend* 115(1–2):150–55.

6. McGregor IS, Callaghan PD, Hunt GE (2008) From Ultrasocial to Antisocial: A Role for Oxytocin in the Acute Reinforcing Effects and Long-Term Adverse Consequences of Drug Use? *Br J Pharmacol* 154(2):358–68.

7. Ruck CA, Bigwood J, Staples D, Ott J, Wasson RG (1979) Entheogens. *J Psychedelic Drugs* 11(1–2):145–46.

8. Nichols DE, Chemel BR (2006) The Neuropharmacology of Religious Experience: Hallucinogens and the Experience of the Divine. In: McNamara P [ed] *Where God and Science Meet: How Brain and Evolutionary Studies Alter Our Understanding of Religion. Vol 3: The Psychology of Religious Experience.* Praeger Perspectives: Westport, CT, pp 1–33.

9. Pahnke WN, Kurland AA, Unger S, Savage C, Grof S (1970) The Experimental Use of Psychedelic (LSD) Psychotherapy. *JAMA* 212(11):1856–63, p 1861.

10. Baggott MJ, Siegrist JD, Galloway GP, Robertson LC, Coyle JR, Mendelson JE (2010) Investigating the Mechanisms of Hallucinogen-Induced Visions Using 3,4-Methylenedioxyamphetamine (MDA): A Randomized Controlled Trial in Humans. *PLoS One* 5(12):e14074. Fisher CM (1991) Visual Hallucinations on Eye Closure Associated with Atropine Toxicity. A Neurological Analysis and Comparison with Other Visual Hallucinations. *Can J Neurol Sci* 18(1):18–27. Fisher CM (1991) Visual Hallucinations and Racing Thoughts on Eye Closure after Minor Surgery. *Arch Neurol* 48(10):1091–92.

11. Hanawalt NG (1954) Recurrent Images: New Instances and a Summary of the Older Ones. *Am J Psychol* 67(1):170–74.

12. Ingle D (2005) Two Kinds of "Memory Images": Experimental Models for Hallucinations? *Behav Brain Sci* 28(6):768. Ingle D (2005) Central Visual Persistences: I. Visual and Kinesthetic Interactions. *Perception* 34(9):1135–51. Ingle D (2006) Central Visual Persistences: II. Effects of Hand and Head Rotations. *Perception* 35(10):1315–29.

13. Behrendt RP, Young C (2004) Hallucinations in Schizophrenia, Sensory Impairment, and Brain Disease: A Unifying Model. *Behav Brain Sci* 27(6):771–87; discussion 787–830. Collerton D, Perry E, McKeith I (2005) Why People See Things That Are Not There: A Novel Perception and Attention Deficit Model for Recurrent Complex Visual Hallucinations. *Behav Brain Sci* 28(6):737–57; discussion 757–94.

14. Wathey JC (2016) *The Illusion of God's Presence: The Biological Origins of Spiritual Longing.* Prometheus Books: Amherst, NY, pp 226–27.

15. Corlett PR, Frith CD, Fletcher PC (2009) From Drugs to Deprivation: A Bayesian Framework for Understanding Models of Psychosis. *Psychopharmacol (Berl)* 206(4):515–30. Friston K (2002) Functional Integration and Inference in the Brain. *Prog Neurobiol* 68(2):113–43. Friston K (2005) A Theory of Cortical Responses. *Philos Trans R Soc Lond B Biol Sci* 360(1456):815–36.

16. Dolgov I, McBeath MK (2005) Signal-Detection-Theory Representation of Normal and Hallucinatory Perception. *Behav Brain Sci* 28(6):761–62.

17. Gill TM, Sarter M, Givens B (2000) Sustained Visual Attention Performance-Associated Prefrontal Neuronal Activity: Evidence for Cholinergic Modulation. *J Neurosci* 20(12):4745–57.

18. McGaughy J, Sarter M (1995) Behavioral Vigilance in Rats: Task Validation and Effects of Age, Amphetamine, and Benzodiazepine Receptor Ligands. *Psychopharmacol (Berl)* 117(3):340–57. Robbins TW (2002) The 5-Choice Serial Reaction Time Task: Behavioural Pharmacology and Functional Neurochemistry. *Psychopharmacology (Berl)* 163(3–4):362–80.

19. Sarter M, Gehring WJ, Kozak R (2006) More Attention Must Be Paid: The Neurobiology of Attentional Effort. *Brain Res Rev* 51(2):145–60.

20. Parikh V, Sarter M (2008) Cholinergic Mediation of Attention: Contributions of Phasic and Tonic Increases in Prefrontal Cholinergic Activity. *Ann N Y Acad Sci* 1129:225–35.

21. Eckenstein FP, Baughman RW, Quinn J (1988) An Anatomical Study of Cholinergic Innervation in Rat Cerebral Cortex. *Neuroscience* 25(2):457–74. Everitt BJ, Robbins TW (1997) Central Cholinergic Systems and Cognition. *Annu Rev Psychol* 48:649–84. Mesulam MM (2004) The Cholinergic Innervation of the Human Cerebral Cortex. *Prog Brain Res* 145:67–78. Varga C, Hartig W, Grosche J, Keijser J, Luiten PG, Seeger J, Brauer K, Harkany T (2003) Rabbit Forebrain Cholinergic System: Morphological Characterization of Nuclei and Distribution of Cholinergic Terminals in the Cerebral Cortex and Hippocampus. *J Comp Neurol* 460(4):597–611.

22. Arnold HM, Burk JA, Hodgson EM, Sarter M, Bruno JP (2002) Differential Cortical Acetylcholine Release in Rats Performing a Sustained Attention Task versus Behavioral Control Tasks That Do Not Explicitly Tax Attention. *Neuroscience* 114(2):451–60.

23. Sarter M, Givens B, Bruno JP (2001) The Cognitive Neuroscience of Sustained Attention: Where Top-Down Meets Bottom-Up. *Brain Res Rev* 35(2):146–60. Rasmusson DD, Smith SA, Semba K (2007) Inactivation of Prefrontal Cortex Abolishes Cortical Acetylcholine Release Evoked by Sensory or Sensory Pathway Stimulation in the Rat. *Neuroscience* 149(1):232–41.

24. Fournier GN, Semba K, Rasmusson DD (2004) Modality- and Region-Specific Acetylcholine Release in the Rat Neocortex. *Neuroscience* 126(2):257–62. Golmayo L, Nunez A, Zaborszky L (2003) Electrophysiological Evidence for the Existence of a Posterior Cortical-Prefrontal-Basal Forebrain Circuitry in Modulating Sensory Responses in Visual and Somatosensory Rat Cortical Areas. *Neuroscience* 119(2):597–609.

25. Ego-Stengel V, Shulz DE, Haidarliu S, Sosnik R, Ahissar E (2001) Acetylcholine-Dependent Induction and Expression of Functional Plasticity in the Barrel Cortex of the Adult Rat. *J Neurophysiol* 86(1):422–37. Herrero JL, Roberts MJ, Delicato LS,

Gieselmann MA, Dayan P, Thiele A (2008) Acetylcholine Contributes through Muscarinic Receptors to Attentional Modulation in V1. *Nature* 454(7208):1110–14. Kilgard MP, Merzenich MM (1998) Cortical Map Reorganization Enabled by Nucleus Basalis Activity. *Science* 279(5357):1714–18. Murphy PC, Sillito AM (1991) Cholinergic Enhancement of Direction Selectivity in the Visual Cortex of the Cat. *Neuroscience* 40(1):13–20. Roberts MJ, Zinke W, Guo K, Robertson R, McDonald JS, Thiele A (2005) Acetylcholine Dynamically Controls Spatial Integration in Marmoset Primary Visual Cortex. *J Neurophysiol* 93(4):2062–72. Sato H, Hata Y, Hagihara K, Tsumoto T (1987) Effects of Cholinergic Depletion on Neuron Activities in the Cat Visual Cortex. *J Neurophysiol* 58(4):781–94. Weinberger NM (2007) Associative Representational Plasticity in the Auditory Cortex: A Synthesis of Two Disciplines. *Learn Mem* 14(1–2):1–16.

26. Verhoog MB, Mansvelder HD (2011) Presynaptic Ionotropic Receptors Controlling and Modulating the Rules for Spike Timing-Dependent Plasticity. *Neural Plast* 2011:870763.

27. Benveniste M, Wilhelm J, Dingledine RJ, Mott DD (2010) Subunit-Dependent Modulation of Kainate Receptors by Muscarinic Acetylcholine Receptors. *Brain Res* 1352:61–69.

28. Schätzle P, Ster J, Verbich D, McKinney RA, Gerber U, Sonderegger P, Mateos JM (2011) Rapid and Reversible Formation of Spine Head Filopodia in Response to Muscarinic Receptor Activation in CA1 Pyramidal Cells. *J Physiol* 589(Pt 17):4353–64.

29. Wang L, Conner JM, Rickert J, Tuszynski MH (2011) Structural Plasticity within Highly Specific Neuronal Populations Identifies a Unique Parcellation of Motor Learning in the Adult Brain. *Proc Natl Acad Sci USA* 108(6):2545–50.

30. Corbetta M, Patel G, Shulman GL (2008) The Reorienting System of the Human Brain: From Environment to Theory of Mind. *Neuron* 58(3):306–24.

31. Carnes KM, Fuller TA, Price JL (1990) Sources of Presumptive Glutamatergic/ Aspartatergic Afferents to the Magnocellular Basal Forebrain in the Rat. *J Comp Neurol* 302(4):824–52. Zaborszky L, Cullinan WE, Braun A (1991) Afferents to Basal Forebrain Cholinergic Projection Neurons: An Update. In: Napier TC, Kalivas PW, Hanin I [eds] *The Basal Forebrain: Anatomy to Function.* Plenum: New York, pp 1–42.

32. Cavada C, Company T, Tejedor J, Cruz-Rizzolo RJ, Reinoso-Suarez F (2000) The Anatomical Connections of the Macaque Monkey Orbitofrontal Cortex. A Review. *Cereb Cortex* 10(3):220–42. Ghashghaei HT, Barbas H (2001) Neural Interaction between the Basal Forebrain and Functionally Distinct Prefrontal Cortices in the Rhesus Monkey. *Neuroscience* 103(3):593–614.

33. See table 3, case AE of Ghashghaei and Barbas (2001).

34. Bermpohl F, Pascual-Leone A, Amedi A, Merabet LB, Fregni F, Gaab N, Alsop D, Schlaug G, Northoff G (2006) Dissociable Networks for the Expectancy and Perception of Emotional Stimuli in the Human Brain. *Neuroimage* 30(2):588–600.

35. Cocker KD, Moseley MJ, Bissenden JG, Fielder AR (1994) Visual Acuity and Pupillary Responses to Spatial Structure in Infants. *Invest Ophthalmol Vis Sci* 35(5):2620–25.

36. Brown AM, Lindsey DT (2009) Contrast Insensitivity: The Critical Immaturity in Infant Visual Performance. *Optom Vis Sci* 86(6):572–76.

37. Adams RJ, Courage ML (1998) Human Newborn Color Vision: Measurement with Chromatic Stimuli Varying in Excitation Purity. *J Exp Child Psychol* 68(1):22–34.

38. Held R, Birch E, Gwiazda J (1980) Stereoacuity of Human Infants. *Proc Natl Acad Sci USA* 77(9):5572–74.

39. Lipsitt LP, Engen T, Kaye H (1963) Developmental Changes in the Olfactory Threshold of the Neonate. *Child Dev* 34:371–76. Tharpe AM, Ashmead DH (2001) A Longitudinal Investigation of Infant Auditory Sensitivity. *Am J Audiol* 10(2):104–12.

40. Kracun I, Rösner H (1986) Early Cytoarchitectonic Development of the Anlage of the Basal Nucleus of Meynert in the Human Fetus. *Int J Dev Neurosci* 4(2):143–49.

41. Perry EK, Smith CJ, Atack JR, Candy JM, Johnson M, Perry RH (1986) Neocortical Cholinergic Enzyme and Receptor Activities in the Human Fetal Brain. *J Neurochem* 47(4):1262–69.

42. Bruel-Jungerman E, Lucassen PJ, Francis F (2011) Cholinergic Influences on Cortical Development and Adult Neurogenesis. *Behav Brain Res* 221(2):379–88. Candy JM, Perry EK, Perry RH, Bloxham CA, Thompson J, Johnson M, Oakley AE, Edwardson JA (1985) Evidence for the Early Prenatal Development of Cortical Cholinergic Afferents from the Nucleus of Meynert in the Human Foetus. *Neurosci Lett* 61(1–2):91–95. Koh S, Santos TC, Cole AJ (2005) Susceptibility to Seizure-Induced Injury and Acquired Microencephaly Following Intraventricular Injection of Saporin-Conjugated 192 IgG in Developing Rat Brain. *Exp Neurol* 194(2):457–66.

43. Schliebs R, Arendt T (2011) The Cholinergic System in Aging and Neuronal Degeneration. *Behav Brain Res* 221(2):555–63.

44. Perry EK, Perry RH (1995) Acetylcholine and Hallucinations: Disease-Related Compared to Drug-Induced Alterations in Human Consciousness. *Brain Cogn* 28(3):240–58.

45. Ketchum JS, Sidell FR, Crowell EB Jr, Aghajanian GK, Hayes AH Jr (1973) Atropine, Scopolamine, and Ditran: Comparative Pharmacology and Antagonists in Man. *Psychopharmacologia* 28(2):121–45.

46. Chan D, Rossor MN (2002) "But Who Is That on the Other Side of You?" Extracampine Hallucinations Revisited. *Lancet* 360:2064–66.

47. González-Maeso J, Weisstaub NV, Zhou M, Chan P, Ivic L, Ang R, Lira A, Bradley-Moore M, Ge Y, Zhou Q, Sealfon SC, Gingrich JA (2007) Hallucinogens Recruit Specific Cortical 5-HT(2A) Receptor-Mediated Signaling Pathways to Affect Behavior. *Neuron* 53(3):439–52.

48. Ohayon MM (2000) Prevalence of Hallucinations and Their Pathological Associations in the General Population. *Psychiatry Res* 97(2–3):153–64.

49. Faure A, Richard JM, Berridge KC (2010) Desire and Dread from the Nucleus Accumbens: Cortical Glutamate and Subcortical GABA Differentially Generate Motivation and Hedonic Impact in the Rat. *PLoS One* 5(6):e11223.

50. Ricceri L, Cutuli D, Venerosi A, Scattoni ML, Calamandrei G (2007) Neonatal Basal Forebrain Cholinergic Hypofunction Affects Ultrasonic Vocalizations and Fear Conditioning Responses in Preweaning Rats. *Behav Brain Res* 183(1):111–17.

51. Khazipov R, Minlebaev M, Valeeva G (2013) Early Gamma Oscillations. *Neuroscience* 250:240–52. Cifuentes Castro VH, López Valenzuela CL, Salazar Sánchez JC, Peña KP, López Pérez SJ, Ibarra JO, Villagrán AM (2014) An Update of the Classical

and Novel Methods Used for Measuring Fast Neurotransmitters during Normal and Brain Altered Function. *Curr Neuropharmacol* 12(6):490–508. Jing M, Li Y, Zeng J, Huang P, Skirzewski M, Kljakic O, Peng W, Qian T, Tan K, Zou J, Trinh S, Wu R, Zhang S, Pan S, Hires SA, Xu M, Li H, Saksida LM, Prado VF, Bussey TJ, Prado MAM, Chen L, Cheng H, Li Y (2020) An Optimized Acetylcholine Sensor for Monitoring in Vivo Cholinergic Activity. *Nat Meth* 17(11):1139–46. Parikh and Sarter (2008).

52. St Peters M, Dmeter E, Lustig C, Bruno JP, Sarter M (2011) Enhanced Control of Attention by Stimulating Mesolimbic-Corticopetal Cholinergic Circuitry. *J Neurosci* 31(26):9760–71. Zmarowski A, Sarter M, Bruno JP (2005) NMDA and Dopamine Interactions in the Nucleus Accumbens Modulate Cortical Acetylcholine Release. *Eur J Neurosci* 22(7):1731–40.

53. Neigh GN, Arnold HM, Rabenstein RL, Sarter M, Bruno JP (2004) Neuronal Activity in the Nucleus Accumbens Is Necessary for Performance-Related Increases in Cortical Acetylcholine Release. *Neuroscience* 123(3):635–45.

54. Zmarowski A, Sarter M, Bruno JP (2007) Glutamate Receptors in Nucleus Accumbens Mediate Regionally Selective Increases in Cortical Acetylcholine Release. *Synapse* 61(3):115–23.

55. Zmarowski et al. (2005).

56. Brooks JM, Sarter M, Bruno JP (2007) D2-Like Receptors in Nucleus Accumbens Negatively Modulate Acetylcholine Release in Prefrontal Cortex. *Neuropharmacology* 53(3):455–63.

57. Muller JM, Moore H, Myers MM, Shair HN (2008) Ventral Striatum Dopamine D2 Receptor Activity Inhibits Rat Pups' Vocalization Response to Loss of Maternal Contact. *Behav Neurosci* 122(1):119–28.

58. Turchi J, Sarter M (2001) Bidirectional Modulation of Basal Forebrain N-Methyl-D-Aspartate Receptor Function Differentially Affects Visual Attention but Not Visual Discrimination Performance. *Neuroscience* 104(2):407–17.

59. Honey GD, Corlett PR, Absalom AR, Lee M, Pomarol-Clotet E, Murray GK, McKenna PJ, Bullmore ET, Menon DK, Fletcher PC (2008) Individual Differences in Psychotic Effects of Ketamine Are Predicted by Brain Function Measured under Placebo. *J Neurosci* 28(25):6295–303. Newcomer JW, Farber NB, Jevtovic-Todorovic V, Selke G, Melson AK, Hershey T, Craft S, Olney JW (1999) Ketamine-Induced NMDA Receptor Hypofunction as a Model of Memory Impairment and Psychosis. *Neuropsychopharmacology* 20(2):106–18.

Chapter 11

1. Campbell J (1998) *The Myths and Masks of God: Joseph Campbell Audio Collection.* Highbridge Audio: Minneapolis, MN. Disc 1: Interpreting Symbolic Forms, Track 11.

2. McNamara P, Durso R, Brown A, Harris E (2006) The Chemistry of Religiosity: Evidence from Patients with Parkinson's Disease. In: McNamara P [ed] *Where God and Science Meet: How Brain and Evolutionary Studies Alter Our Understanding of Religion. Vol 2: The Neurology of Religious Experience.* Praeger Perspectives: Westport, CT, pp

1–14. Previc FH (2006) The Role of the Extrapersonal Brain Systems in Religious Activity. *Conscious Cogn* 15(3):500–539. Rogers SA, Paloutzian RF (2006) Schizophrenia, Neurology, and Religion: What Can Psychosis Teach Us about the Evolutionary Role of Religion? In: McNamara P [ed] *Where God and Science Meet: How Brain and Evolutionary Studies Alter Our Understanding of Religion. Vol 3: The Psychology of Religious Experience.* Praeger Perspectives: Westport, CT, pp 161–86. Schachter SC (2006) Religion and the Brain: Evidence from Temporal Lobe Epilepsy. In: McNamara P [ed] *Where God and Science Meet: How Brain and Evolutionary Studies Alter Our Understanding of Religion. Vol 2: The Neurology of Religious Experience.* Praeger Perspectives: Westport, CT pp 171–88.

3. Cardno AG, Marshall EJ, Coid B, Macdonald AM, Ribchester TR, Davies NJ, Venturi P, Jones LA, Lewis SW, Sham PC, Gottesman II, Farmer AE, McGuffin P, Reveley AM, Murray RM (1999) Heritability Estimates for Psychotic Disorders: The Maudsley Twin Psychosis Series. *Arch Gen Psychiatry* 56(2):162–68.

4. Gearing RE, Alonzo D, Smolak A, McHugh K, Harmon S, Baldwin S (2011) Association of Religion with Delusions and Hallucinations in the Context of Schizophrenia: Implications for Engagement and Adherence. *Schizophr Res* 126(1–3):150–63.

5. Rudalevičienė P, Stompe T, Narbekovas A, Raškauskienė N, Bunevičius R (2008) Are Religious Delusions Related to Religiosity in Schizophrenia? *Medicina (Kaunas, Lithuania)* 44(7):529–35.

6. European Commission (2005) Eurobarometer on Social Values, Science and Technology 2005. www.eurosfaire.prd.fr/bibliotheque/pdf/ebs_225_report_en_06 -2005.pdf (accessed 18 December 2021). Rudalevičienė et al. (2008).

7. Klaf FS, Hamilton JG (1961) Schizophrenia–A Hundred Years Ago and Today. *J Ment Sci* 107:819–27.

8. Picchioni MM, Murray RM (2007) Schizophrenia. *BMJ* 335(7610):91–95.

9. Synofzik M, Lindner A, Thier P (2008) The Cerebellum Updates Predictions about the Visual Consequences of One's Behavior. *Curr Biol* 18(11):814–18. Wolpert DM, Miall RC, Kawato M (1998) Internal Models in the Cerebellum. *Trends Cogn Sci* 2(9):338–47.

10. Kontaris I, Wiggett AJ, Downing PE (2009) Dissociation of Extrastriate Body and Biological-Motion Selective Areas by Manipulation of Visual-Motor Congruency. *Neuropsychologia* 47(14):3118–24. Leube DT, Knoblich G, Erb M, Grodd W, Bartels M, Kircher TT (2003) The Neural Correlates of Perceiving One's Own Movements. *Neuroimage* 20(4):2084–90. Farrer C, Frey SH, Van Horn JD, Tunik E, Turk D, Inati S, Grafton ST (2008) The Angular Gyrus Computes Action Awareness Representations. *Cereb Cortex* 18(2):254–61.

11. Feinberg I (2011) Corollary Discharge, Hallucinations, and Dreaming. *Schizophr Bull* 37(1):1–3.

12. Ford JM, Mathalon DH (2004) Electrophysiological Evidence of Corollary Discharge Dysfunction in Schizophrenia during Talking and Thinking. *J Psychiatr Res* 38(1):37–46.

13. Botvinick M, Cohen J (1998) Rubber Hands "Feel" Touch That Eyes See. *Nature* 391(6669):756.

14. Peled A, Ritsner M, Hirschmann S, Geva AB, Modai I (2000) Touch Feel Illusion in Schizophrenic Patients. *Biol Psychiatry* 48(11):1105–8. Thakkar KN, Nichols HS, McIntosh LG, Park S (2011) Disturbances in Body Ownership in Schizophrenia: Evidence from the Rubber Hand Illusion and Case Study of a Spontaneous Out-of-Body Experience. *PLoS One* 6(10):e27089.

15. Thakkar et al. (2011).

16. Lenggenhager B, Smith ST, Blanke O (2006) Functional and Neural Mechanisms of Embodiment: Importance of the Vestibular System and the Temporal Parietal Junction. *Rev Neurosci* 17(6):643–57.

17. Shaqiri A, Roinishvili M, Kaliuzhna M, Favrod O, Chkonia E, Herzog MH, Blanke O, Salomon R (2018) Rethinking Body Ownership in Schizophrenia: Experimental and Meta-Analytical Approaches Show No Evidence for Deficits. *Schizophr Bull* 44(3):643–52.

18. Farrer C, Frith CD (2002) Experiencing Oneself vs Another Person as Being the Cause of an Action: The Neural Correlates of the Experience of Agency. *Neuroimage* 15(3):596–603. Kontaris I, Wiggett AJ, Downing PE (2009) Dissociation of Extrastriate Body and Biological-Motion Selective Areas by Manipulation of Visual-Motor Congruency. *Neuropsychologia* 47(14):3118–24. Leube DT, Knoblich G, Erb M, Grodd W, Bartels M, Kircher TT (2003) The Neural Correlates of Perceiving One's Own Movements. *Neuroimage* 20(4):2084–90.

19. Synofzik M, Thier P, Leube DT, Schlotterbeck P, Lindner A (2010) Misattributions of Agency in Schizophrenia Are Based on Imprecise Predictions about the Sensory Consequences of One's Actions. *Brain* 133(Pt 1):262–71.

20. Franck N, Farrer C, Georgieff N, Marie-Cardine M, Dalery J, d'Amato T, Jeannerod M (2001) Defective Recognition of One's Own Actions in Patients with Schizophrenia. *Am J Psychiatry* 158(3):454–59.

21. Farrer C, Franck N, Frith CD, Decety J, Georgieff N, d'Amato T, Jeannerod M (2004) Neural Correlates of Action Attribution in Schizophrenia. *Psychiatry Res* 131(1):31–44.

22. Brunelin J, d'Amato T, Brun P, Bediou B, Kallel L, Senn M, Poulet E, Saoud M (2007) Impaired Verbal Source Monitoring in Schizophrenia: An Intermediate Trait Vulnerability Marker? *Schizophr Res* 89(1–3):287–92.

23. Brunelin J, Poulet E, Bediou B, Kallel L, Dalery J, d'Amato T, Saoud M (2006) Low Frequency Repetitive Transcranial Magnetic Stimulation Improves Source Monitoring Deficit in Hallucinating Patients with Schizophrenia. *Schizophr Res* 81(1):41–45.

24. Grossman ED, Battelli L, Pascual-Leone A (2005) Repetitive TMS over Posterior STS Disrupts Perception of Biological Motion. *Vision Res* 45(22):2847–53.

25. Farrer et al. (2004).

26. Brunelin et al. (2006). Fitzgerald PB, Daskalakis ZJ (2008) A Review of Repetitive Transcranial Magnetic Stimulation Use in the Treatment of Schizophrenia. *Can J Psychiatry* 53(9):567–76. Hoffman RE, Hampson M, Wu K, Anderson AW, Gore JC, Buchanan RJ, Constable RT, Hawkins KA, Sahay N, Krystal JH (2007) Probing the Pathophysiology of Auditory/Verbal Hallucinations by Combining Functional Magnetic Resonance Imaging and Transcranial Magnetic Stimulation. *Cereb Cortex* 17(11):2733–43. Tranulis C, Sepehry AA, Galinowski A, Stip E (2008) Should We Treat Auditory

Hallucinations with Repetitive Transcranial Magnetic Stimulation? A Metaanalysis. *Can J Psychiatry* 53(9):577–86.

27. Kammers MP, Verhagen L, Dijkerman HC, Hogendoorn H, De Vignemont F, Schutter DJ (2009) Is This Hand for Real? Attenuation of the Rubber Hand Illusion by Transcranial Magnetic Stimulation over the Inferior Parietal Lobule. *J Cogn Neurosci* 21(7):1311–20. Tsakiris M, Costantini M, Haggard P (2008) The Role of the Right Temporo-Parietal Junction in Maintaining a Coherent Sense of One's Body. *Neuropsychologia* 46(12):3014–18.

28. Stephan KE, Friston KJ, Frith CD (2009) Dysconnection in Schizophrenia: From Abnormal Synaptic Plasticity to Failures of Self-Monitoring. *Schizophr Bull* 35(3):509–27.

29. Perner J, Aichhorn M, Kronbichler M, Staffen W, Ladurner G (2006) Thinking of Mental and Other Representations: The Roles of Left and Right Temporo-Parietal Junction. *Soc Neurosci* 1(3–4):245–58.

30. Abi-Dargham A (2004) Do We Still Believe in the Dopamine Hypothesis? New Data Bring New Evidence. *Int J Neuropsychopharmacol* 7(Suppl 1):S1–S5.

31. Martin LF, Freedman R (2007) Schizophrenia and the Alpha7 Nicotinic Acetylcholine Receptor. *Int Rev Neurobiol* 78:225–46. Verhoog MB, Mansvelder HD (2011) Presynaptic Ionotropic Receptors Controlling and Modulating the Rules for Spike Timing-Dependent Plasticity. *Neural Plast* 2011:870763.

32. Berman JA, Talmage DA, Role LW (2007) Cholinergic Circuits and Signaling in the Pathophysiology of Schizophrenia. *Int Rev Neurobiol* 78:193–223.

33. Verhoog and Mansvelder (2011).

34. Stephan et al. (2009).

35. Basham ME, Nordeen EJ, Nordeen KW (1996) Blockade of NMDA Receptors in the Anterior Forebrain Impairs Sensory Acquisition in the Zebra Finch *(Poephila guttata)*. *Neurobiol Learn Mem* 66(3):295–304. Nakamori T, Sato K, Atoji Y, Kanamatsu T, Tanaka K, Ohki-Hamazaki H (2010) Demonstration of a Neural Circuit Critical for Imprinting Behavior in Chicks. *J Neurosci* 30(12):4467–80.

36. Kelley AE (2004) Ventral Striatal Control of Appetitive Motivation: Role in Ingestive Behavior and Reward-Related Learning. *Neurosci Biobehav Rev* 27(8):765–76. Parkes SL, Westbrook RF (2011) Role of the Basolateral Amygdala and NMDA Receptors in Higher-Order Conditioned Fear. *Rev Neurosci* 22(3):317–33.

37. Rolls ET (2010) A Computational Theory of Episodic Memory Formation in the Hippocampus. *Behav Brain Res* 215(2):180–96.

38. Oye I, Paulsen O, Maurset A (1992) Effects of Ketamine on Sensory Perception: Evidence for a Role of N-Methyl-D-Aspartate Receptors. *J Pharmacol Exp Ther* 260(3):1209–13.

39. Honey GD, Corlett PR, Absalom AR, Lee M, Pomarol-Clotet E, Murray GK, McKenna PJ, Bullmore ET, Menon DK, Fletcher PC (2008) Individual Differences in Psychotic Effects of Ketamine Are Predicted by Brain Function Measured under Placebo. *J Neurosci* 28(25):6295–303. Krystal JH, Karper LP, Seibyl JP, Freeman GK, Delaney R, Bremner JD, Heninger GR, Bowers MB Jr, Charney DS (1994) Subanesthetic Effects of the Noncompetitive NMDA Antagonist, Ketamine, in Humans. Psychotomimetic, Perceptual, Cognitive, and Neuroendocrine Responses. *Arch Gen Psychiatry*

51(3):199–214. Muetzelfeldt L, Kamboj SK, Rees H, Taylor J, Morgan CJ, Curran HV (2008) Journey through the K-Hole: Phenomenological Aspects of Ketamine Use. *Drug Alcohol Depend* 95(3):219–29. Pomarol-Clotet E, Honey GD, Murray GK, Corlett PR, Absalom AR, Lee M, McKenna PJ, Bullmore ET, Fletcher PC (2006) Psychological Effects of Ketamine in Healthy Volunteers. Phenomenological Study. *Br J Psychiatry* 189:173–79. Wilkins LK, Girard TA, Cheyne JA (2011) Ketamine as a Primary Predictor of Out-of-Body Experiences Associated with Multiple Substance Use. *Conscious Cogn* 20(3):943–50.

40. Morgan HL, Turner DC, Corlett PR, Absalom AR, Adapa R, Arana FS, Pigott J, Gardner J, Everitt J, Haggard P, Fletcher PC (2011) Exploring the Impact of Ketamine on the Experience of Illusory Body Ownership. *Biol Psychiatry* 69(1):35–41.

41. Cahalan S (2009) My Mysterious Lost Month of Madness. *New York Post* [Oct 4] www.nypost.com/p/news/local/item_OseCEXxo6axZ8Uyig17QKL (accessed 18 December 2021).

42. Day GS, High SM, Cot B, Tang-Wai DF (2011) Anti-NMDA-Receptor Encephalitis: Case Report and Literature Review of an Under-Recognized Condition. *J Gen Intern Med* 26(7):811–16.

43. Kohl BK, Dannhardt G (2001) The NMDA Receptor Complex: A Promising Target for Novel Antiepileptic Strategies. *Curr Med Chem* 8(11):1275–89.

44. Blumer D (1999) Evidence Supporting the Temporal Lobe Epilepsy Personality Syndrome. *Neurology* 53(5 Suppl 2):S9–12. Devinsky O, Najjar S (1999) Evidence against the Existence of a Temporal Lobe Epilepsy Personality Syndrome. *Neurology* 53(5 Suppl 2):S13–25.

45. Ramachandran VS, Blakeslee S (1998) *Phantoms in the Brain: Probing the Mysteries of the Human Mind.* William Morrow: New York, chapter 9. Trimble M, Freeman A (2006) An Investigation of Religiosity and the Gastaut-Geschwind Syndrome in Patients with Temporal Lobe Epilepsy. *Epilepsy Behav* 9(3):407–14. Van Elst LT, Krishnamoorthy ES, Baumer D, Selai C, von Gunten A, Gene-Cos N, Ebert D, Trimble MR (2003) Psychopathological Profile in Patients with Severe Bilateral Hippocampal Atrophy and Temporal Lobe Epilepsy: Evidence in Support of the Geschwind Syndrome? *Epilepsy Behav* 4(3):291–97. Waxman SG, Geschwind N (1975) The Interictal Behavior Syndrome of Temporal Lobe Epilepsy. *Arch Gen Psychiatry* 32:1580–86.

46. Devinsky O, Lai G (2008) Spirituality and Religion in Epilepsy. *Epilepsy Behav* 12(4):636–43.

47. Devinsky and Lai (2008). Saver JL, Rabin J (1997) The Neural Substrates of Religious Experience. *J Neuropsychiatry Clin Neurosci* 9(3):498–510.

48. Beauregard M, O'Leary D (2007) *The Spiritual Brain: A Neuroscientist's Case for the Existence of the Soul.* HarperOne: New York, chapter 3.

49. Devinsky and Lai (2008).

50. Ogata A, Miyakawa T (1998) Religious Experiences in Epileptic Patients with a Focus on Ictus-Related Episodes. *Psychiatry Clin Neurosci* 52(3):321–25.

51. Wuerfel J, Krishnamoorthy ES, Brown RJ, Lemieux L, Koepp M, Tebartz van Elst L, Trimble MR (2004) Religiosity Is Associated with Hippocampal but Not Amygdala Volumes in Patients with Refractory Epilepsy. *J Neurol Neurosurg Psychiatry* 75(4):640–42.

52. Brugger P, Regard M, Landis T (1996) Unilaterally Felt "Presences": The Neuropsychiatry of One's Invisible Doppelganger. *Neuropsychiatry Neuropsychol Behav Neurol* 9(2):114–22.

53. Oyebode F, Davison K (1989) Epileptic Schizophrenia: Clinical Features and Outcome. *Acta Psychiatr Scand* 79(4):327–31.

54. Monaco F, Cavanna A, Magli E, Barbagli D, Collimedaglia L, Cantello R, Mula M (2005) Obsessionality, Obsessive-Compulsive Disorder, and Temporal Lobe Epilepsy. *Epilepsy Behav* 7(3):491–96.

55. Fiske AP, Haslam N (1997) Is Obsessive-Compulsive Disorder a Pathology of the Human Disposition to Perform Socially Meaningful Rituals? Evidence of Similar Content. *J Nerv Ment Dis* 185(4):211–22.

56. Huppert JD, Siev J, Kushner ES (2007) When Religion and Obsessive-Compulsive Disorder Collide: Treating Scrupulosity in Ultra-Orthodox Jews. *J Clin Psychol* 63(10):925–41. Tek C, Ulug B (2001) Religiosity and Religious Obsessions in Obsessive-Compulsive Disorder. *Psychiatry Res* 104(2):99–108.

57. Zhong CB, Liljenquist K (2006) Washing Away Your Sins: Threatened Morality and Physical Cleansing. *Science* 313:1451–52.

58. Lamm C, Singer T (2010) The Role of Anterior Insular Cortex in Social Emotions. *Brain Struct Funct* 214(5–6):579–91.

59. Fusar-Poli P, Placentino A, Carletti F, Landi P, Allen P, Surguladze S, Benedetti F, Abbamonte M, Gasparotti R, Barale F, Perez J, McGuire P, Politi P (2009) Functional Atlas of Emotional Faces Processing: A Voxel-Based Meta-Analysis of 105 Functional Magnetic Resonance Imaging Studies. *J Psychiatry Neurosci* 34(6):418–32.

60. Berle D, Phillips ES (2006) Disgust and Obsessive-Compulsive Disorder: An Update. *Psychiatry* 69(3):228–38.

61. Fiddick L (2011) There Is More Than the Amygdala: Potential Threat Assessment in the Cingulate Cortex. *Neurosci Biobehav Rev* 35(4):1007–18.

62. Huey ED, Zahn R, Krueger F, Moll J, Kapogiannis D, Wassermann EM, Grafman J (2008) A Psychological and Neuroanatomical Model of Obsessive-Compulsive Disorder. *J Neuropsychiatry Clin Neurosci* 20(4):390–408.

63. Fitzgerald KD, Welsh RC, Gehring WJ, Abelson JL, Himle JA, Liberzon I, Taylor SF (2005) Error-Related Hyperactivity of the Anterior Cingulate Cortex in Obsessive-Compulsive Disorder. *Biol Psychiatry* 57(3):287–94.

64. Dougherty DD, Baer L, Cosgrove GR, Cassem EH, Price BH, Nierenberg AA, Jenike MA, Rauch SL (2002) Prospective Long-Term Follow-up of 44 Patients Who Received Cingulotomy for Treatment-Refractory Obsessive-Compulsive Disorder. *Am J Psychiatry* 159(2):269–75.

65. Swain JE, Lorberbaum JP, Kose S, Strathearn L (2007) Brain Basis of Early Parent-Infant Interactions: Psychology, Physiology, and in Vivo Functional Neuroimaging Studies. *J Child Psychol Psychiatry* 48(3–4):262–87.

66. Kessler RC, Berglund P, Demler O, Jin R, Merikangas KR, Walters EE (2005) Lifetime Prevalence and Age-of-Onset Distributions of DSM-IV Disorders in the National Comorbidity Survey Replication. *Arch Gen Psychiatry* 62(6):593–602.

67. Feygin DL, Swain JE, Leckman JF (2006) The Normalcy of Neurosis: Evolutionary Origins of Obsessive-Compulsive Disorder and Related Behaviors. *Prog Neuropsychopharmacol Biol Psychiatry* 30(5):854–64.

68. McGregor I, Haji R, Nash KA, Teper R (2008) Religious Zeal and the Uncertain Self. *Basic Appl Soc Psych* 30(2):183–88.

69. McGregor I, Nash K, Prentice M (2012) Religious Zeal after Goal Frustration. In: Hogg MA, Blaylock DL [eds] *Extremism and the Psychology of Uncertainty*. Wiley-Blackwell: Hoboken, NJ, pp 147–64.

70. Baker KC, Aureli F (1997) Behavioural Indicators of Anxiety: An Empirical Test in Chimpanzees. *Behaviour* 134(13/14):1031–50. Luescher AU (2003) Diagnosis and Management of Compulsive Disorders in Dogs and Cats. *Vet Clin North Am Small Anim Pract* 33(2):253–67, vi. Maestripieri D, Schino G, Aureli F, Troisi A (1992) A Modest Proposal: Displacement Activities as an Indicator of Emotions in Primates. *Anim Behav* 44(5):967–79.

71. Wathey JC (2016) *The Illusion of God's Presence: The Biological Origins of Spiritual Longing*. Prometheus Books: Amherst, NY, pp 81–82.

72. Anastasi MW, Newberg AB (2008) A Preliminary Study of the Acute Effects of Religious Ritual on Anxiety. *J Altern Complement Med* 14(2):163–65. Sosis R, Handwerker WP (2011) Psalms and Coping with Uncertainty: Religious Israeli Women's Responses to the 2006 Lebanon War. *Am Anthropol* 113(1):40–55.

73. Kay AC, Gaucher D, McGregor I, Nash K (2010) Religious Belief as Compensatory Control. *Pers Soc Psychol Rev* 14(1):37–48. McGregor et al. (2012) pp 147–64.

74. Vannucci M, Mazzoni G, Cartocci G (2011) Lack of Control Enhances Accurate and Inaccurate Identification Responses to Degraded Visual Objects. *Psychon Bull Rev* 18(3):524–30. Whitson JA, Galinsky AD (2008) Lacking Control Increases Illusory Pattern Perception. *Science* 322(5898):115–17.

75. Perkins SL, Allen R (2006) Childhood Physical Abuse and Differential Development of Paranormal Belief Systems. *J Nerv Ment Dis* 194(5):349–55. Wathey (2016) pp 119–20.

76. Moulding R, Kyrios M (2006) Anxiety Disorders and Control Related Beliefs: The Exemplar of Obsessive-Compulsive Disorder (OCD). *Clin Psychol Rev* 26(5):573–83. Reuven-Magril O, Dar R, Liberman N (2008) Illusion of Control and Behavioral Control Attempts in Obsessive-Compulsive Disorder. *J Abnorm Psychol* 117(2):334–41. Tolin DF, Worhunsky P, Maltby N (2004) Sympathetic Magic in Contamination-Related OCD. *J Behav Ther Exp Psychiatry* 35(2):193–205.

77. Menzies L, Chamberlain SR, Laird AR, Thelen SM, Sahakian BJ, Bullmore ET (2008) Integrating Evidence from Neuroimaging and Neuropsychological Studies of Obsessive-Compulsive Disorder: The Orbitofronto-Striatal Model Revisited. *Neurosci Biobehav Rev* 32(3):525–49.

78. Kwon JS, Jang JH, Choi JS, Kang DH (2009) Neuroimaging in Obsessive-Compulsive Disorder. *Expert Rev Neurother* 9(2):255–69. Menzies et al. (2008). Milad MR, Rauch SL (2007) The Role of the Orbitofrontal Cortex in Anxiety Disorders. *Ann N Y Acad Sci* 1121:546–61. Whiteside SP, Port JD, Abramowitz JS (2004) A Meta-Analysis of Functional Neuroimaging in Obsessive-Compulsive Disorder. *Psychiatry Res* 132(1):69–79.

79. Rolls ET (2004) The Functions of the Orbitofrontal Cortex. *Brain Cogn* 55(1):11–29. Thorpe SJ, Rolls ET, Maddison S (1983) The Orbitofrontal Cortex: Neuronal Activity in the Behaving Monkey. *Exp Brain Res* 49(1):93–115.

80. Chamberlain SR, Menzies L, Hampshire A, Suckling J, Fineberg NA, del Campo N, Aitken M, Craig K, Owen AM, Bullmore ET, Robbins TW, Sahakian BJ (2008) Orbitofrontal Dysfunction in Patients with Obsessive-Compulsive Disorder and Their Unaffected Relatives. *Science* 321(5887):421–22. Remijnse PL, Nielen MM, van Balkom AJ, Cath DC, van Oppen P, Uylings HB, Veltman DJ (2006) Reduced Orbitofrontal-Striatal Activity on a Reversal Learning Task in Obsessive-Compulsive Disorder. *Arch Gen Psychiatry* 63(11):1225–36.

81. Da Rocha FF, Alvarenga NB, Malloy-Diniz L, Correa H (2011) Decision-Making Impairment in Obsessive-Compulsive Disorder as Measured by the Iowa Gambling Task. *Arq Neuropsiquiatr* 69(4):642–47.

82. Bechara A, Tranel D, Damasio H (2000) Characterization of the Decision-Making Deficit of Patients with Ventromedial Prefrontal Cortex Lesions. *Brain* 123(11):2189–202.

83. Chamberlain et al. (2008).

84. Clarke HF, Walker SC, Dalley JW, Robbins TW, Roberts AC (2007) Cognitive Inflexibility after Prefrontal Serotonin Depletion Is Behaviorally and Neurochemically Specific. *Cereb Cortex* 17(1):18–27. Da Rocha FF, Malloy-Diniz L, Lage NV, Romano-Silva MA, de Marco LA, Correa H (2008) Decision-Making Impairment Is Related to Serotonin Transporter Promoter Polymorphism in a Sample of Patients with Obsessive-Compulsive Disorder. *Behav Brain Res* 195(1):159–63.

85. Kellner M (2010) Drug Treatment of Obsessive-Compulsive Disorder. *Dialogues Clin Neurosci* 12(2):187–97.

86. Bloch MH, Landeros-Weisenberger A, Kelmendi B, Coric V, Bracken MB, Leckman JF (2006) A Systematic Review: Antipsychotic Augmentation with Treatment Refractory Obsessive-Compulsive Disorder. *Mol Psychiatry* 11(7):622–32.

87. Clarke HF, Robbins TW, Roberts AC (2008) Lesions of the Medial Striatum in Monkeys Produce Perseverative Impairments during Reversal Learning Similar to Those Produced by Lesions of the Orbitofrontal Cortex. *J Neurosci* 28(43):10972–82. Clarke HF, Hill GJ, Robbins TW, Roberts AC (2011) Dopamine, but Not Serotonin, Regulates Reversal Learning in the Marmoset Caudate Nucleus. *J Neurosci* 31(11):4290–97.

88. Schjoedt U, Stodkilde-Jorgensen H, Geertz AW, Roepstorff A (2008) Rewarding Prayers. *Neurosci Lett* 443(3):165–68.

89. Everitt BJ, Robbins TW (2005) Neural Systems of Reinforcement for Drug Addiction: From Actions to Habits to Compulsion. *Nat Neurosci* 8(11):1481–89.

90. Fineberg NA, Potenza MN, Chamberlain SR, Berlin HA, Menzies L, Bechara A, Sahakian BJ, Robbins TW, Bullmore ET, Hollander E (2010) Probing Compulsive and Impulsive Behaviors, from Animal Models to Endophenotypes: A Narrative Review. *Neuropsychopharmacology* 35(3):591–604. Grant JE, Brewer JA, Potenza MN (2006) The Neurobiology of Substance and Behavioral Addictions. *CNS Spectr* 11(12):924–30. Saxena S (2008) Neurobiology and Treatment of Compulsive Hoarding. *CNS Spectr* 13(9 Suppl 14):29–36.

91. Buxton JA, Dove NA (2008) The Burden and Management of Crystal Meth Use. *CMAJ* 178(12):1537–39.

92. Jankovic J (2008) Parkinson's Disease: Clinical Features and Diagnosis. *J Neurol Neurosurg Psychiatry* 79(4):368–76.

93. Cotzias GC, Papavasiliou PS, Gellene R (1969) Modification of Parkinson-ism–Chronic Treatment with L-Dopa. *N Engl J Med* 280(7):337–45. Fahn S (1999) Parkinson Disease, the Effect of Levodopa, and the ELLDOPA Trial. Earlier vs Later L-DOPA. *Arch Neurol* 56(5):529–35.

94. Fasano A, Petrovic I (2010) Insights into Pathophysiology of Punding Reveal Possible Treatment Strategies. *Mol Psychiatry* 15(6):560–73.

95. Fénelon G, Soulas T, Zenasni F, de Langavant LC (2010) The Changing Face of Parkinson's Disease–Associated Psychosis: A Cross-Sectional Study Based on the New NINDS-NIMH Criteria. *Mov Disord* 25(6):763–66. O'Sullivan SS, Evans AH, Lees AJ (2009) Dopamine Dysregulation Syndrome: An Overview of Its Epidemiol-ogy, Mechanisms and Management. *CNS Drugs* 23(2):157–70. O'Sullivan SS, Wu K, Politis M, Lawrence AD, Evans AH, Bose SK, Djamshidian A, Lees AJ, Piccini P (2011) Cue-Induced Striatal Dopamine Release in Parkinson's Disease–Associated Impulsive-Compulsive Behaviours. *Brain* 134(Pt 4):969–78. Voon V, Mehta AR, Hallett M (2011) Impulse Control Disorders in Parkinson's Disease: Recent Advances. *Curr Opin Neurol* 24(4):324–30.

96. Previc FH (2006) The Role of the Extrapersonal Brain Systems in Religious Activity. *Conscious Cogn* 15(3):500–539.

Chapter 12

1. Wansbrough H [ed] (1990) *The New Jerusalem Bible.* Doubleday: New York, p 875.

2. Weiland IH, Sperber Z (1970) Patterns of Mother-Infant Contact: The Signifi-cance of Lateral Preference. *J Genet Psychol* 117(2):157–65.

3. Jordan LC, Hillis AE (2005) Aphasia and Right Hemisphere Syndromes in Stroke. *Curr Neurol Neurosci Rep* 5(6):458–64.

4. Nicholls ME, Ellis BE, Clement JG, Yoshino M (2004) Detecting Hemifacial Asymmetries in Emotional Expression with Three-Dimensional Computerized Image Analysis. *Proc R Soc Lond B Biol Sci* 271(1540):663–68.

5. Coolican J, Eskes GA, McMullen PA, Lecky E (2008) Perceptual Biases in Process-ing Facial Identity and Emotion. *Brain Cogn* 66(2):176–87.

6. Devinsky O, Lai G (2008) Spirituality and Religion in Epilepsy. *Epilepsy Behav* 12(4):636–43. Roberts JK, Robertson MM, Trimble MR (1982) The Lateralising Sig-nificance of Hypergraphia in Temporal Lobe Epilepsy. *J Neurol Neurosurg Psychiatry* 45(2):131–38. Wuerfel J, Krishnamoorthy ES, Brown RJ, Lemieux L, Koepp M, Tebartz van Elst L, Trimble MR (2004) Religiosity Is Associated with Hippocampal but Not Amygdala Volumes in Patients with Refractory Epilepsy. *J Neurol Neurosurg Psychiatry* 75(4):640–42.

7. Owen AD, Hayward RD, Koenig HG, Steffens DC, Payne ME (2011) Religious Factors and Hippocampal Atrophy in Late Life. *PLoS One* 6(3):e17006. Trimble M, Freeman A (2006) An Investigation of Religiosity and the Gastaut-Geschwind Syndrome in Patients with Temporal Lobe Epilepsy. *Epilepsy Behav* 9(3):407–14.

8. Allen P, Laroi F, McGuire PK, Aleman A (2008) The Hallucinating Brain: A Review of Structural and Functional Neuroimaging Studies of Hallucinations. *Neurosci Biobehav Rev* 32(1):175–91, table 2.

9. Lutz A, Brefczynski-Lewis J, Johnstone T, Davidson RJ (2008) Regulation of the Neural Circuitry of Emotion by Compassion Meditation: Effects of Meditative Expertise. *PLoS One* 3(3):e1897.

10. Schjoedt U, Stødkilde-Jørgensen H, Geertz AW, Roepstorff A (2009) Highly Religious Participants Recruit Areas of Social Cognition in Personal Prayer. *Soc Cogn Affect Neurosci* 4(2):199–207.

11. Previc FH (2006) The Role of the Extrapersonal Brain Systems in Religious Activity. *Conscious Cogn* 15(3):500–539, p 515.

12. Broca P (1861) Remarques Sur Le Siége de la Faculté Du Langage Articulé, Suivies D'une Observation d'Aphémie (Perte de la Parole) [Remarks on the Seat of the Faculty of Articulated Language, Following an Observation of Aphemia (Loss of Speech)]. *Bull Société Anatomique* 6:330–57.

13. Corballis MC (1991) *The Lopsided Ape: Evolution of the Generative Mind.* Oxford University Press: New York.

14. De Veer MW, Gallup GG Jr, Theall LA, van den Bos R, Povinelli DJ (2003) An 8-Year Longitudinal Study of Mirror Self-Recognition in Chimpanzees *(Pan troglodytes). Neuropsychologia* 41(2):229–34. Plotnik JM, de Waal FB, Reiss D (2006) Self-Recognition in an Asian Elephant. *Proc Natl Acad Sci USA* 103(45):17053–57. Prior H, Schwarz A, Gunturkun O (2008) Mirror-Induced Behavior in the Magpie *(Pica pica):* Evidence of Self-Recognition. *PLoS Biol* 6(8):e202. Reiss D, Marino L (2001) Mirror Self-Recognition in the Bottlenose Dolphin: A Case of Cognitive Convergence. *Proc Natl Acad Sci USA* 98(10):5937–42.

15. Hixson MD (1998) Ape Language Research: A Review and Behavioral Perspective. *Anal Verbal Behav* 15:17–39.

16. Gentner TQ, Fenn KM, Margoliash D, Nusbaum HC (2006) Recursive Syntactic Pattern Learning by Songbirds. *Nature* 440(7088):1204–7.

17. Rogers LJ, Andrew M [eds] (2002) *Comparative Vertebrate Lateralization.* Cambridge University Press: Cambridge, UK.

18. Lippolis G, Bisazza A, Rogers LJ, Vallortigara G (2002) Lateralisation of Predator Avoidance Responses in Three Species of Toads. *Laterality* 7(2):163–83.

19. Rogers LJ, Zucca P, Vallortigara G (2004) Advantages of Having a Lateralized Brain. *Proc R Soc Lond B Biol Sci* 271 Suppl 6:S420–22.

20. Deckel AW (1995) Laterality of Aggressive Responses in *Anolis. J Exp Zool* 272(3):194–200.

21. Bisazza A, Cantalupo C, Capocchiano M, Vallortigara G (2000) Population Lateralisation and Social Behaviour: A Study with 16 Species of Fish. *Laterality* 5(3):269–84.

22. O'Connor KN, Roitblat HL, Bever TG (1993) Auditory Sequence Complexity and Hemispheric Asymmetry of Function in Rats. In: Roitblat HL, Herman LM,

Nachtigall PE [eds] *Language and Communication: Comparative Perspectives.* Erlbaum: Hillsdale, NJ, pp 277–92.

23. Hopkins WD, Wesley MJ, Izard MK, Hook M, Schapiro SJ (2004) Chimpanzees *(Pan troglodytes)* Are Predominantly Right-Handed: Replication in Three Populations of Apes. *Behav Neurosci* 118(3):659–63.

24. Vallortigara G, Rogers LJ (2005) Survival with an Asymmetrical Brain: Advantages and Disadvantages of Cerebral Lateralization. *Behav Brain Sci* 28(4):575–89; discussion 589–633.

25. Levin M (2005) Left-Right Asymmetry in Embryonic Development: A Comprehensive Review. *Mech Dev* 122(1):3–25. Schaafsma SM, Riedstra BJ, Pfannkuche KA, Bouma A, Groothuis TG (2009) Epigenesis of Behavioural Lateralization in Humans and Other Animals. *Philos Trans R Soc Lond B Biol Sci* 364(1519):915–27.

26. Rogers LJ (2002) Advantages and Disadvantages of Lateralization. In: Rogers LJ, Andrew M [eds] *Comparative Vertebrate Lateralization.* Cambridge University Press: Cambridge, UK, pp 126–53.

27. Hellige JB (1996) Hemispheric Asymmetry for Visual Information Processing. *Acta Neurobiol Exp (Wars)* 56(1):485–97.

28. See Vallortigara and Rogers (2005) and the associated commentary.

29. Hori M (1993) Frequency-Dependent Natural Selection in the Handedness of Scale-Eating Cichlid Fish. *Science* 260:216–19.

30. Ghirlanda S, Vallortigara G (2004) The Evolution of Brain Lateralization: A Game-Theoretical Analysis of Population Structure. *Proc R Soc Lond B Biol Sci* 271(1541):853–57.

31. Ghirlanda S, Frasnelli E, Vallortigara G (2009) Intraspecific Competition and Coordination in the Evolution of Lateralization. *Philos Trans R Soc Lond B Biol Sci* 364(1519):861–66.

32. Maynard Smith J (1982) *Evolution and the Theory of Games.* Cambridge University Press: Cambridge, UK.

33. Coren S, Porac C (1977) Fifty Centuries of Right-Handedness: The Historical Record. *Science* 198(4317):631–32.

34. Raymond M, Pontier D, Dufour AB, Moller AP (1996) Frequency-Dependent Maintenance of Left Handedness in Humans. *Proc R Soc Lond B Biol Sci* 263(1377):1627–33.

35. Faurie C, Raymond M (2005) Handedness, Homicide and Negative Frequency-Dependent Selection. *Proc R Soc Lond B Biol Sci* 272(1558):25–28.

36. Schaafsma SM, Geuze RH, Riedstra B, Schiefenhövel W, Bouma A, Groothuis TG (2012) Handedness in a Nonindustrial Society Challenges the Fighting Hypothesis as an Evolutionary Explanation for Left-Handedness. *Evol Hum Behav* 33(2):94–99.

37. Llaurens V, Raymond M, Faurie C (2009) Why Are Some People Left-Handed? An Evolutionary Perspective. *Philos Trans R Soc Lond B Biol Sci* 364(1519):881–94.

38. This is also a major theme of my previous book, especially chapter 4: Wathey JC (2016) *The Illusion of God's Presence: The Biological Origins of Spiritual Longing.* Prometheus Books: Amherst, NY.

39. Decety J, Chaminade T, Grezes J, Meltzoff AN (2002) A PET Exploration of the Neural Mechanisms Involved in Reciprocal Imitation. *Neuroimage* 15(1):265–72.

Dieterich M, Bense S, Lutz S, Drzezga A, Stephan T, Bartenstein P, Brandt T (2003) Dominance for Vestibular Cortical Function in the Non-Dominant Hemisphere. *Cereb Cortex* 13(9):994–1007. Farrer C, Frith CD (2002) Experiencing Oneself vs Another Person as Being the Cause of an Action: The Neural Correlates of the Experience of Agency. *Neuroimage* 15(3):596–603. Grossman ED, Blake R (2001) Brain Activity Evoked by Inverted and Imagined Biological Motion. *Vision Res* 41(10–11):1475–82. Grossman ED, Battelli L, Pascual-Leone A (2005) Repetitive TMS over Posterior STS Disrupts Perception of Biological Motion. *Vision Res* 45(22):2847–53. Noordzij ML, Newman-Norlund SE, de Ruiter JP, Hagoort P, Levinson SC, Toni I (2009) Brain Mechanisms Underlying Human Communication. *Front Hum Neurosci* 3:14. Vander Wyk BC, Hudac CM, Carter EJ, Sobel DM, Pelphrey KA (2009) Action Understanding in the Superior Temporal Sulcus Region. *Psychol Sci* 20(6):771–77. Young L, Camprodon JA, Hauser M, Pascual-Leone A, Saxe R (2010) Disruption of the Right Temporoparietal Junction with Transcranial Magnetic Stimulation Reduces the Role of Beliefs in Moral Judgments. *Proc Natl Acad Sci USA* 107(15):6753–58.

40. Fasold O, von Brevern M, Kuhberg M, Ploner CJ, Villringer A, Lempert T, Wenzel R (2002) Human Vestibular Cortex as Identified with Caloric Stimulation in Functional Magnetic Resonance Imaging. *Neuroimage* 17(3):1384–93.

41. Yovel G, Tambini A, Brandman T (2008) The Asymmetry of the Fusiform Face Area Is a Stable Individual Characteristic That Underlies the Left-Visual-Field Superiority for Faces. *Neuropsychologia* 46(13):3061–68.

42. Baas D, Aleman A, Kahn RS (2004) Lateralization of Amygdala Activation: A Systematic Review of Functional Neuroimaging Studies. *Brain Res Rev* 45(2):96–103.

43. Sergerie K, Chochol C, Armony JL (2008) The Role of the Amygdala in Emotional Processing: A Quantitative Meta-Analysis of Functional Neuroimaging Studies. *Neurosci Biobehav Rev* 32(4):811–30.

44. Adolphs R, Tranel D, Damasio H (2001) Emotion Recognition from Faces and Prosody Following Temporal Lobectomy. *Neuropsychology* 15(3):396–404.

45. Pujol J, Lopez A, Deus J, Cardoner N, Vallejo J, Capdevila A, Paus T (2002) Anatomical Variability of the Anterior Cingulate Gyrus and Basic Dimensions of Human Personality. *Neuroimage* 15(4):847–55.

46. Eisenberger NI, Lieberman MD (2004) Why Rejection Hurts: A Common Neural Alarm System for Physical and Social Pain. *Trends Cogn Sci* 8(7):294–300.

47. Fiddick L (2011) There Is More Than the Amygdala: Potential Threat Assessment in the Cingulate Cortex. *Neurosci Biobehav Rev* 35(4):1007–18. Huey ED, Zahn R, Krueger F, Moll J, Kapogiannis D, Wassermann EM, Grafman J (2008) A Psychological and Neuroanatomical Model of Obsessive-Compulsive Disorder. *J Neuropsychiatry Clin Neurosci* 20(4):390–408.

48. Murphy MR, MacLean PD, Hamilton SC (1981) Species-Typical Behavior of Hamsters Deprived from Birth of the Neocortex. *Science* 213(4506):459–61. Swain JE, Lorberbaum JP, Kose S, Strathearn L (2007) Brain Basis of Early Parent-Infant Interactions: Psychology, Physiology, and in Vivo Functional Neuroimaging Studies. *J Child Psychol Psychiatry* 48(3–4):262–87.

49. Hill SY, Wang S, Kostelnik B, Carter H, Holmes B, McDermott M, Zezza N, Stiffler S, Keshavan MS (2009) Disruption of Orbitofrontal Cortex Laterality in Offspring from Multiplex Alcohol Dependence Families. *Biol Psychiatry* 65(2):129–36.

50. Minagawa-Kawai Y, Matsuoka S, Dan I, Naoi N, Nakamura K, Kojima S (2009) Prefrontal Activation Associated with Social Attachment: Facial-Emotion Recognition in Mothers and Infants. *Cereb Cortex* 19(2):284–92. Swain et al. (2007).

51. Nitschke JB, Nelson EE, Rusch BD, Fox AS, Oakes TR, Davidson RJ (2004) Orbitofrontal Cortex Tracks Positive Mood in Mothers Viewing Pictures of Their Newborn Infants. *Neuroimage* 21(2):583–92.

52. Hill et al. (2009).

53. Shamay-Tsoory SG, Tomer R, Berger BD, Aharon-Peretz J (2003) Characterization of Empathy Deficits Following Prefrontal Brain Damage: The Role of the Right Ventromedial Prefrontal Cortex. *J Cogn Neurosci* 15(3):324–37.

54. Nagy E, Compagne H, Orvos H, Pal A, Molnar P, Janszky I, Loveland KA, Bardos G (2005) Index Finger Movement Imitation by Human Neonates: Motivation, Learning, and Left-Hand Preference. *Pediatr Res* 58(4):749–53.

55. Chiron C, Jambaque I, Nabbout R, Lounes R, Syrota A, Dulac O (1997) The Right Brain Hemisphere Is Dominant in Human Infants. *Brain* 120(Pt 6):1057–65.

56. Bracco L, Tiezzi A, Ginanneschi A, Campanella C, Amaducci L (1984) Lateralization of Choline Acetyltransferase (ChAT) Activity in Fetus and Adult Human Brain. *Neurosci Lett* 50(1–3):301–5.

57. Grossmann T, Johnson MH, Lloyd-Fox S, Blasi A, Deligianni F, Elwell C, Csibra G (2008) Early Cortical Specialization for Face-to-Face Communication in Human Infants. *Proc R Soc Lond B Biol Sci* 275(1653):2803–11. Honda Y, Nakato E, Otsuka Y, Kanazawa S, Kojima S, Yamaguchi MK, Kakigi R (2010) How Do Infants Perceive Scrambled Face? A Near-Infrared Spectroscopic Study. *Brain Res* 1308:137–46. Lloyd-Fox S, Blasi A, Volein A, Everdell N, Elwell CE, Johnson MH (2009) Social Perception in Infancy: A Near Infrared Spectroscopy Study. *Child Dev* 80(4):986–99. Nakato E, Otsuka Y, Kanazawa S, Yamaguchi MK, Watanabe S, Kakigi R (2009) When Do Infants Differentiate Profile Face from Frontal Face? A Near-Infrared Spectroscopic Study. *Hum Brain Mapp* 30(2):462–72. Otsuka Y, Nakato E, Kanazawa S, Yamaguchi MK, Watanabe S, Kakigi R (2007) Neural Activation to Upright and Inverted Faces in Infants Measured by Near Infrared Spectroscopy. *Neuroimage* 34(1):399–406. Senju A, Johnson MH, Csibra G (2006) The Development and Neural Basis of Referential Gaze Perception. *Soc Neurosci* 1(3–4):220–34.

58. Sheridan CJ, Matuz T, Draganova R, Eswaran H, Preissl H (2010) Fetal Magnetoencephalography—Achievements and Challenges in the Study of Prenatal and Early Postnatal Brain Responses: A Review. *Infant Child Dev* 19(1):80–93.

59. Schleussner E, Schneider U, Arnscheidt C, Kahler C, Haueisen J, Seewald HJ (2004) Prenatal Evidence of Left-Right Asymmetries in Auditory Evoked Responses Using Fetal Magnetoencephalography. *Early Hum Dev* 78(2):133–36.

60. Mento G, Suppiej A, Altoe G, Bisiacchi PS (2010) Functional Hemispheric Asymmetries in Humans: Electrophysiological Evidence from Preterm Infants. *Eur J Neurosci* 31(3):565–74.

61. Homae F, Watanabe H, Nakano T, Asakawa K, Taga G (2006) The Right Hemisphere of Sleeping Infant Perceives Sentential Prosody. *Neurosci Res* 54(4):276–80.

62. Arimitsu T, Uchida-Ota M, Yagihashi T, Kojima S, Watanabe S, Hokuto I, Ikeda K, Takahashi T, Minagawa-Kawai Y (2011) Functional Hemispheric Specialization in Processing Phonemic and Prosodic Auditory Changes in Neonates. *Front Psychol* 2:202.

63. Rönnqvist L, Hopkins B (1998) Head Position Preference in the Human Newborn: A New Look. *Child Dev* 69(1):13–23.

64. Previc FH (1991) A General Theory Concerning the Prenatal Origins of Cerebral Lateralization in Humans. *Psychol Rev* 98(3):299–334. Rönnqvist L, Hopkins B, van Emmerik R, de Groot L (1998) Lateral Biases in Head Turning and the Moro Response in the Human Newborn: Are They Both Vestibular in Origin? *Dev Psychobiol* 33(4):339–49.

65. Richards JL, Finger S (1975) Mother-Child Holding Patterns: A Cross-Cultural Photographic Survey. *Child Dev* 46(4):1001–4. Harris LJ, Cardenas RA, Spradlin MP Jr, Almerigi JB (2010) Why Are Infants Held on the Left? A Test of the Attention Hypothesis with a Doll, a Book, and a Bag. *Laterality* 15(5):548–71.

66. Abel EL (2010) Human Left-Sided Cradling Preferences for Dogs. *Psychol Rep* 107(1):336–38. Almerigi JB, Carbary TJ, Harris LJ (2002) Most Adults Show Opposite-Side Biases in the Imagined Holding of Infants and Objects. *Brain Cogn* 48(2–3):258–63. Harris et al. (2010).

67. Damerose E, Vauclair J (2002) Posture and Laterality in Human and Nonhuman Primates: Asymmetries in Maternal Handling and the Infant's Early Motor Asymmetries. In: Rogers LJ, Andrew M [eds] *Comparative Vertebrate Lateralization.* Cambridge University Press: Cambridge, UK, pp 306–62.

68. Salk L (1973) The Role of the Heartbeat in the Relations between Mother and Infant. *Sci Am* 228(5):24–29.

69. Harris et al. (2010). Manning JT, Chamberlain AT (1991) Left-Side Cradling and Brain Lateralization. *Ethol Sociobiol* 12(3):237–44. Sieratzki JS, Woll B (2002) Neuropsychological and Neuropsychiatric Perspectives on Maternal Cradling Preferences. *Epidemiol Psichiatr Soc* 11(3):170–76.

70. Borod JC, Haywood CS, Koff E (1997) Neuropsychological Aspects of Facial Asymmetry during Emotional Expression: A Review of the Normal Adult Literature. *Neuropsychol Rev* 7(1):41–60.

71. Hendriks AW, van Rijswijk M, Omtzigt D (2011) Holding-Side Influences on Infant's View of Mother's Face. *Laterality* 16(6):641–55.

72. Blasi A, Mercure E, Lloyd-Fox S, Thomson A, Brammer M, Sauter D, Deeley Q, Barker GJ, Renvall V, Deoni S, Gasston D, Williams SC, Johnson MH, Simmons A, Murphy DG (2011) Early Specialization for Voice and Emotion Processing in the Infant Brain. *Curr Biol* 21(14):1220–24. Homae et al. (2006).

73. Le Grand R, Mondloch CJ, Maurer D, Brent HP (2003) Expert Face Processing Requires Visual Input to the Right Hemisphere during Infancy. *Nat Neurosci* 6(10):1108–12.

74. Parente R, Tommasi L (2008) A Bias for the Female Face in the Right Hemisphere. *Laterality* 13(4):374–86.

75. Best CT, Womer JS, Queen HF (1994) Hemispheric Asymmetries in Adults' Perception of Infant Emotional Expressions. *J Exp Psychol Hum Percept Perform* 20(4):751–65. Hoptman MJ, Levy J (1988) Perceptual Asymmetries in Left- and Right-Handers for Cartoon and Real Faces. *Brain Cogn* 8(2):178–88.

76. Vervloed MP, Hendriks AW, van den Eijnde E (2011) The Effects of Mothers' Past Infant-Holding Preferences on Their Adult Children's Face Processing Lateralisation. *Brain Cogn* 75(3):248–54.

77. Bryden MP, Free T, Gagné S, Groff P (1991) Handedness Effects in the Detection of Dichotically-Presented Words and Emotions. *Cortex* 27(2):229–35.

78. Bourne VJ, Todd BK (2004) When Left Means Right: An Explanation of the Left Cradling Bias in Terms of Right Hemisphere Specializations. *Dev Sci* 7(1):19–24. Donnot J, Vauclair J (2007) Infant Holding Preferences in Maternity Hospitals: Testing the Hypothesis of the Lateralized Perception of Emotions. *Dev Neuropsychol* 32(3):881–90. Harris LJ, Almerigi JB, Carbary TJ, Fogel TG (2001) Left-Side Infant Holding: A Test of the Hemispheric Arousal-Attentional Hypothesis. *Brain Cogn* 46(1–2):159–65. Harris et al. (2010). Lucas MD, Turnbull OH, Kaplan-Solms KL (1993) Laterality of Cradling in Relation to Perception and Expression of Facial Affect. *J Genet Psychol* 154(3):347–52. Vauclair J, Donnot J (2005) Infant Holding Biases and Their Relations to Hemispheric Specializations for Perceiving Facial Emotions. *Neuropsychologia* 43(4):564–71.

79. Damerose E, Hopkins WD (2002) Scan and Focal Sampling: Reliability in the Laterality for Maternal Cradling and Infant Nipple Preferences in Olive Baboons, *Papio anubis*. *Anim Behav* 63(3):511–18.

80. Staddon JER (2003) *Adaptive Behavior and Learning: Internet Edition*. Cambridge University Press: Cambridge, http://citeseerx.ist.psu.edu/viewdoc/download?doi=10.1.1.730.7231&rep=rep1&type=pdf (accessed 19 December 2021), chapter 8.

81. Yellott JI Jr (1969) Probability Learning with Noncontingent Success. *J Math Psychol* 6(3):541–75.

82. Wolford G, Miller MB, Gazzaniga M (2000) The Left Hemisphere's Role in Hypothesis Formation. *J Neurosci* 20(6):RC64.

83. Wathey (2016) pp 278–79. De Haan EHF, Corballis PM, Hillyard SA, Marzi CA, Seth A, Lamme VAF, Volz L, Fabri M, Schechter E, Bayne T, Corballis M, Pinto Y (2020) Split-Brain: What We Know Now and Why This Is Important for Understanding Consciousness. *Neuropsychol Rev* 30(2):224–33.

84. O'Connor KN, Roitblat HL, Bever TG (1993) Auditory Sequence Complexity and Hemispheric Asymmetry of Function in Rats. In: Roitblat HL, Herman LM, Nachtigall PE [eds] *Language and Communication: Comparative Perspectives*. Erlbaum: Hillsdale, NJ, pp 277–92.

85. Bell V, Reddy V, Halligan P, Kirov G, Ellis H (2007) Relative Suppression of Magical Thinking: A Transcranial Magnetic Stimulation Study. *Cortex* 43(4):551–57.

86. Lindeman M, Svedholm AM, Riekki T, Raij T, Hari R (2013) Is It Just a Brick Wall or a Sign from the Universe? An fMRI Study of Supernatural Believers and Skeptics. *Soc Cogn Affect Neurosci* 8(8):943–49.

87. Aron AR, Robbins TW, Poldrack RA (2014) Inhibition and the Right Inferior Frontal Cortex: One Decade On. *Trends Cogn Sci* 18(4):177–85. Volkow ND, Fowler JS, Wang GJ, Telang F, Logan J, Jayne M, Ma Y, Pradhan K, Wong C, Swanson JM

(2010) Cognitive Control of Drug Craving Inhibits Brain Reward Regions in Cocaine Abusers. *Neuroimage* 49(3):2536–43.

88. Shyamalan MN (2002) *Signs.* Touchstone Pictures.

89. Morgan T (1988) *Literary Outlaw: The Life and Times of William S. Burroughs.* New York: Henry Holt, p 236.

90. Erard M (2012) *Babel No More: The Search for the World's Most Extraordinary Language Learners.* Free Press: New York.

91. Sacks O (1985) *The Man Who Mistook His Wife for a Hat, and Other Clinical Tales.* Summit Books: New York, chapter 9.

92. Etcoff NL, Ekman P, Magee JJ, Frank MG (2000) Lie Detection and Language Comprehension. *Nature* 405(6783):139.

93. Ramachandran VS (2004) *A Brief Tour of Human Consciousness: From Impostor Poodles to Purple Numbers.* Pi Press: New York, p 156.

Chapter 13

1. Twain M (1904) [1996] *The Diary of Adam and Eve.* Random House: New York, p 47.

2. Panksepp J, Nelson E, Bekkedal M (1997) Brain Systems for the Mediation of Social Separation-Distress and Social-Reward. Evolutionary Antecedents and Neuropeptide Intermediaries. *Ann N Y Acad Sci* 807:78–100. Newman JD (2007) Neural Circuits Underlying Crying and Cry Responding in Mammals. *Behav Brain Res* 182(2):155–65.

3. Levine S, Wiener SG (1988) Psychoendocrine Aspects of Mother-Infant Relationships in Nonhuman Primates. *Psychoneuroendocrinology* 13(1–2):143–54.

4. Muller JM, Moore H, Myers MM, Shair HN (2008) Ventral Striatum Dopamine D2 Receptor Activity Inhibits Rat Pups' Vocalization Response to Loss of Maternal Contact. *Behav Neurosci* 122(1):119–28.

5. Khazipov R, Minlebaev M, Valeeva G (2013) Early Gamma Oscillations. *Neuroscience* 250:240–45.

6. Cifuentes Castro VH, López Valenzuela CL, Salazar Sánchez JC, Peña KP, López Pérez SJ, Ibarra JO, Villagrán AM (2014) An Update of the Classical and Novel Methods Used for Measuring Fast Neurotransmitters during Normal and Brain Altered Function. *Curr Neuropharmacol* 12(6):490–508. Jing M, Li Y, Zeng J, Huang P, Skirzewski M, Kljakic O, Peng W, Qian T, Tan K, Zou J, Trinh S, Wu R, Zhang S, Pan S, Hires SA, Xu M, Li H, Saksida LM, Prado VF, Bussey TJ, Prado MAM, Chen L, Cheng H, Li Y (2020) An Optimized Acetylcholine Sensor for Monitoring in Vivo Cholinergic Activity. *Nat Meth* 17(11):1139–46. Parikh V, Sarter M (2008) Cholinergic Mediation of Attention: Contributions of Phasic and Tonic Increases in Prefrontal Cholinergic Activity. *Ann N Y Acad Sci* 1129:225–35.

7. Holley LA, Wiley RG, Lappi DA, Sarter M (1994) Cortical Cholinergic Deafferentation Following the Intracortical Infusion of 192 IgG-Saporin: A Quantitative Histochemical Study. *Brain Res* 663(2):277–86.

8. Neigh GN, Arnold HM, Rabenstein RL, Sarter M, Bruno JP (2004) Neuronal Activity in the Nucleus Accumbens Is Necessary for Performance-Related Increases in Cortical Acetylcholine Release. *Neuroscience* 123(3):635–45.

9. Kehoe P, Callahan M, Daigle A, Mallinson K, Brudzynski S (2001) The Effect of Cholinergic Stimulation on Rat Pup Ultrasonic Vocalizations. *Dev Psychobiol* 38(2):92–100.

10. Ricceri L (2003) Behavioral Patterns under Cholinergic Control during Development: Lessons Learned from the Selective Immunotoxin 192 IgG Saporin. *Neurosci Biobehav Rev* 27(4):377–84. Scattoni ML, Puopolo M, Calamandrei G, Ricceri L (2005) Basal Forebrain Cholinergic Lesions in 7-Day-Old Rats Alter Ultrasound Vocalisations and Homing Behaviour. *Behav Brain Res* 161(1):169–72.

11. Ricceri L, Cutuli D, Venerosi A, Scattoni ML, Calamandrei G (2007) Neonatal Basal Forebrain Cholinergic Hypofunction Affects Ultrasonic Vocalizations and Fear Conditioning Responses in Preweaning Rats. *Behav Brain Res* 183(1):111–17.

12. Ricceri et al. (2007).

13. Brudzynski SM (2009) Medial Cholinoceptive Vocalization Strip in the Cat and Rat Brains: Initiation of Defensive Vocalizations. In: Brudzynski SM [ed] *Handbook of Mammalian Vocalization*. Academic Press: Oxford, pp 265–79.

14. Hofer MA (2002) Unexplained Infant Crying: An Evolutionary Perspective. *Acta Paediatr* 91(5):491–96, p 494.

15. Shair HN (2007) Acquisition and Expression of a Socially Mediated Separation Response. *Behav Brain Res* 182(2):180–92.

16. Soltis J (2004) The Signal Functions of Early Infant Crying. *Behav Brain Sci* 27(4):443–58.

17. Ainsworth MD, Blehar MC, Waters E, Wall S (1978) *Patterns of Attachment: A Psychological Study of the Strange Situation*. Erlbaum: Hillsdale, NJ.

18. Young L, Camprodon JA, Hauser M, Pascual-Leone A, Saxe R (2010) Disruption of the Right Temporoparietal Junction with Transcranial Magnetic Stimulation Reduces the Role of Beliefs in Moral Judgments. *Proc Natl Acad Sci USA* 107(15):6753–58.

19. Andersen M, Schjoedt U, Nielbo K, Sorensen J (2014) Mystical Experience in the Lab. *Method Theory Study Relig* 26:217–45.

20. Boucher PO, Ozdemir RA, Momi D, Burke MJ, Jannati A, Fried PJ, Pascual-Leone A, Shafi MM, Santarnecchi E (2021) Sham-Derived Effects and the Minimal Reliability of Theta Burst Stimulation. *Sci Rep* 11(1):21170.

21. Arzy S, Seeck M, Ortigue S, Spinelli L, Blanke O (2006) Induction of an Illusory Shadow Person. *Nature* 443(7109):287. Cazzato V, Mian E, Serino A, Mele S, Urgesi C (2015) Distinct Contributions of Extrastriate Body Area and Temporoparietal Junction in Perceiving One's Own and Others' Body. *Cogn Affect Behav Neurosci* 15(1):211–28.

22. Schurz M, Radua J, Aichhorn M, Richlan F, Perner J (2014) Fractionating Theory of Mind: A Meta-Analysis of Functional Brain Imaging Studies. *Neurosci Biobehav Rev* 42:9–34.

23. Andrews-Hanna JR (2012) The Brain's Default Network and Its Adaptive Role in Internal Mentation. *Neuroscientist* 18(3):251–70.

24. Bentley P, Husain M, Dolan RJ (2004) Effects of Cholinergic Enhancement on Visual Stimulation, Spatial Attention, and Spatial Working Memory. *Neuron* 41(6):969–82.

25. Buchheim A, Heinrichs M, George C, Pokorny D, Koops E, Henningsen P, O'Connor MF, Gundel H (2009) Oxytocin Enhances the Experience of Attachment Security. *Psychoneuroendocrinology* 34(9):1417–22. Feldman R (2012) Oxytocin and Social Affiliation in Humans. *Horm Behav* 61(3):380–91. Insel TR (2010) The Challenge of Translation in Social Neuroscience: A Review of Oxytocin, Vasopressin, and Affiliative Behavior. *Neuron* 65(6):768–79. Panksepp J, Nelson E, Bekkedal M (1997) Brain Systems for the Mediation of Social Separation-Distress and Social-Reward. Evolutionary Antecedents and Neuropeptide Intermediaries. *Ann N Y Acad Sci* 807:78–100.

26. Mesulam MM (1999) Spatial Attention and Neglect: Parietal, Frontal and Cingulate Contributions to the Mental Representation and Attentional Targeting of Salient Extrapersonal Events. *Philos Trans R Soc Lond B Biol Sci* 354(1387):1325–46.

27. Van der Kamp J, Canal-Bruland R (2011) Kissing Right? On the Consistency of the Head-Turning Bias in Kissing. *Laterality* 16(3):257–67.

28. Brugger P, Regard M, Landis T (1996) Unilaterally Felt "Presences": The Neuropsychiatry of One's Invisible Doppelganger. *Neuropsychiatry Neuropsychol Behav Neurol* 9(2):114–22.

29. Arzy S, Seeck M, Ortigue S, Spinelli L, Blanke O (2006) Induction of an Illusory Shadow Person. *Nature* 443(7109):287. Zijlmans M, van Eijsden P, Ferrier CH, Kho KH, van Rijen PC, Leijten FS (2009) Illusory Shadow Person Causing Paradoxical Gaze Deviations during Temporal Lobe Seizures. *J Neurol Neurosurg Psychiatry* 80(6):686–88.

30. Alister Hardy Religious Experience Research Centre (2021) Alister Hardy RERC Archive Database. https://uwtsd.ac.uk/library/alister-hardy-religious-experience-research -centre/online-archive/ (accessed 19 December 2021).

31. Wathey JC (2016) *The Illusion of God's Presence: The Biological Origins of Spiritual Longing.* Prometheus Books: Amherst, NY, pp 75–76.

32. Pew Forum (2009) Faith in Flux: Changes in Religious Affiliation in the U.S. The Pew Forum on Religion and Public Life. www.pewforum.org/2009/04/27/faith-in-flux/ (accessed 14 September 2021).

33. Ring RH, Schechter LE, Leonard SK, Dwyer JM, Platt BJ, Graf R, Grauer S, Pulicicchio C, Resnick L, Rahman Z, Sukoff Rizzo SJ, Luo B, Beyer CE, Logue SF, Marquis KL, Hughes ZA, Rosenzweig-Lipson S (2010) Receptor and Behavioral Pharmacology of WAY-267464, a Non-Peptide Oxytocin Receptor Agonist. *Neuropharmacology* 58(1):69–77.

34. Adolphs R (2010) What Does the Amygdala Contribute to Social Cognition? *Ann N Y Acad Sci* 1191:42–61.

35. Schneider B, Koenigs M (2017) Human Lesion Studies of Ventromedial Prefrontal Cortex. *Neuropsychologia* 107:84–93.

36. Haidt J (2007) The New Synthesis in Moral Psychology. *Science* 316(5827): 998–1002.

37. Van Bavel JJ, Packer DJ, Cunningham WA (2008) The Neural Substrates of In-Group Bias: A Functional Magnetic Resonance Imaging Investigation. *Psychol Sci* 19(11):1131–39. Westen D, Blagov PS, Harenski K, Kilts C, Hamann S (2006)

Neural Bases of Motivated Reasoning: An FMRI Study of Emotional Constraints on Partisan Political Judgment in the 2004 U.S. Presidential Election. *J Cogn Neurosci* 18(11):1947–58.

38. Stone VE, Cosmides L, Tooby J, Kroll N, Knight RT (2002) Selective Impairment of Reasoning about Social Exchange in a Patient with Bilateral Limbic System Damage. *Proc Natl Acad Sci USA* 99(17):11531–36.

39. Livesley WJ, Jang KL, Jackson DN, Vemon PA (1993) Genetic and Environmental Contributions to Dimensions of Personality Disorder. *Am J Psychiatry* 150:1826–31.

40. Stinson FS, Dawson DA, Goldstein RB, Chou SP, Huang B, Smith SM, Ruan WJ, Pulay AJ, Saha TD, Pickering RP, Grant BF (2008) Prevalence, Correlates, Disability, and Comorbidity of DSM-IV Narcissistic Personality Disorder: Results from the Wave 2 National Epidemiologic Survey on Alcohol and Related Conditions. *J Clin Psychiatry* 69(7):1033–45.

41. Maynard Smith J (1982) *Evolution and the Theory of Games.* Cambridge University Press: Cambridge, UK.

42. Baskin-Sommers A, Krusemark E, Ronningstam E (2014) Empathy in Narcissistic Personality Disorder: From Clinical and Empirical Perspectives. *Personal Disord* 5(3):323–33.

43. Shamay-Tsoory S (2015) The Neuropsychology of Empathy: Evidence from Lesion Studies. *Revue Neuropsychologie* 7(4):237–43, https://doi.org/10.3917/rne.074.0237.

44. Fan Y, Wonneberger C, Enzi B, de Greck M, Ulrich C, Tempelmann C, Bogerts B, Doering S, Northoff G (2011) The Narcissistic Self and Its Psychological and Neural Correlates: An Exploratory fMRI Study. *Psychol Med* 41(8):1641–50.

45. Nenadić I, Lorenz C, Gaser C (2021) Narcissistic Personality Traits and Prefrontal Brain Structure. *Sci Rep* 11(1):15707.

46. Schulze L, Dziobek I, Vater A, Heekeren HR, Bajbouj M, Renneberg B, Heuser I, Roepke S (2013) Gray Matter Abnormalities in Patients with Narcissistic Personality Disorder. *J Psychiatr Res* 47(10):1363–69.

47. Chester DS, Lynam DR, Powell DK, DeWall CN (2016) Narcissism Is Associated with Weakened Frontostriatal Connectivity: A DTI Study. *Soc Cogn Affect Neurosci* 11(7):1036–40.

48. Eisenberger NI, Lieberman MD (2004) Why Rejection Hurts: A Common Neural Alarm System for Physical and Social Pain. *Trends Cogn Sci* 8(7):294–300.

49. Altemeyer B, Hunsberger B (1992) Authoritarianism, Religious Fundamentalism, Quest, and Prejudice. *Int J Psychol Relig* 2(2):113–33, doi: 10.1207/s15327582ijpr0202_5. Altemeyer B, Hunsberger B (2004) A Revised Religious Fundamentalism Scale: The Short and Sweet of It. *Int J Psychol Relig* 14(1):47–54, doi: 10.1207/s15327582ijpr1401_4.

50. Asp E, Ramchandran K, Tranel D (2012) Authoritarianism, Religious Fundamentalism, and the Human Prefrontal Cortex. *Neuropsychology* 26(4):414–21.

51. Asp E, Manzel K, Koestner B, Cole CA, Denburg NL, Tranel D (2012) A Neuropsychological Test of Belief and Doubt: Damage to Ventromedial Prefrontal Cortex Increases Credulity for Misleading Advertising. *Front Neurosci* 6:100.

52. Prado J, Chadha A, Booth JR (2011) The Brain Network for Deductive Reasoning: A Quantitative Meta-Analysis of 28 Neuroimaging Studies. *J Cogn Neurosci*

23(11):3483–97. Vartanian O, Beatty EL, Smith I, Blackler K, Lam Q, Forbes S, De Neys W (2018) The Reflective Mind: Examining Individual Differences in Susceptibility to Base Rate Neglect with fMRI. *J Cogn Neurosci* 30(7):1011–22.

53. Harris S, Sheth SA, Cohen MS (2008) Functional Neuroimaging of Belief, Disbelief, and Uncertainty. *Ann Neurol* 63(2):141–47. Harris S, Kaplan JT, Curiel A, Bookheimer SY, Iacoboni M, Cohen MS (2009) The Neural Correlates of Religious and Nonreligious Belief. *PLoS One* 4(10):e0007272. Howlett JR, Paulus MP (2015) The Neural Basis of Testable and Non-Testable Beliefs. *PLoS One* 10(5):e0124596.

54. Haidt J (2012) *The Righteous Mind: Why Good People Are Divided by Politics and Religion.* Pantheon Books: New York.

55. Jasmin KM, McGettigan C, Agnew ZK, Lavan N, Josephs O, Cummins F, Scott SK (2016) Cohesion and Joint Speech: Right Hemisphere Contributions to Synchronized Vocal Production. *J Neurosci* 36(17):4669–80.

56. Chang EF, Niziolek CA, Knight RT, Nagarajan SS, Houde JF (2013) Human Cortical Sensorimotor Network Underlying Feedback Control of Vocal Pitch. *Proc Natl Acad Sci USA* 110(7):2653–58.

Epilogue

1. Hagerty BB (2009b) The God Choice. *USA Today* [22 June]:9A.

2. Hagerty BB (2009a) *Fingerprints of God: The Search for the Science of Spirituality.* Riverhead: New York.

Appendix A

1. Bullock TH, Orkand R, Grinnell A (1977) *Introduction to Nervous Systems.* WH Freeman: San Francisco, p xiii.

2. Azevedo FA, Carvalho LR, Grinberg LT, Farfel JM, Ferretti RE, Leite RE, Jacob Filho W, Lent R, Herculano-Houzel S (2009) Equal Numbers of Neuronal and Non-neuronal Cells Make the Human Brain an Isometrically Scaled-Up Primate Brain. *J Comp Neurol* 513(5):532–41.

3. Martin AR, Brown DA, Diamond ME, Cattaneo A, De-Miguel FF (2020) *From Neuron to Brain (Sixth Edition).* Sinauer Associates: Sunderland, MA. Fields RD (2009) *The Other Brain.* Simon & Schuster: New York.

4. Kasai H, Fukuda M, Watanabe S, Hayashi-Takagi A, Noguchi J (2010) Structural Dynamics of Dendritic Spines in Memory and Cognition. *Trends Neurosci* 33(3):121–29.

5. Martin et al. (2020).

6. Mu Y, Poo MM (2006) Spike Timing-Dependent LTP/LTD Mediates Visual Experience-Dependent Plasticity in a Developing Retinotectal System. *Neuron* 50(1):115–25.

7. Viriyopase A, Bojak I, Zeitler M, Gielen S (2012) When Long-Range Zero-Lag Synchronization Is Feasible in Cortical Networks. *Front Comput Neurosci* 6:49.

8. Axmacher N, Mormann F, Fernandez G, Elger CE, Fell J (2006) Memory Formation by Neuronal Synchronization. *Brain Res Rev* 52(1):170–82. Buzsáki G, Draguhn A (2004) Neuronal Oscillations in Cortical Networks. *Science* 304(5679):1926–29. Engel AK, Fries P, Singer W (2001) Dynamic Predictions: Oscillations and Synchrony in Top-Down Processing. *Nat Rev Neurosci* 2(10):704–16. Rutishauser U, Ross IB, Mamelak AN, Schuman EM (2010) Human Memory Strength Is Predicted by Theta-Frequency Phase-Locking of Single Neurons. *Nature* 464(7290):903–7. Senkowski D, Schneider TR, Foxe JJ, Engel AK (2008) Crossmodal Binding through Neural Coherence: Implications for Multisensory Processing. *Trends Neurosci* 31(8):401–9. Summerfield C, Mangels JA (2005) Functional Coupling between Frontal and Parietal Lobes during Recognition Memory. *Neuroreport* 16(2):117–22. Womelsdorf T, Fries P (2007) The Role of Neuronal Synchronization in Selective Attention. *Curr Opin Neurobiol* 17(2):154–60.

9. Blumenfeld H (2002) *Neuroanatomy through Clinical Cases.* Sinauer: Sunderland, MA.

10. Allison T, Puce A, McCarthy G (2000) Social Perception from Visual Cues: Role of the STS Region. *Trends Cogn Sci* 4(7):267–78. Haxby JV, Hoffman EA, Gobbini MI (2000) The Distributed Human Neural System for Face Perception. *Trends Cogn Sci* 4(6):223–33.

11. Mountcastle V (1978) An Organizing Principle for Cerebral Function: The Unit Model and the Distributed System. In: Edelman G, Mountcastle V [eds] *The Mindful Brain.* MIT Press: Cambridge, MA.

12. Hawkins J, Blakeslee S (2004) *On Intelligence.* Times Books: New York.

13. Casey BJ, Davidson M, Rosen B (2002) Functional Magnetic Resonance Imaging: Basic Principles of and Application to Developmental Science. *Dev Sci* 5(3):301–9. Huettel SA, Song AW, McCarthy G (2009) *Functional Magnetic Resonance Imaging (Second Edition).* Sinauer Associates: Sunderland, MA. Smith SM (2004) Overview of fMRI Analysis. *Br J Radiol* 77 Spec No 2:S167–75. Ward J (2020) Students Guide to Cognitive Neuroscience: Basics of fMRI. www.youtube.com/watch?v=-C84RFgyzuE (accessed 30 November 2021).

14. Owen AM, Coleman MR, Boly M, Davis MH, Laureys S, Pickard JD (2006) Detecting Awareness in the Vegetative State. *Science* 313(5792):1402.

15. I present some examples of this in chapter 14 of Wathey JC (2016) *The Illusion of God's Presence: The Biological Origins of Spiritual Longing.* Prometheus Books: Amherst, NY.

16. Ellis CT, Skalaban LJ, Yates TS, Bejjanki VR, Córdova NI, Turk-Browne NB (2020) Re-Imagining fMRI for Awake Behaving Infants. *Nat Commun* 11(1):4523.

17. Sheridan CJ, Matuz T, Draganova R, Eswaran H, Preissl H (2010) Fetal Magnetoencephalography—Achievements and Challenges in the Study of Prenatal and Early Postnatal Brain Responses: A Review. *Infant Child Dev* 19(1):80–93.

18. Morgan EU, van der Meer A, Vulchanova M, Blasi DE, Baggio G (2020) Meaning before Grammar: A Review of ERP Experiments on the Neurodevelopmental Origins of Semantic Processing. *Psychon Bull Rev* 27(3):441–64. Sur S, Sinha VK (2009) Event-Related Potential: An Overview. *Ind Psychiatry J* 18(1):70–73.

19. Grossmann T, Johnson MH, Lloyd-Fox S, Blasi A, Deligianni F, Elwell C, Csibra G (2008) Early Cortical Specialization for Face-to-Face Communication in Human Infants. *Proc R Soc Lond B Biol Sci* 275(1653):2803–11. Sharon D, Hamalainen MS, Tootell RB, Halgren E, Belliveau JW (2007) The Advantage of Combining MEG and EEG: Comparison to fMRI in Focally Stimulated Visual Cortex. *Neuroimage* 36(4):1225–35.

20. Abdelnour F, Schmidt B, Huppert TJ (2009) Topographic Localization of Brain Activation in Diffuse Optical Imaging Using Spherical Wavelets. *Phys Med Biol* 54(20):6383–413. Liu Z, He B (2008) fMRI-EEG Integrated Cortical Source Imaging by Use of Time-Variant Spatial Constraints. *Neuroimage* 39(3):1198–214.

21. Pascual-Leone A, Bartres-Faz D, Keenan JP (1999) Transcranial Magnetic Stimulation: Studying the Brain-Behaviour Relationship by Induction of "Virtual Lesions." *Philos Trans R Soc Lond B Biol Sci* 354(1387):1229–38.

22. Boucher PO, Ozdemir RA, Momi D, Burke MJ, Jannati A, Fried PJ, Pascual-Leone A, Shafi MM, Santarnecchi E (2021) Sham-Derived Effects and the Minimal Reliability of Theta Burst Stimulation. *Sci Rep* 11(1):21170.

23. See figure 5 of Pascual-Leone et al. (1999).

24. Fenno L, Yizhar O, Deisseroth K (2011) The Development and Application of Optogenetics. *Annu Rev Neurosci* 34:389–412.

25. Hawkins and Blakeslee (2004).

26. Schwarzlose R (2021) *Brainscapes: The Warped, Wondrous Maps Written into Your Brain—And How They Guide You.* Houghton Mifflin Harcourt: Boston.

27. Chudler EH, Bergsman KC (2014) Explain the Brain: Websites to Help Scientists Teach Neuroscience to the General Public. *CBE Life Sci Educ* 13(4):577–83. Dingman M (2021) Neuroscientfically Challenged. www.neuroscientificallychallenged.com/ (accessed 10 September 2021). Dubuc B (2021) The Brain from Top to Bottom. https://thebrain.mcgill.ca/index.php (accessed 10 September 2021). Krebs C, Fejtek M (2021) Functional Neuroanatomy. University of British Columbia: www.neuroanatomy.ca/index.html (accessed 10 September 2021). Rose C, Kandel E (2017) *The Charlie Rose Brain Series.* https://charlierose.com/collections/3 (accessed 10 September 2021).

28. Eagleman D (2015) The Brain. https://weta.org/watch/shows/brain-david-eagleman (accessed 30 November 2021).

Appendix B

1. Anderson F, Salamo L, Stein BL [eds] (1975) *Mark Twain's Notebooks & Journals: Volume II (1877–1883).* University of California Press: Berkeley, CA, p 305.

2. Hill PC, Hood RW [eds] (1999) *Measures of Religiosity.* Religious Education Press: Birmingham, AL.

3. Persinger MA, Healey F (2002) Experimental Facilitation of the Sensed Presence: Possible Intercalation between the Hemispheres Induced by Complex Magnetic Fields. *J Nerv Ment Dis* 190(8):533–41.

4. Persinger MA (2000) The UFO Experience: A Normal Correlate of Human Brain Function. In: Jacobs DM [ed] *UFOs and Abduction: Challenging the Borders of Knowledge*. University Press: Lawrence, KS, pp 262–302.

5. Crescentini C, Aglioti SM, Fabbro F, Urgesi C (2014) Virtual Lesions of the Inferior Parietal Cortex Induce Fast Changes of Implicit Religiousness/Spirituality. *Cortex* 54:1–15.

6. Cohen-Zimerman S, Cristofori I, Zhong W, Bulbulia J, Krueger F, Gordon B, Grafman J (2020) Neural Underpinning of a Personal Relationship with God and Sense of Control: A Lesion-Mapping Study. *Cogn Affect Behav Neurosci* 20(3):575–87.

7. Cloninger CR, Svrakic DM, Przybeck TR (1993) A Psychobiological Model of Temperament and Character. *Arch Gen Psychiatry* 50(12):975–90.

Index

acetylcholine, 139–41, 143–44;
  in schizophrenia, 153
acquired sociopathy, 62
action potential, 201
adaptiveness:
  of innate neural maternal model, 17–23;
  of lateralization, 163–68;
  of motherese, 29
addiction:
  to drugs, 74–75;
  insula and, 87;
  religion as, 65
agency, sense of, 78, 86, 90;
  in schizophrenia, 148, 150–52
Ainsworth, Mary, 13, 182
alien abduction, illusion of, 120
alien hand syndrome, 79, 86
Allman, John, 49
altruism, reciprocal, 4–5, 188
amphetamines, 153, 161
amygdala, 58–60, 141;
  hemispheric dominance in, 168–69;
  research directions for, 184
anterior, term, 202
anterior cingulate cortex, 45, 53–54, 87, 141, 156;
  and belief, 190;
  functions of, 47–49;

hemispheric dominance in, 169;
  mechanism of, 49–51;
  and religiosity, 121–23
anticipatory signals, 148–49, 153
anxiety:
  lack of control and, 158–59;
  OCD and, 157–58;
  religious rituals and, 158
Asp, Erik, 125
attachment:
  to mother/God, 13–17, 14f;
  periaqueductal gray and, 131
attention:
  and acetylcholine, 139–41;
  emotion and, 141–45;
  and perception, 138–39
attraction, 58–60
auditory system:
  fetal, 171–72;
  neonatal, 94–100
authoritarianism, lesions and, 124–29, 190
autism spectrum disorder, 49
aversion, 58–60
axons, 201

babbling, 29, 36, 68, 98
Barker, Dan, iv, vii

D1 receptors, 69, 71–73, 145
D2 receptors, 66, 71–73, 145
Darwin, Charles, 46, 53
Dawkins, Richard, 4
Deese-Roediger-McDermott paradigm, 91
déjà vu, 89
delayed reciprocal altruism, 4–5
delusions, 147–61;
  neurochemistry of, 152–54
dendritic spines, 141, 200–201
depression, anterior cingulate cortex
  and, 48
design, argument from, 12, 37
disgust:
  and incest avoidance system, 39;
  in OCD, 156
DMT, 136
dopamine, 67, 72–73;
  blockers, 145, 153, 160;
  drug addiction and, 74;
  in OCD, 160–61;
  and pair-bonding, 71;
  research directions for, 184–85;
  in schizophrenia, 152–53;
  side effects of, 161
dorsal, term, 202
dorsolateral prefrontal cortex, and
  religiosity, 121
doubt:
  OCD and, 157–58;
  ventromedial prefrontal cortex and,
  125–26

Eisenberger, Naomi, 50
electroencephalography (EEG), 205
Eliot, T. S., 117
embodiment, 26, 78–80;
  and certainty, 102;
  gamma oscillation and, 94;
  NMDA blockers and, 153–54;
  in schizophrenia, 149–50;
  in transsexuals, 105
emotion:
  and expectation, 141–45;

hemispheric dominance and, 172–74;
  nucleus accumbens and, 69–70;
  perception of, 49;
  sensed presence and, 119
emotional empathy, 189
empathetic pain, anterior cingulate cortex
  and, 48
empathy, research directions for, 188–89
entheogens, 136
epilepsy:
  and OCD, 155;
  temporal lobe, and religiosity, 154–55
error detection, anterior cingulate cortex
  and, 47
ethology, vii–viii
event-related field (ERF), 205
event-related potential (ERP), 205
evoked field (EF), 205
evoked potential (EP), 205
expectation:
  emotion and, 141–45;
  and perception, 138–39
extrastriate body area, 86–87;
  research directions for, 183

faces:
  expressions, infant imitation of, 34;
  inversion effect, 30;
  midbrain and, 93;
  orbitofrontal neurons and, 55–56;
  perception of, 29–31, 172–74
fear, amygdala and, 58
Ferguson, Michael, 114, 130–31
first-person perspective, 78
follower type, 7;
  research directions for, 190
frequency-dependent selection, 6
Freud, Sigmund, 13
frontal lobe, 98, 202
frontotemporal dementia, 61–62;
  and religiousness, 124–29
functional magnetic resonance imaging
  (fMRI), 203–5
fundamentalism, lesions and, 124–29, 190